Seed Technology

The Authors

Dhirendra Khare is Dean, Faculty of Agriculture, JNKVV, Jabalpur. Dr. Khare has obtained B.Sc. (Ag) [Honours] in 1981, M.Sc. (Ag) [Honours] Plant Breeding and Genetics in 1983 and Ph.D. [Honours] Plant Breeding and Genetics in 1987-88. He is teaching the courses on Seed Technology at Undergraduate and Postgraduate levels since 1986. He has published more than 89 refereed research papers, six books and nine manuals. **The Bharat Ratna Dr. C. Subramaniam Award of ICAR for Outstanding Teacher** in Agriculture and Allied Sciences for the biennium 2007-2008 was conferred on Dr. Dhirendra Khare, for excellence in teaching. **The Outstanding Team Research Award of ICAR** was conferred on Dr. D. Khare alongwith the team of JNKVV in 2007-2008 for the excellent research work done in the field of crop improvement and seed availability of soybean. He is life and Fellow member of Indian Society of Seed Technology and Indian Society of Genetics and Plant Breeding. He has organized several national training programmes on aspects on seed science and technology and performed the duty of resource person in many training programmes across the nation.

Mohan S. Bhale is the Associate Professor (Plant Pathology) at Seed Technology Research, Department of Plant Breeding & Genetics, Jawaharlal Nehru Krishi Vishwa Vidyalaya, Jabalpur (Madhya Pradesh). He received Master of Science in Agricultural Plant Pathology in 1978 and Ph.D. in 1992, both from Jawaharlal Nehru Krishi Vishwa Vidyalaya, Jabalpur. He started his professional career in 1979 and involved with management of diseases of crops. Presently, he is dealing with the designated objectionable seedborne diseases of field crops.

His publications include research and review articles, book chapters, manuals and bulletins, working sheets and edited proceedings. He has educated personnel engaged with seed through mass and group trainings on Seed Health and Seed Production Management. At post graduate level, he is involved with teaching of Seed Health Management.

He has an active long association with several professional societies including Indian Society of Seed Technology, Association of Plant Pathologists of India and Indian Mycological Society. He has been honored as a Fellow of Indian Phytopathological Society, Indian Society of Mycology & Plant Pathology and Indian Society of Oilseeds Research. He is the recipient of ICAR Award for Outstanding Interdisciplinary Team Research in Crop Improvement for contributions in seed pathological aspects of soybean in 2007-2008. He has developed the detection techniques and approaches for the management of major seed associated pathogens of soybean.

He is the author of five other books: Seed Technology; A Handbook of Seed Certification; Seed borne Diseases Objectionable in Seed Production and their Management; Seed borne Diseases of Field Crops and their Management and Seed Pathology in Modern Agriculture and Climate Change and Quality Seed Production and Field Crops, Seed Production Technology : Principles and Practices. He is the Chief Editor of JNKVV Research Journal (ISSN: 0021-3721), an official publication of the University since 2002.

Seed Technology

Dhirendra Khare

Mohan S. Bhale

SCIENTIFIC PUBLISHERS

Published by:

SCIENTIFIC PUBLISHERS (INDIA)

Jodhpur-
5A, New Pali Road, P.O. Box 91
Jodhpur 342 001, India

Delhi-
4806/24, Ansari Road
Daryaganj
New Delhi-110 002 India

E-mail: info@scientificpub.com Website:
www.scientificpubonline.com

Print 2026

First Edition: 2000
Second Enlarged Edition: 2014
Succint Edition : 2021

ISBN: 978-93-88043-58-8
eISBN: 978-93-88172-38-7

Printed in India

ACKNOWLEDGEMENT

Seed is the most important viable component of sustainable agriculture. Availability of quality seed in sufficient quantity is one of the most imperative targets for the scientists and officials engaged in agriculture. On high demand of the book after first edition in the year 2000 and second in 2014, now a concise edition is presented. This book is the outcome of several years of teaching on seed at Bachelor, Master and Doctorate levels; research on almost all the aspects under AICRP; production of Nucleus and Breeder seed and testing and certification of the seed of all the important field and horticultural crops and concurrently training to officials engaged in seed production and certification. This book places a clear emphasis on all the relevant aspects of seed and also ensures that those aspects are penned based upon rigorous and current research.

At this juncture, we thank Jawaharlal Nehru Krishi Vishwa Vidyalaya, Jabalpur, authors of different books on seed technology, web sites of all over the world for providing information, students for questioning their doubts that result in deep thinking on the subject. Discussion on the subject with colleagues and friends helped us to present the views in the form wider accepted by the reader. We were fortunate enough to have a group of excellent reviewers and readers of the previous editions who pointed out problems, identified difficulties and informed us to improve it in this edition.

Affection, inspiration, moral support and devotion of the family members Mrs Anuradha Khare, Sonal + Amit, Nehal + Mayank including Kavya ; and Mrs Usha Bhale, Manu + Kaushal and Tanu + Meghana are highly appreciable to complete the assignment.

We are convinced that the information provided in the book shall be of practical use for researchers, teachers, students, officials, policy makers and seed producer and quality control officers.

Dhirendra Khare
Mohan S. Bhale

Jabalpur
19.11.2020

CONTENTS

Chapter - 1

SEED MORPHOLOGY

1.1. OBJECTIVE

To study the development, structure and genetics of seed and its parts

1.2. DEFINITION

Botanically, seed is a ripened ovule containing an embryo in arrested state of development, usually with food reserve and a protective coat. In seed technological term, the part of the plant use for sowing purpose to raise the crop is considered as *seed.*

1.3. SEED FORMATION

It consists of four stages

- Gametogenesis
- Pollination
- Fertilization
- Development of seed

Gametogenesis

Formation of male and female gametes with haploid chromosome number for fertilization is known as gametogenesis. Gamete formation takes place separately in male (anther) and female (ovule) part of the flower. It involves two steps i.e., sporogenesis followed by gametogenesis in both male and female reproductive parts (Fig. 1).

Sporogenesis: It is the formation of spore in reproductive part i.e., spores of male (microsporogenesis) and of female (megasporogenesis).

Gametogenesis: It is the formation of gamete in reproductive part i.e., gamete of male (microgametogenesis) and of female (megagametogenesis).

Formation of male gamete

It has two steps microsporogenesis and microgametogenesis. Androecium (stamen) is the male part of the flower with anther and filament as its parts. Anther is a four chambered structure placed on a filament. Each chamber is known as pollen sac (microsporangia).

Microsporogenesis: In the pollen sac, pollen grains are formed. Inner most layer of pollen sac in known as Tapetum that provides nutrients for development of pollen mother cells (2n) in each pollen sac. The diploid pollen mother cell undergoes meiosis (reduction) cell division, first to form dyad and then four haploid cells (tetrad). Each one is known as microspore.

Microgametogenesis: Haploid nucleus of the microspore forms two haploid nuclei by mitotic (equational) cell division. Out of two haploid nuclei only one nucleus again divides by mitotic cell division and forms two haploid nuclei. Second nucleus remains as such (no further division). In this way total *three haploid nuclei.* Each one is known as **microgamete**.

Pollen grain is a double walled structure, with hard outer cover (**exine**) and thin inner layer (**intine**). 3-5 germpores are present on the **exine** of the pollen grain. After pollination, pollen grain germinates on the stigma and forms a germ tube which protrudes out from any one germ pore. Through germ tube all the three male gametes enter in the female reproductive part. The gamete present at the tip of the pollen tube is known as **tube nucleus**, whereas the remaining two are known as **generative nuclei** or **sperm cells**. Tube nucleus is responsible for growth and direction of pollen tube whereas generative cells take part in fertilization.

Genetic constitution of each microspore formed in a pollen sac differs from the other because of meiosis cell division that involves crossing over and recombination. But all the three micro-gametes are of same genetic constitution due to mitotic cell division. Study of pollen grain is termed as **Palynology.**

Formation of female gamete

Gynoecium (pistil) is the female part of the flower with stigma, style and ovary as its parts. Ovary contains ovule (megasporangium) with embryosac surrounded by two layers of integuments. In the embryosac female gametes are formed by megasporogenesis and megagametogenesis in the nucellus.

Megasporogenesis: A diploid nucellus cell differentiates into sporogenous cell towards micropylar end with nutrients from other nucellus tissues. It works as **Megaspore mother cell**. Megaspore mother cell undergoes meiosis (reduction) cell division to form four haploid cells. Each haploid cell is considered as **Megaspore.** Out of four megaspores, three are degenerated only one remains functional.

Megagametogenesis: The functional megaspore develops into female gametophyte by three mitotic (equational) cell divisions. By first mitotic cell division two, by second four and by third eight haploid nuclei are formed.

These eight archisporium haploid cells are arranged in three-two-three fashion in the embryosac, all surrounded by nucellus. Three towards chalaza end are termed as *Antipodal*, two at the centre as *Polar nuclei* whereas out of three arranged at the micropylar end, the one present in the middle is termed as *egg cell* and remaining two as *synergids*. The nucellus (plural: nucelli) is the central portion of the ovule inside the integuments. It consists of diploid maternal tissue and has the function of a megasporangium.

All the eight megagametes have same genetic constitution as they all originated from one spore by mitotic cell division.

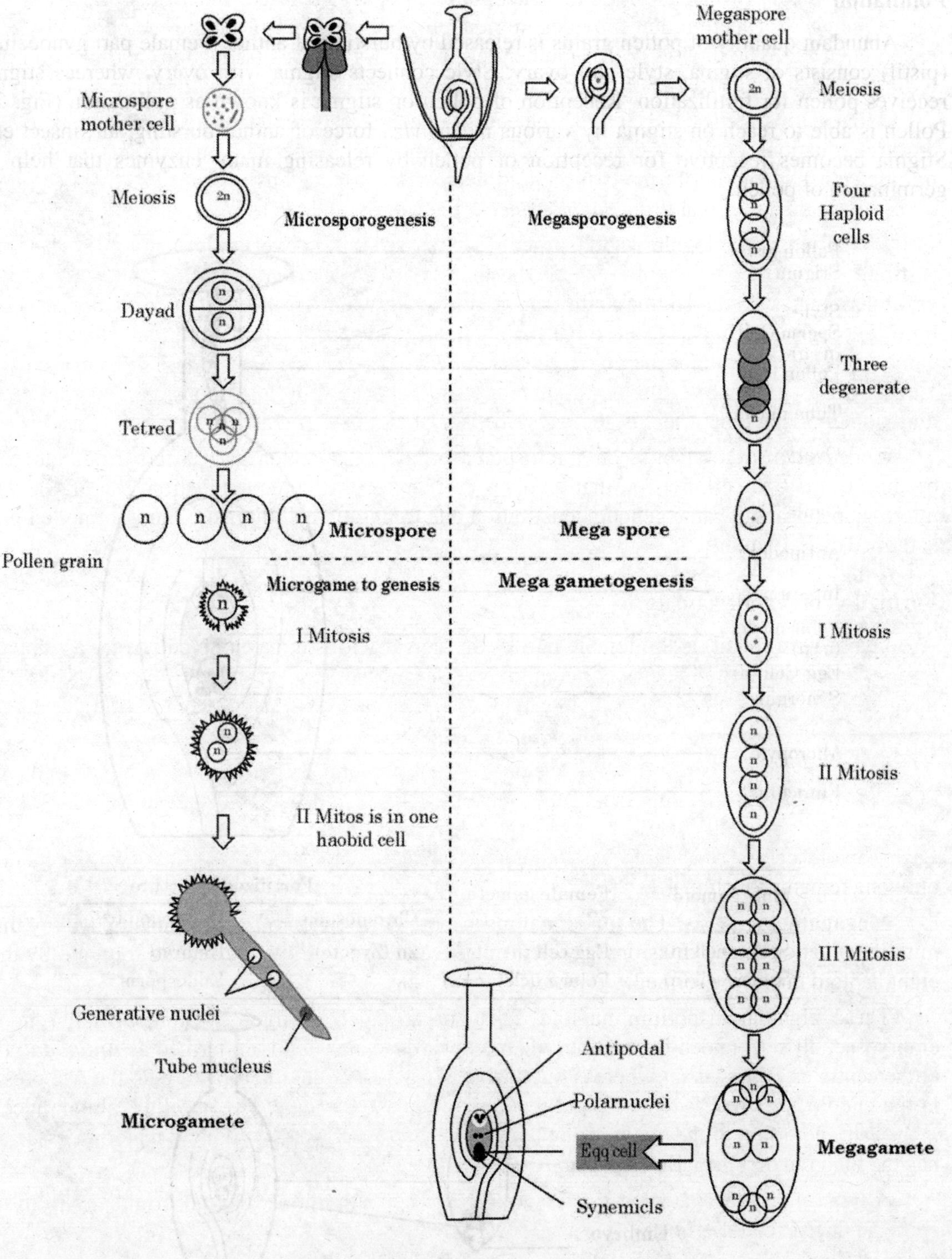

Fig. 1. Sporo and gameto-genesis

Pollination

Abundant quantity of pollen grains is released by bursting of anther. Female part gynoecium (pistil) consists of stigma, style and ovary. Style connects stigma with ovary, whereas stigma receives pollen for fertilization. Reception of pollen on stigma is known as pollination (Fig. 2). Pollen is able to reach on stigma by various means *viz.,* force of anther bursting, air, insect etc. Stigma becomes receptive for reception of pollen by releasing many enzymes that help in germination of pollen.

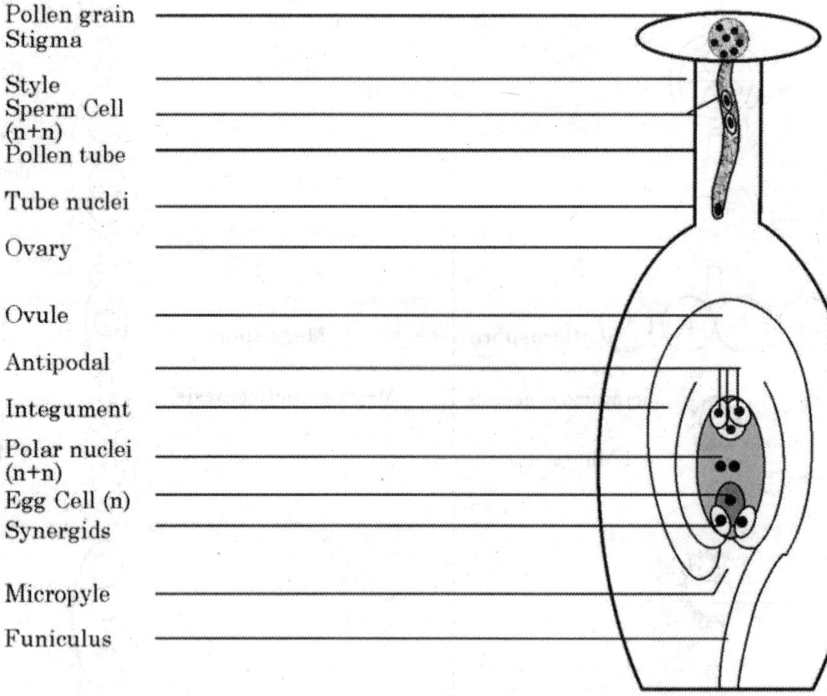

Male gamete	Female gamete	Fertilized structure	
		Chromosome	Developed to
Sperm cell (n)	Egg cell (n)	2n (Zygote)	Embryo
Sperm cell (n)	Polar nuclei (n+n)	3n	Endosperm

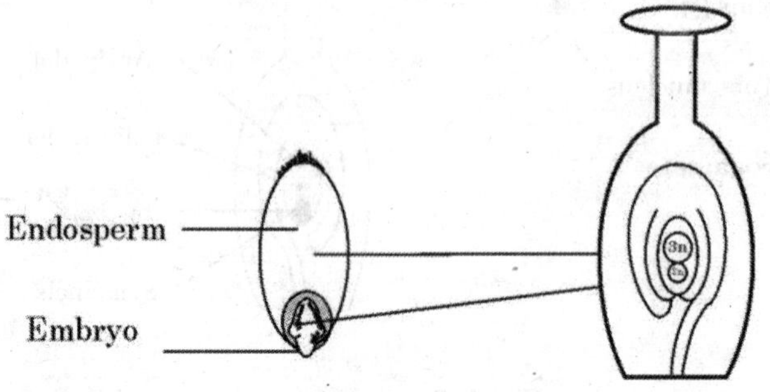

Fig. 2. Pollination and fertilization

Pollen germinates on stigma and forms a pollen tube which comes out through any one germpore. The pollen tube enters in the stigma and travels through style upto ovule. The tube nucleus present at the tip of the germ tube directs the tube towards ovule. Whereas the generative cell present behind the tube nucleus undergoes mitotic cell division and forms two haploid generative nuclei.

Fertilization

In female embryosac eight archisporium cells are present in the ovule inside the ovary. The end of embryosac towards stigma or opposite to the opening (micropyle) is known as chalaza end (non micropylar end). Three haploid cells are arranged at this end known as Antipodal in the embryosac. Two haploid cells are arranged at the center known as Polar nuclei (*sl.* nucleus). Three haploid cells are arranged towards micropylar end of the embryosac. It is known as egg apparatus. The haploid cells of the egg apparatus present on either side are known as synergids, whereas the middle larger one as egg cell.

The pollen tube enters in the embryo sac normally through micropylar end. At the time of peneteration in the embryo sac the tube nucleus enters in one of the synergids and gets burst. In chalazogamous plants, the pollen tube enters the ovule through the chalaza instead of the micropyle opening. During the process of bursting, both the generative haploid cells are released in the embryosac.

During fertilization one sperm nucleus combines with the egg cell and forms a zygote (2n). It further develops into an embryo. The second sperm cell combines with two polar nuclei (n+n) and forms the store of food known as endosperm (3n).

1.4 SEED DEVELOPMENT

After fertilization seed develops through following physiological stages

Histo-differentiation: Formation of embryo from egg cell and endosperm from polar nuclei after fertilization are considered as histo-differentiation.

Cell expansion: After formation of embryo and endosperm the photosynthates are transferred in simple form (sucrose and glucose) from mother plant (source) and stored in complex form i.e., carbohydrate, lipid and protein in the cells of storage (sink) organs i.e., cotyledon or endosperm. It results in expansion of cell size.

Dehydration: After completion of photosynthate transfer, the seed reaches at **physiological maturity** which is defined with no active connection of seed with mother plant. From physiological maturity dehydration starts as physical process in the seed and its completion is considered as **physical maturity**.

Development of a monocot seed

The formation of monocot seed consists of three stage i.e., pro-embryo, globular and scutellar (Fig. 3).

The fertilized egg cell (n+n) undergoes mitotic cell division to form a structure of two diploid cells via transverse section known as **pro-embryo.** The upper cell is known as **apical** (distal or axial), while lower one as **basal** cell.

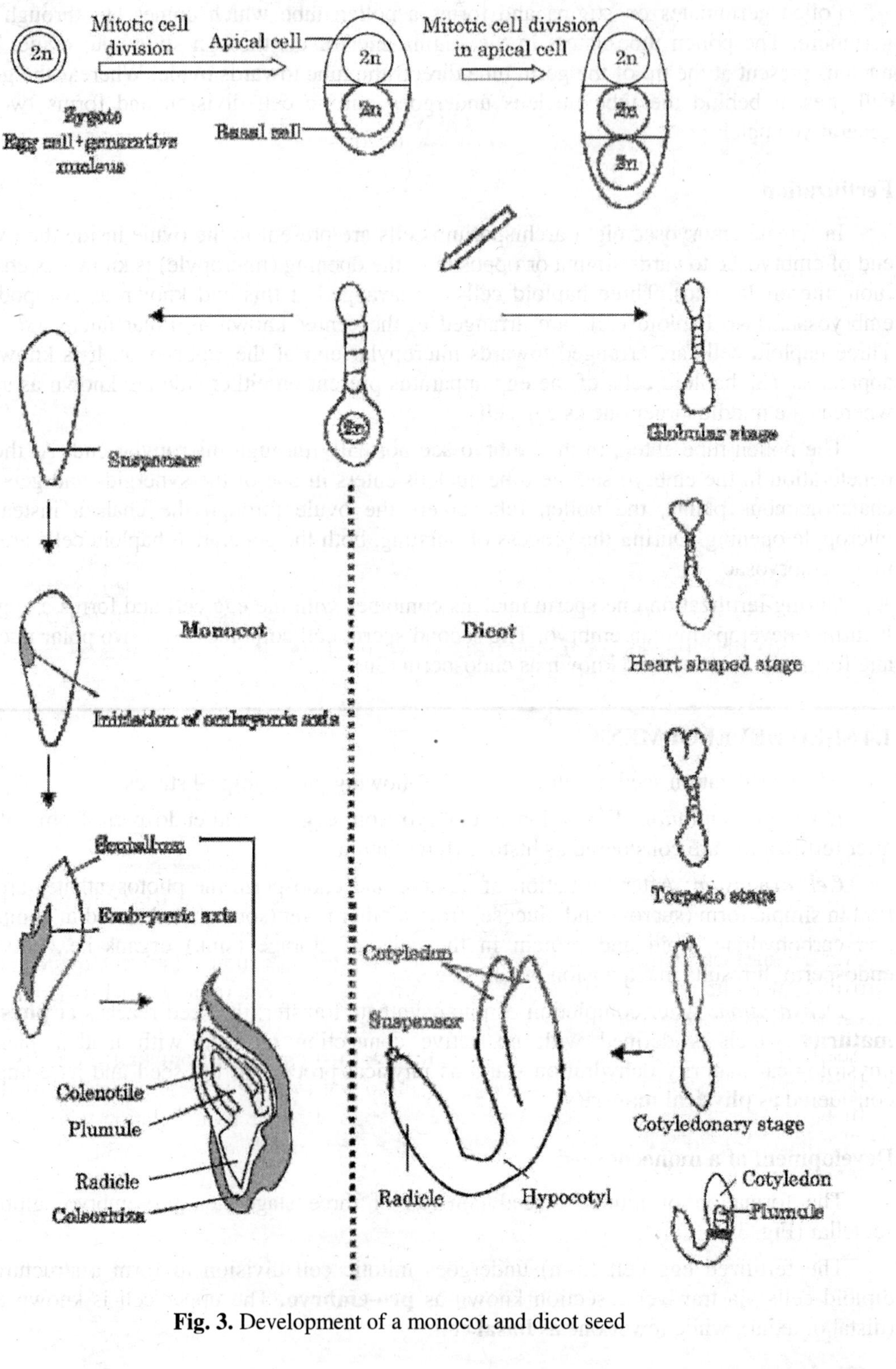

Fig. 3. Development of a monocot and dicot seed

The apical cell divides mitotically and forms a globular structure of 16 diploid cells. At *globular stage* one side of the globular structure divides more rapidly than the other cells and forms embryonic axis. Whereas, other cells form the single cotyledon is known as **scutellum**. The stage is known as *scutellar stage*.

In the seed, scutellum is formed in between endosperm and embryonic axis. It is formed by the diploid apical cell of pro-embryo therefore its chromosomal constitution is same as of embryo. The embryonic axis differentiates into **plumule** (shoot) and **radicle** (root) with covering of specialized tissues **coleoptile** on plumule and **coleorhiza** on radicle.

The basal cell of the pro-embryo forms the structure known as suspensor, but the structure is not well developed in monocot seeds.

The initial triploid endosperm cell undergoes several mitotic cell divisions prior to cell wall formation. The photosynthates are transferred from sink via *transfer cells* to store predominantly as a complex form (starch) in triploid cells of endosperm (sink). The structure of funicle is not well developed and functional in monocot seed. With the accumulation of food material the size of cell gets enlarged. Many enzymes and hormones are stored in scutellum.

Development of a dicot seed

The development of monocot and dicot seed is similar upto globular stage. In dicot the basal cell of proembryo develops into a suspensor that pushes the globular structure deep into the embryosac cavity and absorbs and transfers the food material for storage in globular cells (Fig. 3).

Formation of cotyledon in the globular structure starts with a depression at the tip (heart shaped stage). Elongation of cotyledon by deepening of depression starts at torpedo stage. At cotyledonary stage radicle and hypocotyl are well defined with rudimentary suspensor. Removal of one cotyledon shows presence of plumule in between both the cotyledons.

The food material is absorbed by the suspensor from the surrounding tissue and transferred to the cotyledon. In dicot seeds the food material is predominantly stored in the form of protein and/or lipids. To transfer the photosynthate a vascular strand runs from mother plant through funicle upto one part of seed coat. From the seed coat it is diffused in the nucellus tissues and then absorbed by suspensor. It shows that mother plant is not directly connected with the developing seed.

At physiological maturity the bridge present in the form of funicle between ovule and ovary is broken down with a scar on the seed coat known as **hilum**. The rapid water loss from the seed starts as physical process upto physiological maturity.

Seed ripening: After transfer of photosynthates (food material) from source (leaves) to sink (seed) abscission layer is formed at the base of the ovule at the connection of ovule with ovary *i.e.,* funicle. It shows that now seed with new plant (embryo) has no relation with the mother plant. Formation of abscission layer cut down the supply of water and photosynthates. Seed starts conversion of food material in complex structure and minimize metabolic activities to remain viable for longer period of time (upto favorable conditions of germination). To achieve both the objectives the best option is to minimize presence of water by physical process. The colour of the seed before ripening is usually green due to presence of active chlorophyll. With the gradual loss of moisture, the colour of the testa changes generally to yellow-brown-black or variegated according to the crop.

1.5. SEED STRUCTURE AND HISTOLOGY

Seed Coat: It is a protective coat of diploid maternal tissue made up of two layers testa (outer thick layer) and tegmen (inner thin membrane). The seed coat is present as an envelope to protect the embryo and endosperm from desiccation, mechanical injury, effect of environmental fluctuations and damage due to insects and microorganisms. It also helps in dispersal of seed. Ridge on seed coat is formed by adnate funiculus.

Seed coat develops from the diploid maternal tissue of the integuments and consists of vacuolated thin walled cells. During the process of maturation, it undergoes varied degree of structural alteration.

Embryo: It is a rudimentary plant made up of an axis bearing one or more cotyledons. Embryo is present in axis form with one tip known as plumule (Latin *little feather*) responsible to form shoot portion and the other axis known as radicle (embryonic root) to form the root. The portion of embryonic axis extended above the cotyledon is known as epicotyl and below the cotyledon as hypocotyl (Fig. 4). The shape of the embryo and their position within the seed are variable between species. In the dicot species that have a substantial endosperm (endospermic seeds) the embryo occupies proportionately less of the seed than when the endosperm is rudimentary or absent.

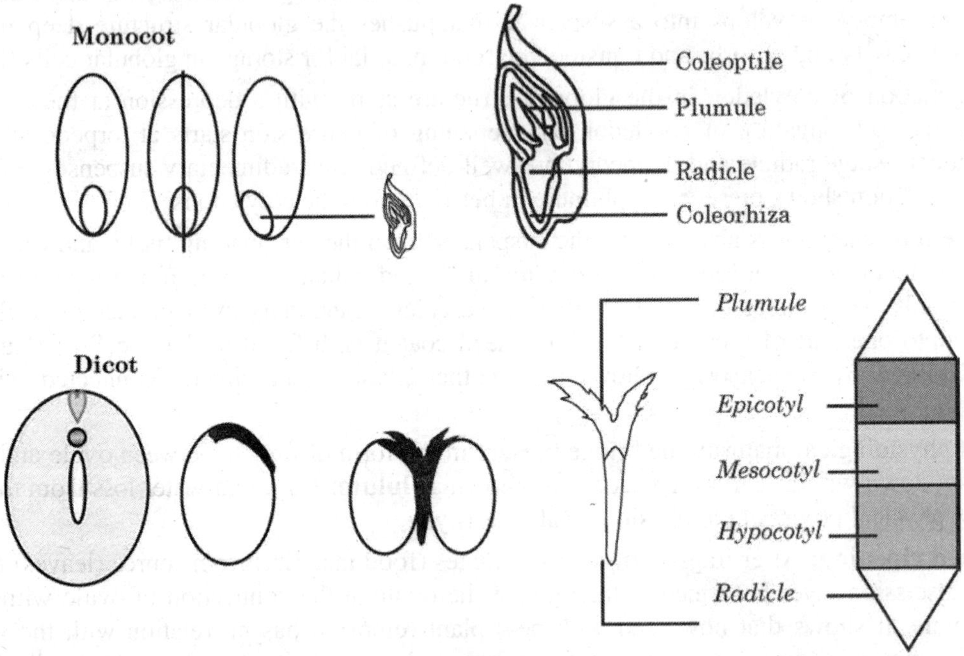

Fig. 4. Embryo of monocot and dicot seed

Embryo of monocot seed is a small structure which lies at one end of the seed. Organs of the monocot embryo are coleoptile (shoot sheath), the plumule, the radicle and coleorhiza (root sheath). The coleoptile is the leaf sheath of the cotyledon that protects the plumule during field emergence. The coleorhiza is a protecting cover on radicle to protect radicle during germination.

The portion in between plumue and radicle is mesocotyl. It is responsible for growth of embryo during germination.

Store of food

Endosperm: (from the Greek words *within* and *seed*) It is a thick and massive structure made up of elongated cells containing abundant starch present in the mature seed and serves as food storage organ. Cells are triploid (3n), result of double fertilization. Two out of three genomes are of maternal origin. Testa and endosperm are the two covering layers of the embryo.

Seeds are categorized as endospermic or non-endospermic in relation to the presence or absence of a well developed endosperm within the mature seed. The relatively massive endosperm is the major source of stored seed reserves in cereals. As a rule endosperm lack intercellular space. In cereal, the storage cells of the endosperm are nonliving with replacement of cytoplasmic contents by starch and protein in cereals.

(i) *Albuminous* (From the Latin *albumin*, the white of an egg): It means well developed endosperm is present as store of food. The amount of endosperm in mature seeds is highly species-dependent and varies from an abundant endosperm layer (*Nicotiana tabaccum*) to a single layer (*Arabidopsis thaliana*).

(ii) *Exalbuminous*: Fusion among polar nuclei and sperm cell occurs and 3n endosperm is initially formed but during the process of development the cotyledons absorb the food reserves from the endosperm. The cotyledons serve as sole food storage organs. The endosperm is almost degraded in the mature seed and the embryo is enclosed by the testa. The endosperm was initially present but now it is not visible therefore, the seed is designated as exalbuminous.

On the outside of the endosperm one to few layers of diploid living cells known as aleurone layer are present. Aleurone layer is responsible to synthesize **enzymes** to convert food material in available form during germination.

Cotyledon: It is the extension of the embryo originates from zygote as embryonic axis with similar genetic constitution. On the basis of number of cotyledons, the crop species are divided into two groups (I) monocotyledonous – presence of one cotyledon (ii) dicotyledonous - two cotyledons are present. Some gymnosperms have 2-18 cotyledons (polycotyledonous). In monocot the scutellum situated in between the endosperm and the embryo is the single cotyledon. The cotyledons of non-endospermic seeds are much bulkier and are the storage tissues, and in peas and beans account for over 90 percent of the mass of the seed in comparison to monocot endospermic seeds.

1.6. FRUIT STRUCTURES CONSIDERED AS SEED

Fruits are mature, ripened ovaries containing seeds. The **pericarp (fruit wall)** is made up of diploid maternal tissue. Following single seeded fruits are usually considered as seed:

Caryopsis: Single seeded fruit with fused fruit and seed wall e.g. wheat, sorghum, pearl millet etc. The fruit of cereals use for sowing as seed are of following types

Caryopsis without husk: Maize, pearl millet, sorghum, wheat etc.

Caryopsis with husk: Rice, barley, oat etc.

Husk is not the part of fruit/seed. It is a part of flower present as an attachment i.e., lemma and palea that covers the caryopsis.

Achene: Single seeded indehiscent fruit where pericarp can be removed from the mature seed, e.g., sunflower, safflower etc.

Schizocarp: A dry fruit, which is separated into two or more units at maturity, e.g. coriander, carrot etc.

Nut : Fruit with thick and hard shell like fruit wall, e.g. ground nut.

Table 1. Examples of different categories of seeds

	Albuminous	**Ex-albuminous**
Monocot	Cereals	Onion
Dicot	Castor, Sunflower	Pulses

1.7. EXPRESSION OF QUALITATIVE GENES IN DIFFERENT PARTS OF A SEED

Seed coat: The seed coat colour of the "hybrid seed" is similar to the female parent irrespective of dominant gene present in the male parent (Table 2). It is termed as non-cytoplasmic maternal inheritance. Example In a crop black seed coat colour (BB) is completely dominant over yellow seed coat colour (bb). In the event hybrid seed is produced by the following combination of parents

Table 2. Genetics of seed coat colour

Parent	**Female**	**Male**	**Female**	**Male**
Seed coat colour	Yellow	Black	Black	Yellow
Genotype	bb	BB	BB	bb
Hybrid (F_1)	Bb		Bb	
Seed coat colour	Yellow		Black	
Genes in seed coat of hybrid	bb		BB	

As per rule of genetics the dominant gene has to express in F_1 generation. But the seed coat of a hybrid will be as of female parent. It never means the principles of genetics are wrong. In the hybrid seed the genes present in male parent (**n**) are able to reach in endosperm (**n**+n+n) and embryo (**n**+n). Whereas, the seed coat is formed by integument and in the integument of hybrid, genes from male parent are not present. The expression of the cell depends on the gene present in it. Therefore, cell will express as of female parent. It shows that in a hybrid seed the seed coat is of female parent. To work out the inheritance of colour, seed coat present on F_2 should be considered as the seed coat of F_1. The seed is a hybrid or not, may not be verified based on seed coat colour. Similarly genetic purity affected by cross-pollination may not be verified based on morphology of seed coat.

Table 3. Contribution of male and female parent in the formation of a hybrid seed structure

Part of seed	Female	Male
Embryo (2n)	Egg cell (n)	Sperm nucleus (n)
Cotyledon (2n)	Egg cell (n)	Sperm nucleus (n)
Endosperm (3n)	Polar nuclei (n+n)	Sperm nucleus (n)
Testa (2n)	Outer integument (2n)	-
Hilum	Scar of funiculus	-
Micropyle	Micropyle	-
Caruncle (2n)	Outer integument (2n)	-
Strophiole (2n)	Wart like outgrowth (2n)	-
Coleoptile (2n)	Egg cell (n)	Sperm nucleus (n)
Coleorhiza (2n)	Egg cell (n)	Sperm nucleus (n)
Scutellum (2n)	Egg cell (n)	Sperm nucleus (n)
Pericarp (2n)	Fruit wall (2n)	-
Caryopsis	Fruit wall + seed wall	-

Endosperm : In endosperm, genes are present in triplicate (3n). The colour of endosperm may depend on number of dominant or recessive genes present (Table 4). As number of dominant genes increases colour intensity increases. The expression is visible in hybrid seed because the expression is in the cells of endosperm that contain genes from male parent also. This phenomenon is known as xenia effect, which is prominent in maize. Female will always contribute genes in homozygous condition (YY or yy) never in heterozygous condition, as polar nuclei (n+n) are formed by mitotic division in **megaspore**.

Table 4. Xenia effect on maize seed

Colour of endosperm			
Polar nuclei	**Sperm nucleus**	**Endosperm**	**Xenia effect**
YY	Y	YYY	Yellow
YY	y	YYy	Medium yellow
yy	Y	Yyy	Light yellow
yy	y	yyy	White

Where, yellow colour of endosperm is controlled by dominant gene Y, while, white colour of endosperm by recessive allele 'y'.

Embryo: The cells of embryo are formed by mitotic cell division in zygote (2n) containing genes from both male and female parents. Therefore, the colour of embryo of a hybrid seed depends on the presence of dominant gene contributed by female or male parent.

Cotyledon: It is the extension of embryo, therefore, the genetic constitution of each cell of cotyledon is same as of embryo. Thus, hybrid seed or genetic contamination may be verified based on cotyledon colour.

■■■■

Chapter 2

VARIETY

Objective: Explanation and definitions of plant/crop varieties

2.1. AS PER BOTANY

In botanical nomenclature, a variety is a sub-group of a species *e.g., Brassica oleracea* var. *capitata* (cabbage) and *B. oleracea* var. *botrytis* (cauliflower); *B. campestris* var. *dichotoma (Kali sarson), B. juncia* var. *rugosa (Pahadi rai), B. napus* var. *napus (Toria)* etc. Variety is a rank below species, as such, it gets a ternary name. A variety has an appearance distinction from other varieties, but may hybridize freely with other varieties of the same species. Usually varieties are separated geographically from each other.

A taxonomic subdivision of a species consisting of naturally occurring or selectively bred populations or individuals that differ from the remainder of the species in certain minor characteristics is considered as a variety.

2.2. AS PER SEED ACT

Seed Act (1966, sub section 16 of section 2) defined a sub division of a kind identified by its growth, yield, plant, fruit, seed or other characters as a variety.

Identified variety: The strain that has been identified by the workshop as better performer than the existing varieties over the location and year for particular zone or state that can be released for cultivation is termed as identified variety.

Pre-released variety: The identified variety whose proposal has been submitted for release to the State or Central Variety Release Committee.

Released variety: Based on better performance of the newly developed strain in comparison to existing variety, the state government can recommend cultivation of that strain in the state. Such strain after recommendation for cultivation in the state is considered as recommended variety. Central Government may release a variety for cultivation in a particular zone constituted based on agro-climatic zone at national level as per recommendation of Central Variety Release Committee (CVRC). After release the proposal for notification of the variety is submitted. The seed of the variety which is recommended but not notified is not multiplied by seed multiplication chain.

Notified variety: A variety notified under section 5 of the Seeds Act, 1966 (54 of 1966) by Government of India is considered as notified variety.

2.3. AS PER SEED TECHNOLOGY

Seed technology considers a group of plants uniform in their morphological, physiological, biochemical and other characters without any variation from generation to generation which can be differentiated from other groups of plants of the same species by expression of some distinguishing characteristics as variety after its release and notification.

2.4. AS PER PLANT BREEDING

Plant Breeding considers the end product of a crop improvement method as variety after its release and notification. It is of different types as per the behaviour of crop *viz.* variety (self pollinated crops), open pollinated variety, synthetic variety, composite variety (cross pollinated crops) and hybrids (commercial exploitation of heterosis).

Cultivar: Name given to all cultivated variants of plant varieties produced by breeding procedure after its notification is designated as *Cultivar* that distinguish it from otherwise similar plants of the same species and its seed is made available by the system of seed multiplication chain. When propagated it retains those characteristics. The word *cultivar* is a portmanteau coined by Liberty Hyde Bailey from "cultivated" and "variety".

Cultigen: Any genera with only one known species that to under cultivation designated as *Cultigen*. Genera of a cultivated plant, not known to have a wild or uncultivated counterpart such as banana and magic mint because either its wild forms are now extinct, or its species was domesticated (grown and selected by humankind) from so far back in history and has undergone such drastic transformation under prehistoric human selection, that the plant's ancestry is essentially unknown.

Variety: a defined strain of a crop plant, selected on the basis of phenotypic (sometimes genotypic) homogeneity from a population developed by the principles of plant breeding and notified for cultivation.

Hybrid varieties: Notified varieties where seed of the variety is produced by hybridization between two selected parents every year to exploit the phenomenon of heterosis. The F_1 seed of the hybrid varieties is made available every year for cultivation. It is a homogeneous population of heterozygous plants. Cultivation of hybrid varieties is popular in maize, pearl millet, sorghum, rice, pigeon pea, sunflower, sugarcane etc.

Open pollinated variety: An improved heterogeneous population of homozygous and heterozygous plants of a cross-pollinated species viz., maize, sorghum etc., is known as open pollinated variety after its notification. It is a plant population in which at least 70% of the progeny result from the crossing of the parent lines every year. It may be a synthetic or composite variety.

Synthetic variety: A variety which originates by random mating among the known mixture of cross fertilizing genotypes is considered as synthetic variety. Lines used as parents for synthetic varieties are selected based on the results of phenotypic selection alongwith good vigour level measured by general combining ability. The crossed seed is harvested as bulk and replanted in successive generations. After synthesis it is maintained by open pollination in isolation. The parents are considered as SYN_0 while successive generations as SYN_1, SYN_2 and so on.

Composite varieties: A variety that originates by random mating among a number of outstanding strains of open pollinated genotypes without testing of general combining ability is considered as composite variety. Composite varieties are maintained by open pollination in isolation.

Mutant: A variant or variety developed by mutation.

Strain: A homogeneous population of homozygous plants of self pollinated crops developed based on the principles of plant breeding to compare with the existing varieties of the crop under cultivation to promote as variety is termed as strain.

Introduction: Established variety of one environment introduced for cultivation in new environment is considered as introduction.

Genotype: The homozygous variant of any crop species with known hereditary constitution is considered as genotype.

Accession: Serial number given to the individual characterized and documented hereditary material of a crop species available with the plant breeder as germplasm.

Culture: *see* genotype

Germplam: Sum total of the characterized and documented hereditary material of a crop species available with the plant breeder.

Advanced breeding lines: The line introduced for evaluation in Initial or Advance Varietal Trail in the All India Coordinated Research Project of the Crop.

Genetically modified variety (GM): A crop variety developed by application of biotechnology with the insertion of specific gene from other genus by genetic manipulation. e.g., Cry1 AC gene from a soil bacteria *Bacillus thirugenensis* is incorporated in the cotton to make it resistant against the attack of boll worm. These varieties are known as Bt. Cotton. Permission for cultivation of GM varieties for commercial cultivation in India is given by GEAC, MOEF.

2.5. AS PER PPV&FR

PPV&FR Act (2001) and submitted proposal of New Seed Bill (2004) consider variety as a plant grouping except microorganism within a single botanical taxon of the lowest known rank, which can be

(i) Defined by the expression of the characteristics resulting from a given genotype of that plant grouping;

(ii) Distinguished from any other plant grouping by expression of at least one of the said characteristics; and

(iii) Considered as a unit with regard to its suitability for being propagated, which remains unchanged after such propagation, and includes propagating material of such variety, extant variety, transgenic variety, farmers' variety and essentially derived variety.

Essentially derived variety: a variety predominantly derived from any protected or non-protected variety that confirms to the initial variety in all aspects except for the differences which result from the act of derivation, and yet is clearly distinguishable from such initial variety.

Example, susceptible extant variety converted to resistant one by back crossing without any change in the phenological traits, however, some variation in the expression of distinguishing morphological trait(s) for differentiation from the original variety is considered as essentially derive variety.

A variety is designated as essentially derived in respect to the initial variety when it is predominantly derived from such initial variety, or from a variety that itself is predominantly derived from such initial variety, while retaining the expression of the essential characteristics that result from the genotype or combination of genotypes of such initial variety. It is clearly distinguishable from such initial variety. It confirms to such initial variety in the expression of the essential characteristics that result from the genotype or combination of genotype of such initial variety.

Extant variety : a notified variety under section 5 of the Seeds Act, 1966 (54 of 1966); or a farmers' variety; or a variety about which there is common knowledge; or any other variety which is in the public domain is considered as Extant variety.

Variety under cultivation in the public domain as of common knowledge is considered as extant variety. All the notified and Farmers' varieties are extant varieties. The candidate variety required for protection is compared with the extant varieties to check the repetition.

Farmers' variety: a variety which- (i) has been traditionally cultivated and evolved by the farmers in their fields; or (ii) is a wild relative or land race of a variety about which the farmers possess the common knowledge; where farmer means any person who (i) cultivates crops by cultivating the land himself; or (ii) cultivates crop by directly supervising the cultivation of land through any other person; or (iii) conserves and preserves, solely or jointly, with any person (iv) any wild species or traditional varieties or adds value to such wild species or traditional varieties through selection and identification of their useful properties.

A variety developed by any person involved in cultivation of crops mainly through selection and has been traditionally cultivated by the farmers as variety of common knowledge is considered as farmers' variety.

Those plant varieties that are homogeneous, traditionally cultivated by farmers, selected by farmers in their own field and are an improvement over the wild relatives and/or landraces, and outlined the possible means to ascertain distinctiveness and uniformity of such varieties (Nagarajan *et al.,* 2008) are considered as farmers' variety.

Farmers' variety means a variety, which has been traditionally cultivated and evolved by the farmers' in their fields. It may be a wild relative or land race of a variety about which farmer possesses the common knowledge.

Landrace: It refers to variety/species domesticated to the natural and cultural environment in which they originated with minimum scientific interference (modern breeding practices) of human beings.

The Biological Diversity Act, 2002 explains the landrace as a primitive cultivar that was grown by ancient farmers and their successors.

Primitive or antique varieties usually associated with traditional agriculture are considered as land race, which is often highly adapted to the environment in which they live or originated. Land races are often evolved naturally with minimal assistance/interference of human (or from human with traditional rather than modern breeding methods). The earmark of a landrace is its adaptation to the environment that results in stable phenotypic expression over time. Land races developed by adaptation to the demands of the particular physical environment. These phenotypic traits are maintained naturally so long as environment is maintained, but are degraded as soon as transplanted to an exotic environment. Very tight control by selective breeding is needed to conserve the ancestral traits that made them so attractive in the first phase.

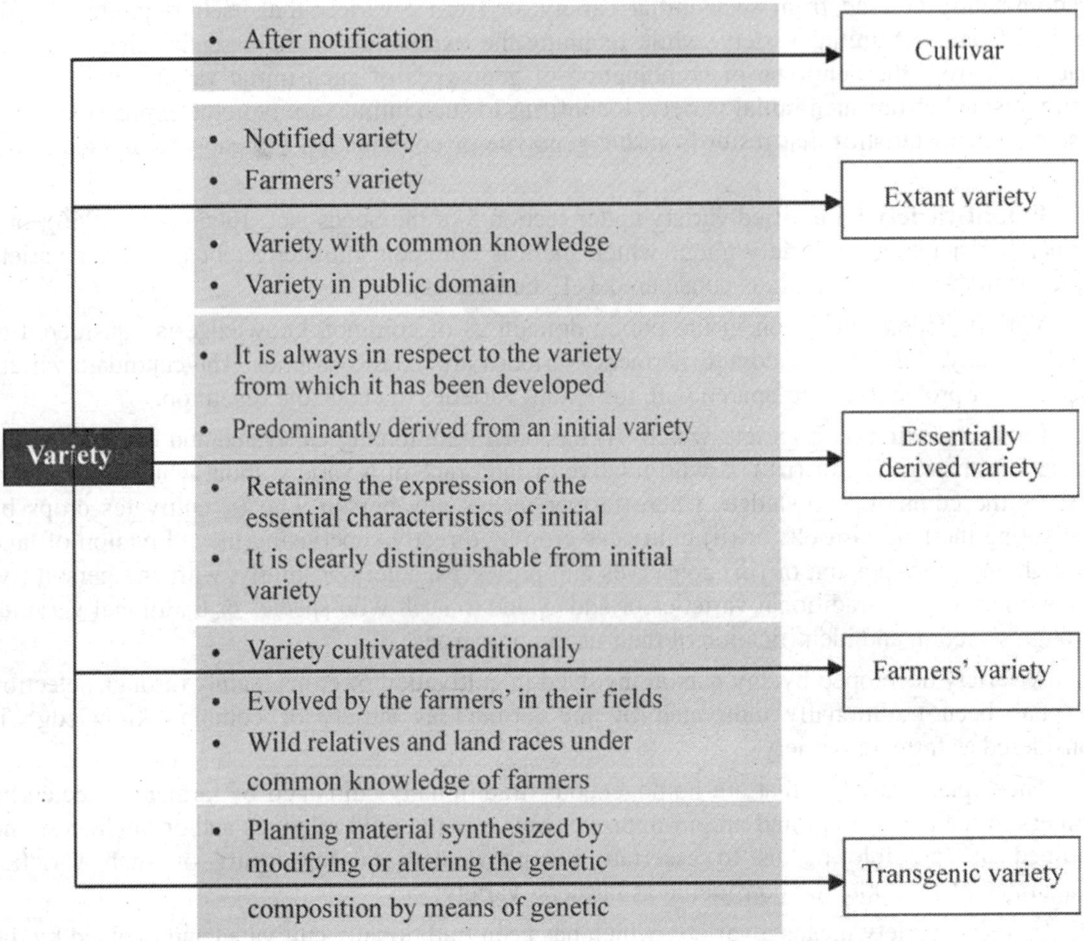

Fig. 1. Classification of varieties based on PPV&FR act

Wild type: The typical form of an organism, strain, gene, or characteristic as it occurs in nature, as distinguished from mutant forms that may result from selective breeding is known as wild type. It is the most common phenotype in the natural population.

Protected variety: A candidate variety passed DUS testing is considered as protected for 15 years after notification. Seed production by any other organization for commerce/business without prior permission is considered as offence.

Variety of common knowledge: A variety is considered as variety of common knowledge when

- The variety is entered in official list/register of varieties in any country granting protection to plant variety.

- Inclusion of the variety in a recognized, publicly accessible collection, including a commercial or botanical reference collection or an accession in a National/International Gene Bank.
- Adequate description of the variety in a publication that may be considered a part of the public technical knowledge (including books, monographs, scientific/ popular journals, theses, technical bulletins and annual reports).
- Variety is freely cultivated by the farmers in a state/region/country.
- Propagating material or harvested material of the variety under authorized marketing or in circulation or trade.
- Shared traditional knowledge of relevant community about the variety (including a landrace or local variety or farmer's variety) that can be substantiated, if needed.

Variety in public domain: The knowledge of a variety that is freely available, commonly shared throughout the world without any access restrictions refers as variety in public domain. It is not restricted by geographical boundaries. Under Protection of Plant Variety Act, a variety of common knowledge can not be protected and will remain freely available for cultivation. On the contrary, a candidate variety should be clearly distinguishable by one or more characteristics from any variety whose existence is a matter of common knowledge at the date of the application for protection. After completion of the period of protection, a protected variety becomes a variety of common knowledge.

Transgenic variety: Seed of planting material synthesized or developed by modifying or altering the genetic composition by means of genetic engineering.

■■■■

Chapter 3

IDENTIFICATION, RELEASE AND NOTIFICATION OF VARIETY

Objective: To know the procedure for Identification, Release and Notification of crop variety

3.1. IDENTIFICATION

In public sector (private sector can also enter) the improved genotypes developed by plant breeders are released after a thorough assessment under multi-disciplinary approach of their Value for Cultivation and Use (VCU test) and comparison with other developed genotypes alongwith at least three checks (national/zonal/local) under All India Coordinated Crop Improvement Project (AICCIP) of respective crop at national level in three tier system (two under special consideration) at multi-location. The entries with high degree of phenotypic uniformity and genotypic stability are included based on superior performance for yield/other traits irrespective of the method followed for its development.

Initial Variety Trial (IVT): Testing of the strain under All India Coordinated Testing in the first year is designated as IVT. Sponsoring agencies supply the pure quality seed. The seed from same lot is distributed to all the testing centers after coding. In each center, plot size and number of replications are kept uniform and the performance is tested under best agronomic management with recommended package of practice. The required data on yield, days to flowering and maturity, plant height, seed characteristic, reaction to biotic and abiotic stresses etc. generated by the testing centres are processed by coordinated unit. Based on the performance, the entries are promoted to Advance Varietal Trial–I.

Advance Variety Trial I (AVT-I): It is constituted for each recognized agro-ecological zone by the entries promoted from IVT for second year testing. In AVT-I the number of entries are restricted to 16 with three checks and each testing centers (not less than 4) well distributed in the particular agro-ecological zone. The entries are grown with same specification and crop husbandry as IVT except, larger plot size. The data generated are processed to promote best entries to Advance Variety Trial II.

Advance Variety Trial II (AVT-II): It is repetition of AVT I to generate some additional data and test stability over the years. Under AVT II the seed technologist characterizes the entries. The data are processed and on the basis of performance of a variety under test in a particular region/zone, the tested variety is **identified** by the annual workshop of that crop.

3.2. RELEASE

The object and scope of the Seed Act is to make quality seed of the varieties in demand available for cultivation by regulating the production, quality and marketing. The Seed Act controls only the seed of notified varieties, i.e., where the determination has been made to control their quality. It requires following information

Uniform Regional Trial (URT) After identification of the variety, testing is organized at a larger number of locations, involving research stations, testing centers and regional stations of the universities/institutes and state department of agriculture located in the claimed agro-climatic zone. The URT is laid out in a plot of larger size with the plot of existing variety under identical crop husbandry.

The data generated at different locations are statistically analyzed to promote the best entry for Adoptive Research Trial.

Adoptive Research Trial: It is conducted on research stations of university and farms of State Governments to test the performance in large fields in the given environment. Based on the performance the variety is promoted for next trial. For release of a variety the breeder is required to supply ten quintals of seed (sufficient to grow ten hectares) to the State Farm Corporation of India for multiplication as test stock seed, which is used for conducting the adoptive trials at the state farm and mini-kit demonstrations at the cultivators fields.

Soon after identification of a variety a release proposal to the State or Central Government is submitted by the institute/organization, which evolved the variety. On the basis of the performance in extensive multi-location testing and requirement, State Variety Release Committee or Central Sub committee on Crop standards, Notification and Release of Varieties, Division of Seed, Ministry of Agriculture, Government of India releases the variety for the state or zone, respectively. A State is free to release any variety (transgenic?) of any crops in the State. A strain tested under AVTII but not identified by the workshop may be released by a state for cultivation after multi-location adaptive trials as per requirement.

The purpose of Release of cultivar is to introduce the newly evolved varieties to the farmers for cultivation and this enables the farmers to choose cultivar for cultivation based on the performance. Based on the performance, farmers show desire for the seed of that variety, so in other words, release of a cultivar is the recommendation of the farmers for its adoption. Therefore, notification of a variety is linked to the release of a variety, though the release process itself does not have statutory cover.

The practice of official release of cultivars started in October, 1964 with the formation of the *Central Variety Release Committee (CVRC)* at the Central level and *State Variety Release Committee* at State level. The Central Variety Release Committee functioned up to 1969, then after its functions were taken over by the Central Seed Committee (CSC) established under the Seeds Act, 1966. The Central Seed Committee constituted a Central Sub-Committee on Crop Standards, Notification & Release of Varieties for Agricultural and Horticultural Crops to discharge the functions of release/notification, provisional notification and de-notification of cultivars at Central level, while State Seed Sub-Committee (SSSC) are to discharge similar functions for release at State level.

The cultivar's performance only in one State is treated as 'State variety', which is considered by the State Seed Sub-committee of that particular State for release. The sponsoring authority submits the proposal for release in the prescribed proforma to the State Seed Sub-committee for consideration and release of cultivar.

3.3. NOTIFICATION

After official release (at State / Central level), the cultivars are notified under the Seed Act, so that the quality of seed can be regulated. The main purpose of notification is to bring the seed of a particular crop/variety under the purview of Seed Law Enforcement, mainly to empower the Seed Inspectors to verify the quality by sampling and analysis.

After release, the proposal of the variety is submitted to Central Sub Committee on Crop standards, Notification and Release of Varieties, Division of Seed, Ministry of Agriculture, Government of India for notification. In the event, the variety is notified under section 5 by the Central Government in consultation with the Central Seed Committee for 10-15 years under the chairmanship of Deputy Director General (Crops) ICAR.

The seed of the notified variety is multiplied following seed multiplication chain. Availability of Breeder seed of the notified variety is the responsibility of the Ministry of Agriculture, Government of India. It is fulfiled by National Seed Project (Crops) of ICAR under the supervision of Directorate of Seed Research Maunath Bhanjan UP.

The Release is not a statutory function. Its main purpose is to make known the details of the newly evolved cultivar to the public and also the areas for which it is found suitable for cultivation. The Notification is a statutory function performed under the Seed Act so that the provisions of the Seed Act could be applied to regulate the quality of seed during sale.

Under the Seed Act, Certified seed can be produced only of notified varieties and thus, notification precedes the certification. Therefore, notification is compulsory for production of Certified category seed. Unless the variety is notified, the seed cannot be certified. The advantages of notification are

- Seed Law Enforcement agency notified under the Seed Act can draw and test seed samples of notified varieties. Therefore, for regulation of the quality of seed, notification is a prerequisite.
- The farmers benefit from the notification of varieties because of its certified quality. As seeds of notified varieties can only be sold after proper labeling and packing indicating the minimum specified standards.
- Once the variety is notified, in general, it becomes asset of Government of India. The morphological characters of notified varieties are documented by the Central Seed Committee.
- The notification of the varieties helps to trace out its genesis.
- Subsidy is considered based on the notification status.
- Seed planning/programmes are being undertaken based on the notification statistical data.

The notification is made by the Central Government on the recommendation of the Central Seed Committee. The proposals for notification of a State-important variety are forwarded in the prescribed format by the State government after its release in that State to the Central Seed Committee for consideration.

3.4. DE-NOTIFICATION

Instructions for de-notification are issued by Ministry of Agriculture, GOI upon completion of the validity period of a variety and/or when, the variety is not under cultivation.

3.5. RE-NOTIFICATION

Instructions for re-notification may be issued by GOI for the variety that remains in demand for cultivation even after completion of the validity period.

Chapter 4

PLANTING VALUE OF SEED

Objective: To achieve optimum population of healthy plants of the selected variety adopting normal recommended seed rate.

Genetic potential of a variety, optimum population of healthy plants and environmental conditions govern the seed yield from a unit area. It all depends upon quality of seed. Quality is defined literally as fitness for the purpose and degree of excellence. Seed should be fit for sowing purpose with higher degree of excellence. Seed determines all the three important factors responsible for determination of yield from a unit area *i.e.,* yielding ability of variety, plant population and ability of soil. Therefore, a seed that has the quality to determine yielding ability, plant population and soil health is considered as quality seed (Fig. 1). This is why the saying ***you reap as you sow*** became well established. Quality seed is the basic input of agriculture and all other inputs are contingent upon it for being optimally effective. Availability of quality planting material is the fundamental right of the farmers. The quality of seed is designated planting values of seed.

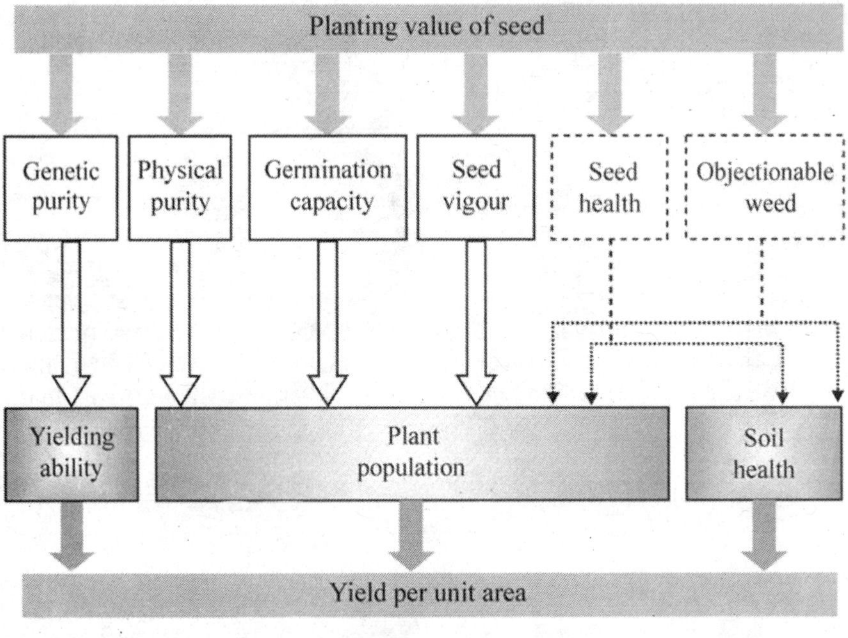

Fig. 1. Seed characteristics determining planting value of seed

The seed use for sowing purpose is considered fit in the event it fulfils following criteria for sowing purpose (Fig. 1).

Genetic purity: Crop variety is an important component of agriculture; exploitation of change in genetic constitution in a plant is governed by variety. Genetic exploitation develops variety for specific condition to exploit inputs, therefore, grower select a variety as per his requirement. The seed is the vehicle of genes of a variety and produces a plant of required characters. Therefore all the seeds used for sowing purpose must have similar genetic constitution for uniform performance and quality of produce. If all the seeds are of similar genetic constitution of the required variety then it is considered as pure at genetic level. All the plants of such seed behave similarly that means synchronous flowering and maturity, same height and response to biotic and abiotic stresses. Action to solve any problem will have common procedure, action plan and time, with same level of correction in all the plants. Apart from quantity, quality of produce is another important aspect. Plants with similar genetic constitution produce the same quality that fetch higher price in market. Therefore, to exploit genetic ability of variety for high quality and quantity of the produce, the quality seed should be purchased from a reliable source and seed of self and cross pollinated crops should be changed after every third year and hybrid every year.

Physical purity: Usually the seed should be 98% pure at physical level. The margin of two percent may be anything other than seed of the crop selected for sowing purpose. In the event the physical purity is less than the upper limit, then the part of impurity will not form plant of that crop it will directly reduce the plant population, may affect working of seed drill and act as source of pathogen and insect pest. Therefore, while selected for sowing purpose seed should be free from inert matter and seeds of other species to maintain optimum plant population and storage of seed safely.

Germination: Plant population is an important key factor to determine seed yield from a unit area. With the use of recommended seed rate one should get optimum plant population. Both low and high plant population affects per unit produce. The seed that has no germinability becomes source for growth of pathogen and insect that may also affect seed with germinability or establishment of seedlings.

Seed vigour: High vigour of seed offers the in establishment of vigorous seedling. Seed vigour helps in establishment of seedling even under abiotic stress. Therefore, use of vigorous seed is a prerequisite to establish optimum plant population even under poor management conditions.

Seed health: Seed free from seed borne pathogens and insects are essential to maintain optimum plant population (ergot of sorghum and pearl millet, ear cockle of wheat etc.), yield/plant (downy mildew of sunflower and pearl millet), further transmission of disease (loose smut of wheat) with the use of recommended seed rate. Continuous use of diseased seed may make soil sick for the disease (Karnal bunt of wheat, downy mildew of sunflower). The quality of produce is affected negatively by the disease for consumption purpose like Karnal bunt of wheat, *Aspergillus* of groundnut etc. Presence of insect in seed deteriorates longevity, shelf life, germinability and vigour of seed.

Weed: Seed lot should not be contaminated with weed seeds, as it is difficult and costly to remove them from the field. Identification of objectionable weed for removal and use of herbicide to control them in associated crop is a challenge. Weeding enhances the cost of cultivation. Many weeds once enter in the field may make soil sick in such a way that it cannot be eradicated e.g., *Convolvulus arvensis* in wheat. Therefore, seed use for sowing purpose should be free from weeds with special reference to objectionable category.

■■■■

Chapter 5

SEED PRODUCTION AND CERTIFICATION SYSTEM IN INDIA

System of Seed Production

In India, seed production of Foundation and Certified class is the responsibility of National or State Seed Corporation. Whereas, maintenance breeding and Breeder seed production are of Agricultural Universities or ICAR institutes under the supervision of plant breeder (Fig. 1).

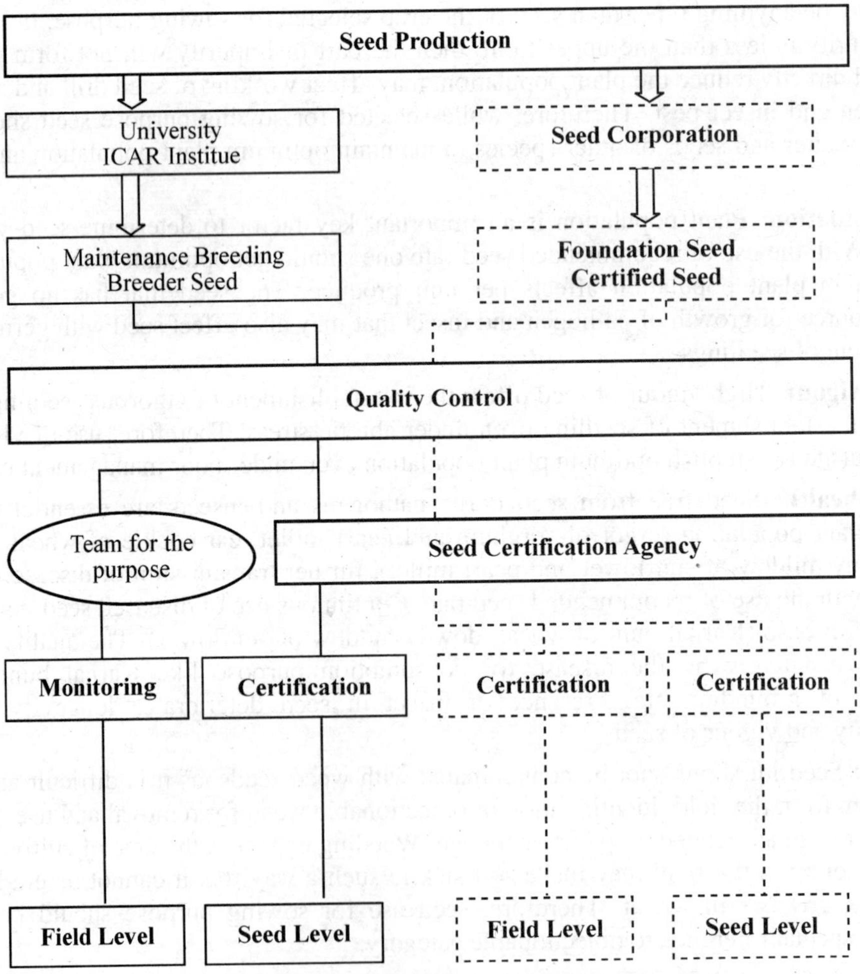

Fig. 1. Seed production and quality control system in India

Seed multiplication chain

Objective: Multiplication of quality seed under vigilant supervision of plant breeder or seed certification agency to distribute seed of notified varieties for sowing purpose.

Improved varieties are developed by plant breeder to fulfil specific objective. The Certified seed of these varieties are made available to the farmers year after year as per demand. The seed of the variety in demand is multiplied with the aim to maintain its genetic constitution as such. For multiplication of seed without altering genetic constitution of the variety, Government of India adopted a three-tier model (Fig. 2) i.e., Breeder, Foundation and Certified seed. It involves State Agricultural Universities/ ICAR Institutes, State/National Seed Corporation and Seed Certification Agencies.

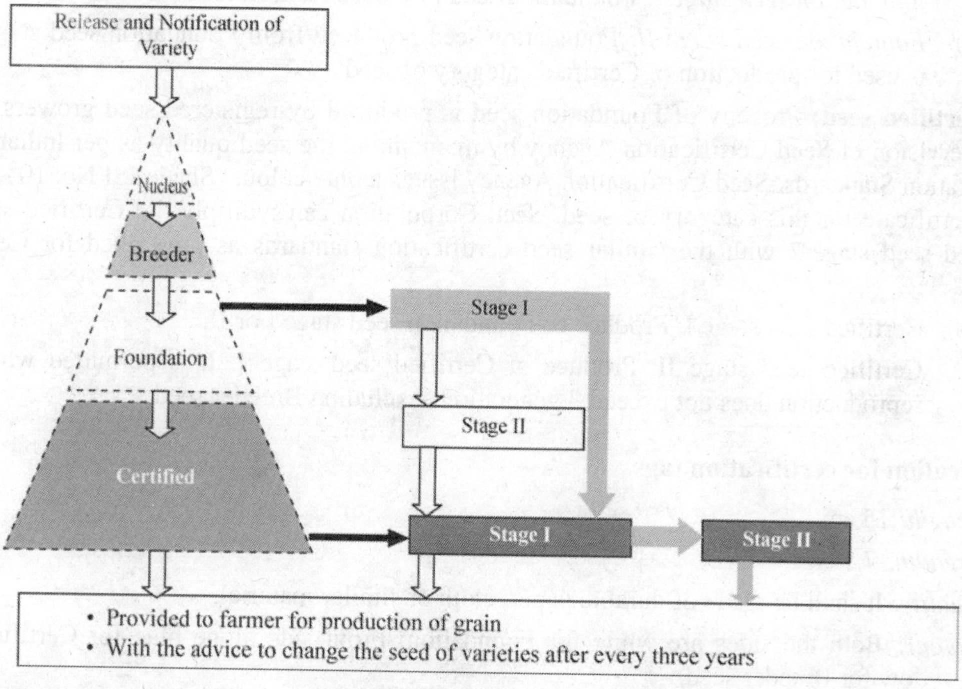

Fig. 2. Seed multiplication chain

Nucleus seed

(a) **Basic Nucleus seed**: This is the cent percent pure seed at genetic and physical levels produced by the plant breeder who evolved the variety. This seed is maintained by the institute that developed the variety. The seed is made available to different agencies on demand to start seed multiplication chain.

(b) **Nucleus seed for seed multiplication chain**: This is the cent percent pure seed at genetic and physical levels produced by a plant breeder from basic nucleus seed stock or produce of the progenies of the plants selected from Breeder seed production plot to start the seed multiplication chain in the university where Breeder seed has to be produced year after year. This seed is not certified by any agency.

Breeder seed: This is the progeny of the Nucleus seed multiplied in large area under the supervision of plant breeder and monitored by a committee. It provides cent percent physical and genetic pure seed for production of Foundation class. Golden yellow (No. 356; IS: 5-1978) coloured certificate is issued for this category by the producing agency.

Foundation seed: Progeny of Breeder seed is handled by recognized seed producing agencies in public and private sector under the supervision of Seed Certification Agency in such a way that its quality is maintained according to the prescribed seed standard. Seed Certification Agency issues a white colour certificate for Foundation class seed. Seed Corporation purchases Foundation seed from seed growers. Foundation seed can again be multiplied in the event of its shortage with similar seed certification standard.

(i) *Foundation seed stage I:* Foundation seed produced from Breeder seed.

(ii) *Foundation seed stage II:* Foundation seed produced from Foundation seed stage I. It is used for production of Certified category of seed.

Certified seed: Progeny of Foundation seed is produced by registered seed growers under the supervision of Seed Certification Agency by maintaining the seed quality as per Indian Seed Certification Standards. Seed Certification Agency issues a blue colour (Shade ISI No. 104, azure blue) certificate for this category of seed. Seed Corporation can multiply the Certified seed as Certified seed stage-2 with the similar seed certification standards as prescribed for Certified seed.

(i) Certified seed stage I: Produce of Foundation seed stage I or II.

(ii) Certified seed stage II: Produce of Certified seed stage I. It is permitted when the reproduction does not exceed 3 generations excluding Breeder seed.

Specification for certification tags

Length: 15 cm.

Breadth: 7.5 cm.

Quality: It shall be made of durable paper, cloth or similar material.

Colour: Both the sides are white for Foundation; Front side alone blue for Certified and golden yellow for Breeder seed.

Contents & layout: The contents shall be as shown:

Preserve this tag for establishing the use of seed to which if relates	Crop : **Serial No.** Variety : Lot No.: Certification No. & Date: Date of test: Producer's Name: Village: District: State: *SCA's emblem* Certified by: (SCA's full address) **Class of seed**	Certification valid up to 8/4 months from the date of issue of certification provided the seed is stored under ideal condition. Use of seed after expiry of the validity period by anyone is entirely at his risk and the SCA shall not be responsible to him for any damage to the seed. No one should accept the seed if the certification on tag or seal is absent or has been tempered with. This tag signifies that seed in the container to which it is attached has been found to confirm to the standards prescribed for certification under the Seed Act, 1966. However, upon the holder of the seed rests the responsibility for delivering seed conforming to the minimum standards for certification.

Specification of labels (15x7.5cm of white colour)

Preserve this label for establishing source of seed	Crop: Lot No.: Variety: Date of test. Genetic purity. 100% Pure seed: Other crop seed: Weed seed: Inert matter: Germination: Moisture % (w/w): Net content: Produced by: **BREEDER SEED**	The seed in this container has been produced in accordance with the production procedure prescribed by ICAR

Poison treated/emblem of certification agency (Known pedigree)	Grown-Inspected-Harvested-Processed-Tested-Packed-Marketed to give assured performance	Class: Kind: Variety: Lot Number: Date of test: Pure seed (minimum): Inert matter (maximum): Other crop seeds (maximum) Total weed seeds (maximum): Germination (minimum): Moisture (maximum): Net Content (when packed): Name and address of Seed Producing Agency **FOUNDATION/CERTIFIED SEED**

Agencies involved in seed multiplication chain: Three autonomous bodies *viz.,* State Agricultural University/ICAR Institutes, Seed Corporation and Seed Certification Agencies are involved in seed multiplication chain apart from ICAR and State Department of Agriculture (Table 1 and Fig. 3).

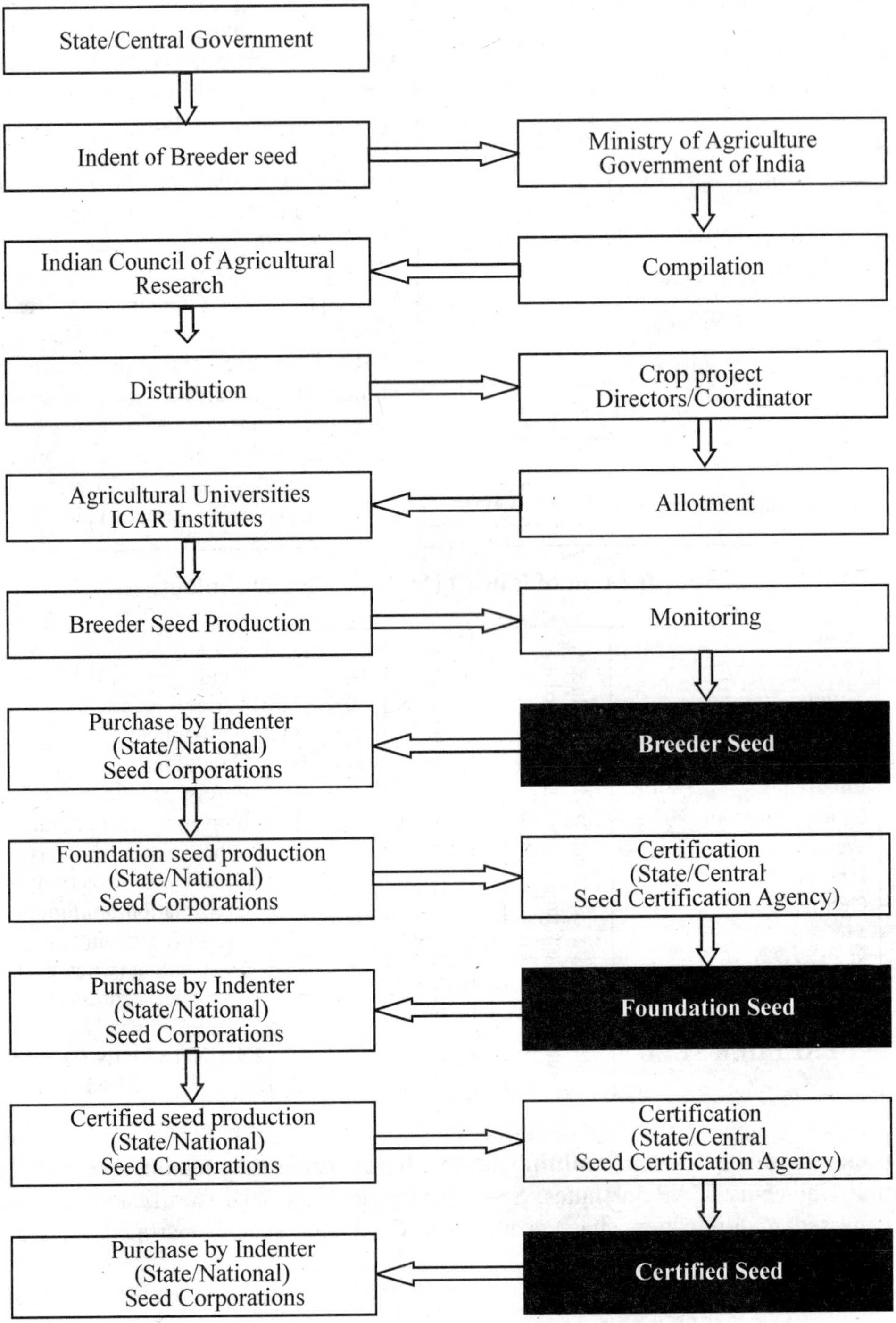

Fig. 3. Role of agencies involved in seed multiplication chain

Table 1. Role of different agencies in seed multiplication

Agency	Role
State Agricultural Universities/ ICAR Institutes	Production
	Nucleus seed (maintenance breeding)
	Breeder seed
	Monitoring
	Nucleus seed
	Breeder seed production programme
National Seed Corporation/ State Seed Corporation	Production
	Foundation seed
	Certified seed
	Procurement
	Breeder seed
	Foundation seed
	Certified seed
Seed Certification Agency	Monitoring
	Breeder seed production programme
	Certification at field and seed levels
	Foundation seed
	Certified seed

Seed certification system

The seed of Foundation and Certified categories produced by respective agencies are tested and certified for sowing seed quality at field and seed levels (Fig. 4). The producers are not recognized as quality control agency by the Government of India. The responsibility of quality control is with seed certification agency an autonomous body. In place of certification, Breeder seed production programme is monitored by a team constituted for the purpose whereas, seed is tested in the notified laboratory for sowing seed quality. No seed certification standard has been fixed at field and seed levels for Breeder seed category.

Other categories of seed

Truthfully labeled seed (TL): When a seed is sold based on the result of the laboratory established by the producer then the seed is considered as TL seed e.g., seed produced and sold by many private agencies. In government sector, many times seed is not able to fulfil seed certification standard for any one parameter, marginally. This seed may be sold by making correction (increase the quantity/ bag if germination percentage is low) after getting certification by the authorized Seed Certification Agency. The price of seed is always lower than the Certified seed in government sector. Seed rejected due to genetic impurity or presence of objectionable disease, pest or weed is not labeled as truthful. Many times some government organizations sell the seed of high standard as TL seed with their own trade name for seed to make variety popular.

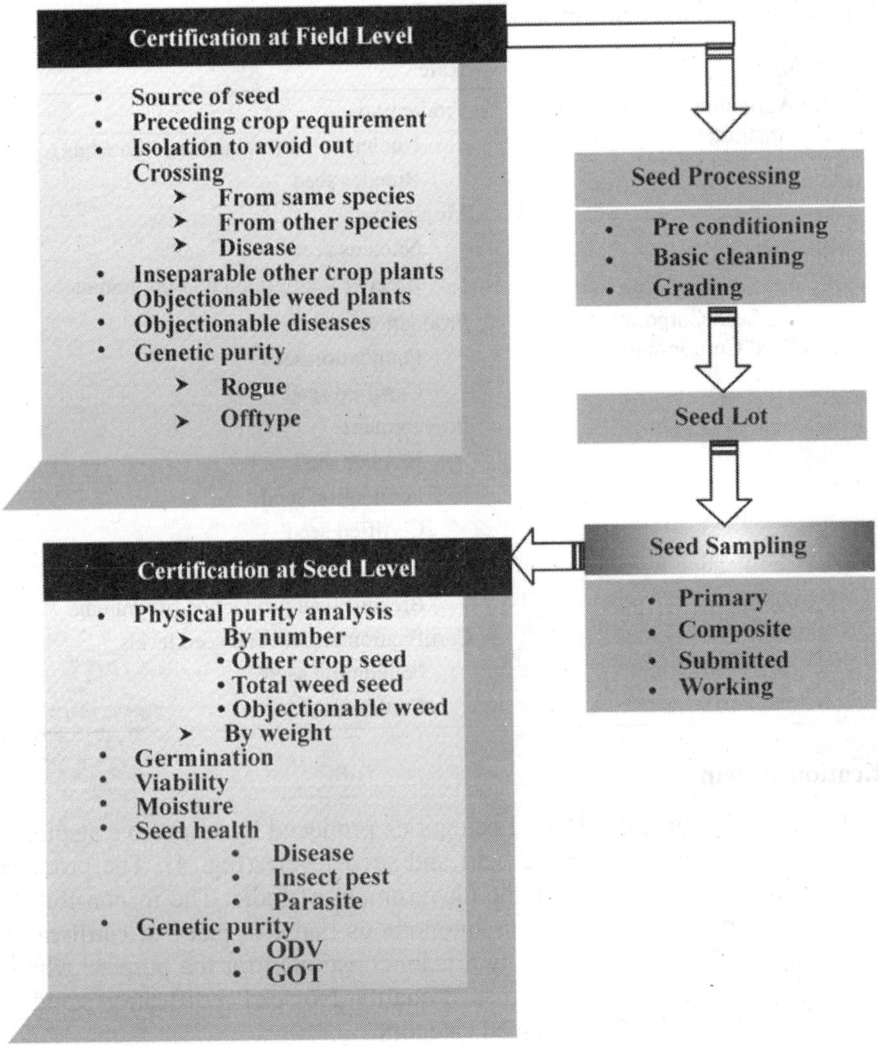

Fig. 4. Seed certification system in India

Registered seed: In United States mainly for autogamous crops the generation between Foundation and Certified seed is considered as Registered seed, which is not a commercial class. Sowing of Foundation seed produces registered seed labeled by purple colour tag. It is not a recognized class in seed multiplication chain in India.

Chapter 6

MAINTENANCE BREEDING

Objective: Maintenance of genetic purity of a variety

Principle

Maintenance breeding refers to the process to maintain uniform expression of all the distinguishable traits of a variety year after year without any genetic drift. It requires negative selection i.e., the plant progeny which deviates in expression has to be rejected but there is no positive selection for performance.

Maintenance breeding is adopted to maintain genetic purity, quality and health of the variety. It produces true to the type, healthy and quality seed sowing material to produce Breeder seed for exploitation of the variety to enhance production and extend the life of the variety.

Theoretically, no emphasis is made on the traits determining the value for cultivation and use of the variety *viz.,* seed yield, reaction against stresses, quality attributes etc., during maintenance breeding. It maintains a population in an organized and systematic way to retain the relationship of small quantity of Nucleus seed with that originally selected by the concerned plant breeder. Hence maintenance breeding has to be under the control of experienced plant breeder and technical staff who know the genetic basis of phenotypic variation.

Seed of a notified variety has to be made available to the farmers for cultivation. The Institute which developed the variety has limited quantity of genetic pure seed of that variety. The initial seed available with the institute which developed the variety is considered as Basic NUCLEUS SEED.

Now maintenance breeding refers to continuous improvement in a variety by overcoming biotic and abiotic stresses, improving grain quality or eliminating other defects that may constrain production and grain marketability and maintaining yielding ability (Richard, 1996). Maintenance breeding plays a critical role in improving adaptation of newly developed varieties to environmental changes that have a negative impact on older varieties.

With limited resources in seed production unit for area, threshing floor processing plant and technical person not more than 5 varieties of one crop should be maintained at one location. As per Harrington's rule (1952) the unit with all the required facilities may maintain 15 varieties of one crop with following procedure

- Sampling of plant of a variety to obtain Nucleus seed.
- Harvesting and threshing of individual plant and table examination of seed from sampled plant.

- Sowing for multiplication as progeny row in the clan field with cropping history to avoid volunteer plants at the location of adaptation.
- Inspection of progeny row for genetic purity and removal of substandard rows before flowering.
- Harvesting and threshing of progeny rows as NUCLEUS SEED.

Basic nucleus seed: The seed of the notified variety collected from the institute developed it with all the passport information is considered as basic nucleus seed.

This seed is sown in prepared healthy field with, more row to row and plant to plant distance than the normal recommendation at required isolation distance with no chance of volunteer plants. Prescribed dose of fertilizer, irrigation and required plant protection measures are provided.

Off-types and diseased plants are rouged out at seedling, before flowering, flowering and at physiological maturity stage. Plants are marked with contrast colour thread and observed regularly. Marked plants are harvested separately and produce of individual plant is stored in separate packet. The expression of distinct traits, physical soundness of seed and its health are observed at seed level. All the seed of a packet with any deviation in expression of distinct trait, physical appearance and seed health are rejected. The seed of remaining packets are bulked to use as Nucleus seed next year. Produce of unmarked plants are used as Nucleus seed for production of Breeder seed.

Maintenance breeding for the varieties of self pollinated crops

- **Positive selection:** Large number of true to the type single plants (>500)/ have to be tagged in a large population representing Breeder seed production plot to avoid genetic drift.
- In place of single plant single inflorescence is selected to avoid chance of contamination in wheat and rice, as all the seed of one ear will be from one plant but it is difficult to confirm that two ears are from the same plant.
- Individual tagged plant has to be visited several times to observe the expression of diagnostic traits and health.
- Tagged plant with any deviation in expression of distinguishing traits or health is rejected.
- Seed of single plant with the claimed expression of distinguishing traits is harvested separately at harvestable maturity.
- Harvested plant is threshed by hand to avoid mechanical damage to seed.
- Produce of one plant is stored in one packet. In this way number of packets will be equal to the number of plants.
- Seed of individual plant is examined for expression of distinguishing traits of the variety at seed level, seed health and number of seed in the laboratory.
- The whole produce of single plant with any variation in the expression of one or more than one seed from the claimed expression is rejected.

- The whole produce of single plant with seed infested by objectionable seed borne disease is rejected.
- The whole produce of single plant with the number of seed less than the required number for designated row length is rejected.
- In the next season seed of one packet is sown in one row of 5m length in most of the crops and 3m in wheat and rice.

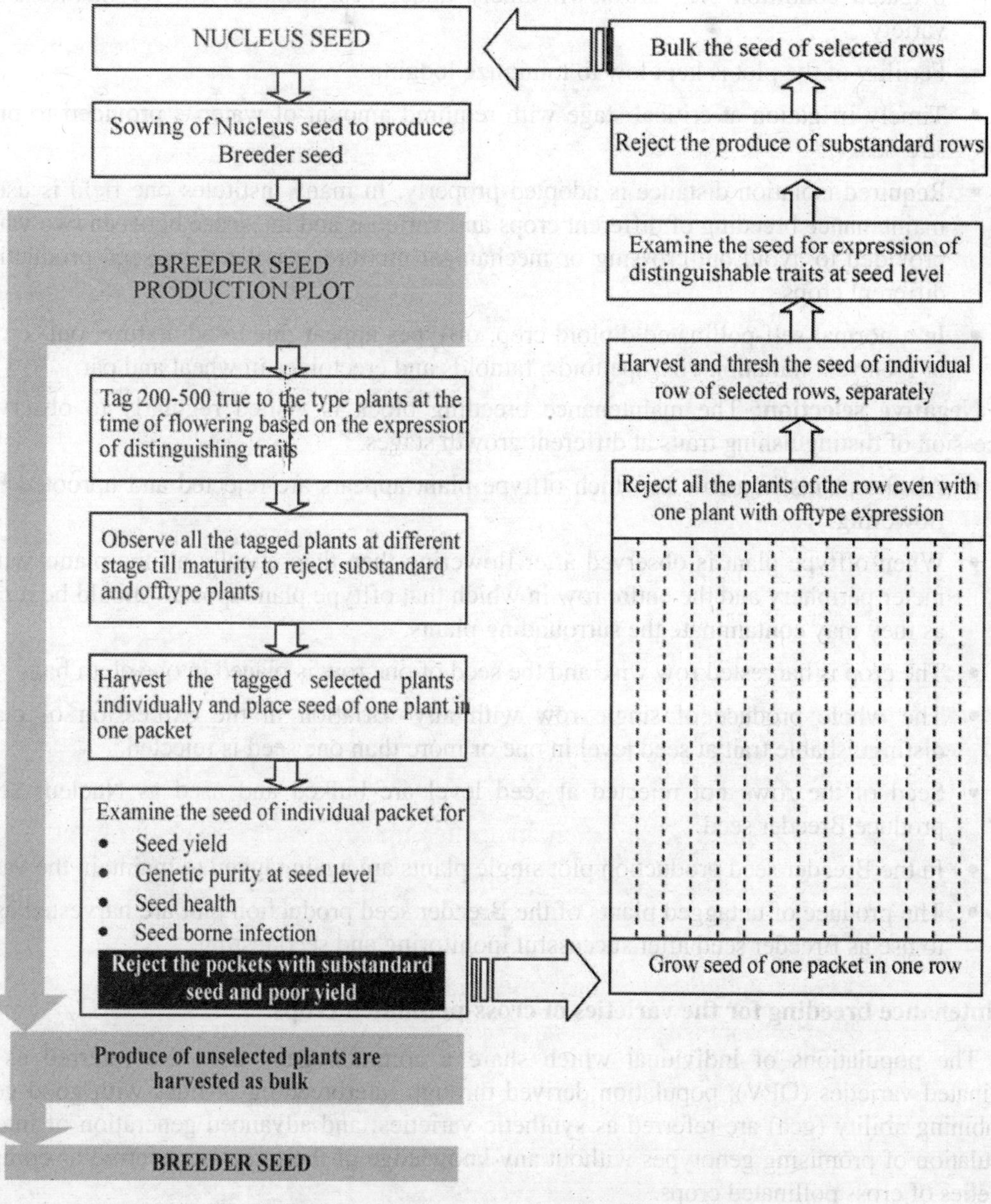

Fig. 1. Maintenance breeding of self-pollinated crops

- The seed of one packet is planted in one row as plant to progeny row with wide row to row and plant to plant distance to examine individual plant and enhance production of individual plant.
- Number of seed more than the required number for one row in one packet should not be used in second row for sowing purpose.
- All the varieties of one crop recommended for early, timely or late sowing; rainfed or irrigated condition etc., are sown timely under best management for maintenance of variety.
- Fertility of the plot is kept low to minimize lodging.
- Timely irrigation at critical stage with required amount of water is provided to prevent stresses.
- Required isolation distance is adopted properly. In many institutes one field is used for maintenance breeding of different crops and varieties and the space between two varieties provided to avoid out-crossing or mechanical mixture is utilized by seed production of different crops.
- In a normal self-pollinated diploid crop, offtypes appear due to admixture, out- crossing or recurrent mutation viz., speltoids, fatuoids and erectoides in wheat and oat.

Negative selection: The maintenance breeding block is visited regularly to observe the expression of distinguishing traits at different growth stages.

- All the plants in a row in which offtype plant appears are rejected and uprooted before flowering.
- When offtype plant is observed after flowering then theoretically all the plants within 3 meter periphery and the entire row in which that offtype plant appears should be removed as they may contaminate the surrounding plants.
- The crop is harvested row wise and the seed of one row is placed in one cloth bag.
- The whole produce of single row with any variation in the expression of claimed distinguishable trait at seed level in one or more than one seed is rejected.
- Seed of the rows not rejected at seed level are bulked and used as Nucleus seed to produce Breeder seed.
- In the Breeder seed production plot single plants are again tagged to maintain the variety.
- The produce of untagged plants of the Breeder seed production plot are harvested as bulk to use as Breeder seed after successful monitoring and seed testing.

Maintenance breeding for the varieties of cross-pollinated crops

The populations of individual which share a common gene pool are referred as open pollinated varieties (OPV); population derived through interbreeding of lines with good general combining ability (gca) are referred as synthetic varieties; and advanced generation of interbred population of promising genotypes without any knowledge of their gca are referred as composite varieties of cross pollinated crops.

Development and certification of OPV is comparatively easy but its maintenance is complicated. Description of distinguishing traits of OPV should always be in range to help the breeder to maintain it comfortably, certification officials to certify it reasonably and producer to enhance the life of variety.

Requirement of experienced plant breeder: For maintenance of varieties of cross pollinated crops, experienced plant breeder is required because

- These populations cannot be uniform as varieties of self-pollinated crops or inbred. The population of these varieties are continuously variable therefore removal of offtype *i.e.,* plant showing non significant variation in plant height, days to flowering, days to maturity etc., from such population is not of any use. But, utmost care to remove rogue *i.e.,* plant of different varieties based on expression of distinguishing traits has to be taken.
- Very high level of uniformity of open pollinated variety may lead to its deterioration.

Number of plants/ears to be used

An important consideration during maintenance of an OPV is the number of plants or ears to be used. The number of plant is determined by

- Quantity of seed require as a source material to produce Breeder seed.
- The number of plants or ears to represent an OPV depends on the genetic variability for expression of quantitative and qualitative traits present within the OPV.
- Theoretical considerations, experiences and practices indicate that >500 plant/ ear are required to represent genetic variability within an OPV and to produce adequate Breeder seed. The number of plants/ears may be enhanced to produce large quantity of Nucleus seed but reduction may lead to genetic drift.
- Inbreeding during maintenance of OPV is inevitable but its effect should be minimized. Any selection for representation of within OPV variability during maintenance breeding should focus on the identification of plants and ears that best represent the OPV.

Effective population size to maintain OPV

As per Hardy-Weinberg law, in a random mating large population gene and genotype frequencies remain constant in the absence of mutation, migration and selection. Therefore, to ensure true to the type population of OPV, multiplication of a large population of random mating individuals under optimum conditions has to be raised. Practically, it is not possible due to large number of varieties. A number of alternative approaches have been developed for multiplication of accessions with limited resources and space.

- Selection of small number of plants from Breeder seed production plot may lead to genetic drift that may result in loss of alleles at the start of maintenance breeding and get compounded with the advance of seed multiplication chain.
- Sampling technique to maintain OPV should aim to capture all alleles occurring at a frequency of 0.95% or more. But in such selection loss of alleles with frequency less than 0.05% becomes high. It means the probability of retention of rare alleles during multiplication of OPV is low. Gale and Lawrence (1984) demonstrated that probability of

retention of an allele of frequency 0.02 after five generation is 0.24 in a breeding population which consists of 25 plants; 0.42 in 50 plants and 0.66 in 100 plants.

Maintenance breeding for OPV of crops from family Comopsiteae and Gramineae

- Tag more than 1000 plants in Breeder seed production plot based on expression of distinguishing qualitative traits with uniformity representing required variation of quantitative traits to avoid genetic drift.
- Individual tagged plat has to be visited several times to observe the expression of diagnostic traits and health.
- Tagged plant with any deviation in expression of distinguishing traits or health is rejected.
- Seed of single plant with the claimed expression of distinguishing traits is harvested separately at harvestable maturity.
- Harvested plant is threshed by hand to avoid mechanical damage to seed.
- Produce of one plant is stored in one packet. In this way number of packets will be equal to the number of plants.
- Seed of individual plant is examined for expression of distinguishing traits of the variety at seed level, seed health and number of seed in the laboratory.
- The whole produce of single plant with any variation in the expression of one or more than one seed from the claimed expression is rejected.
- The whole produce of single plant with seed infested by objectionable seed borne disease is rejected.
- The whole produce of single plant with the number of seed less than the required number for designated row length is rejected.
- Prepare two parts of seed from each packet with same serial number on both the packets.
- One part of the seed from individual packet is used to raise plant to progeny row for evaluation.
- The second packet of each row is stored safely as remnant seed.
- Maintenance requires space isolation to avoid out-crossing.
- The maintenance breeding block is visited regularly to observe the expression of distinguishing traits at different growth stages but meticulously before flowering.
- All the plants in a row in which offtype plant appears are rejected and uprooted strictly before flowering.
- The crop is harvested row wise and the seed of one row is placed in one cloth bag.
- The yield performance of individual row is tested. The progeny row gave high yield than population mean with ±2SE are selected.
- The seed of these progeny row are mixed and used as Breeder seed.
- The seed of selected rows stored earlier in the second packet are bulked and used as NUCLEUS SEED.

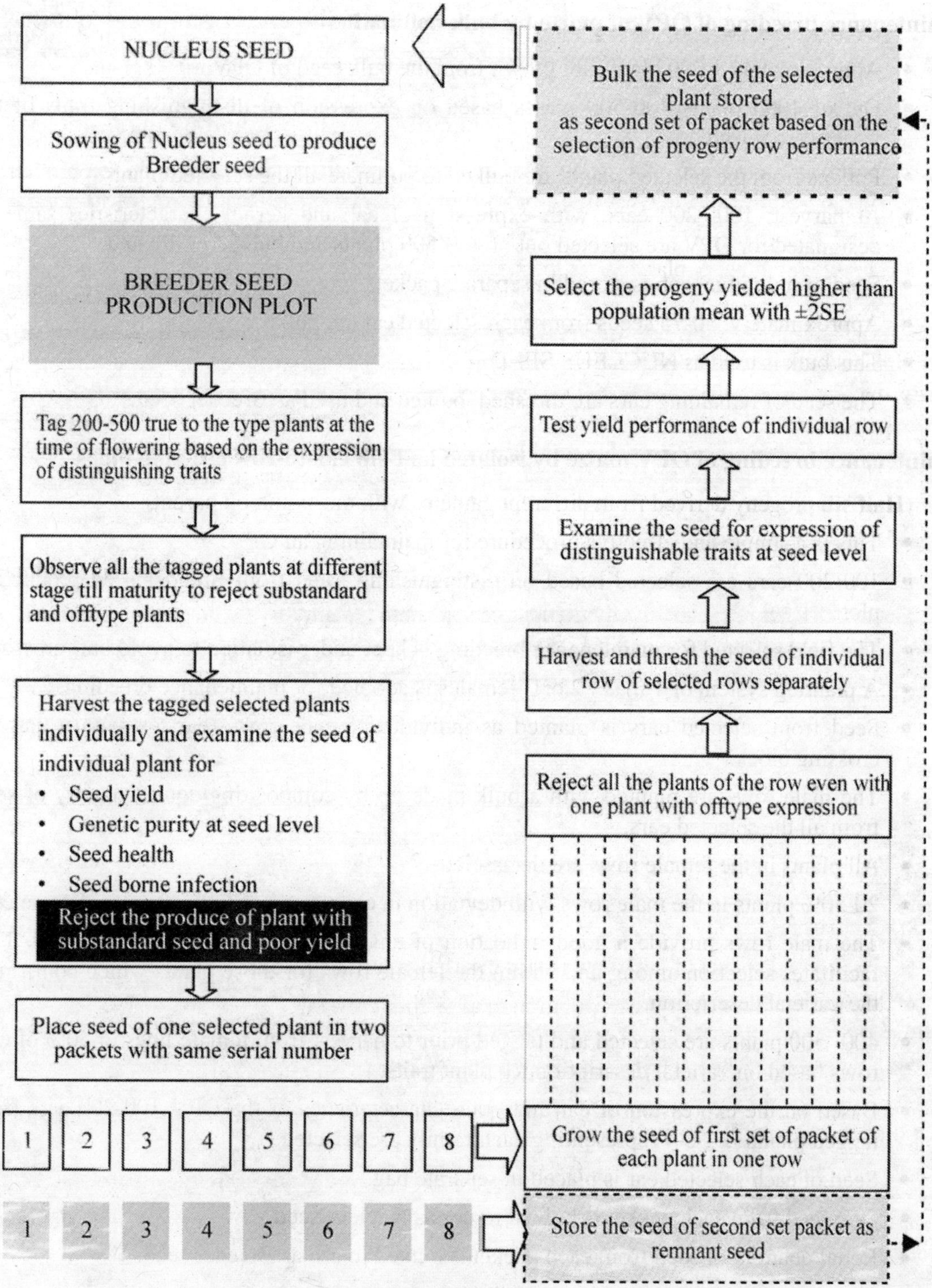

Fig. 2. Maintenance breeding of cross-pollinated crops

Maintenance breeding of OPV of maize by bulk pollination

- Approximately 1,000 plants are grown from the bulk seed of OPV.
- Out of 1000 plants 400-500 plants based on expression of distinguishing traits of the variety are selected.
- Pollens from the selected plants are bulked to pollinate all the selected plants.
- At harvest, 100- 200 ears, with expression of ear and kernel characteristics same as designated for OPV are selected out of 400-500 plants and bulked pollinated.
- Seed of individual cob is stored in separate packets.
- Approximately, 50-75 seeds from each selected ear are bulked.
- This bulk is used as NUCLEUS SEED.
- The seed of remaining ears are threshed, bulked and used as Breeder seed.

Maintenance breeding of OPV maize by isolated half-sib ear-to-row crossing block

(**Half sib** progeny derived from different gametes with one common parent)

- This is a simple and effective procedure for maintaining an OPV.
- 100-200 ears are selected based on distinguishing traits from Breeder seed production plot.
- The field selected for maintenance breeding is kept under isolation to avoid out- crossing.
- A planting system of 1 male: 2 or 3 females is adopted for maintenance breeding.
- Seed from selected ears is planted as individual female rows (ear-rows) in a half-sib crossing block.
- The male rows are planted with a bulk made up by compositing equal quantity of seed from all the selected ears.
- All plants in the female rows are detasseled.
- 20-30% plants in the male rows with deviation in expression of OPV are also detasseled.
- The male rows provide a good indication of environmental variation in the field. This facilitates selection among and within the female rows for those plants which confirm to the varietal description.
- 400- 500 plants are selected and tagged prior to harvest from female lines in 50% of ear-rows based on varietal description for plant traits.
- Based on the expression of ear and grain characteristics of the OPV 100-200 ears from tagged plants i.e., 2-4 representing each family are selected.
- Seed of each selected ear is placed in separate bag.
- One seed from each packet is bulked to use as Nucleus seed.
- Equal quantity of seed from each packet is bulked to use as Breeder seed.

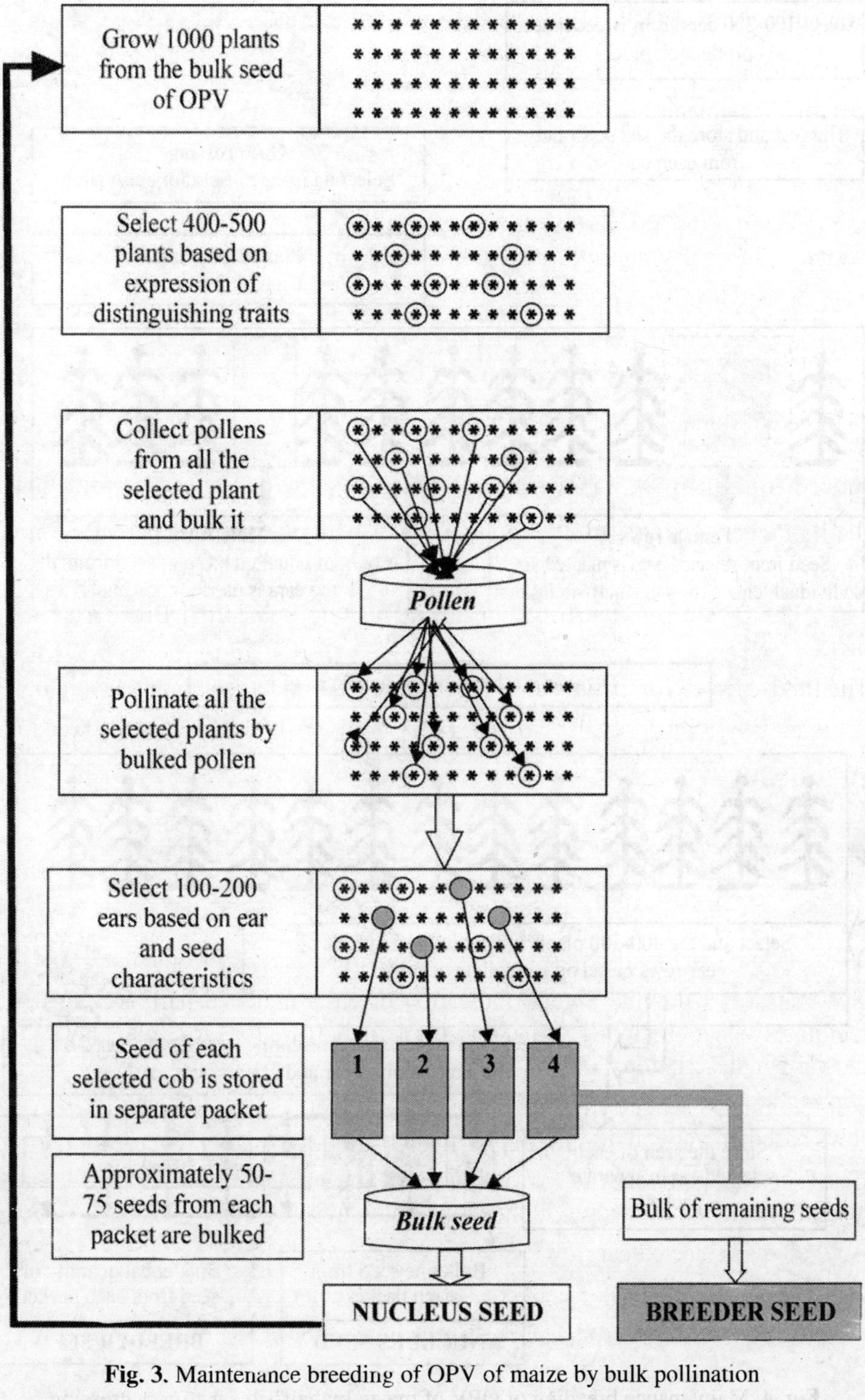

Fig. 3. Maintenance breeding of OPV of maize by bulk pollination

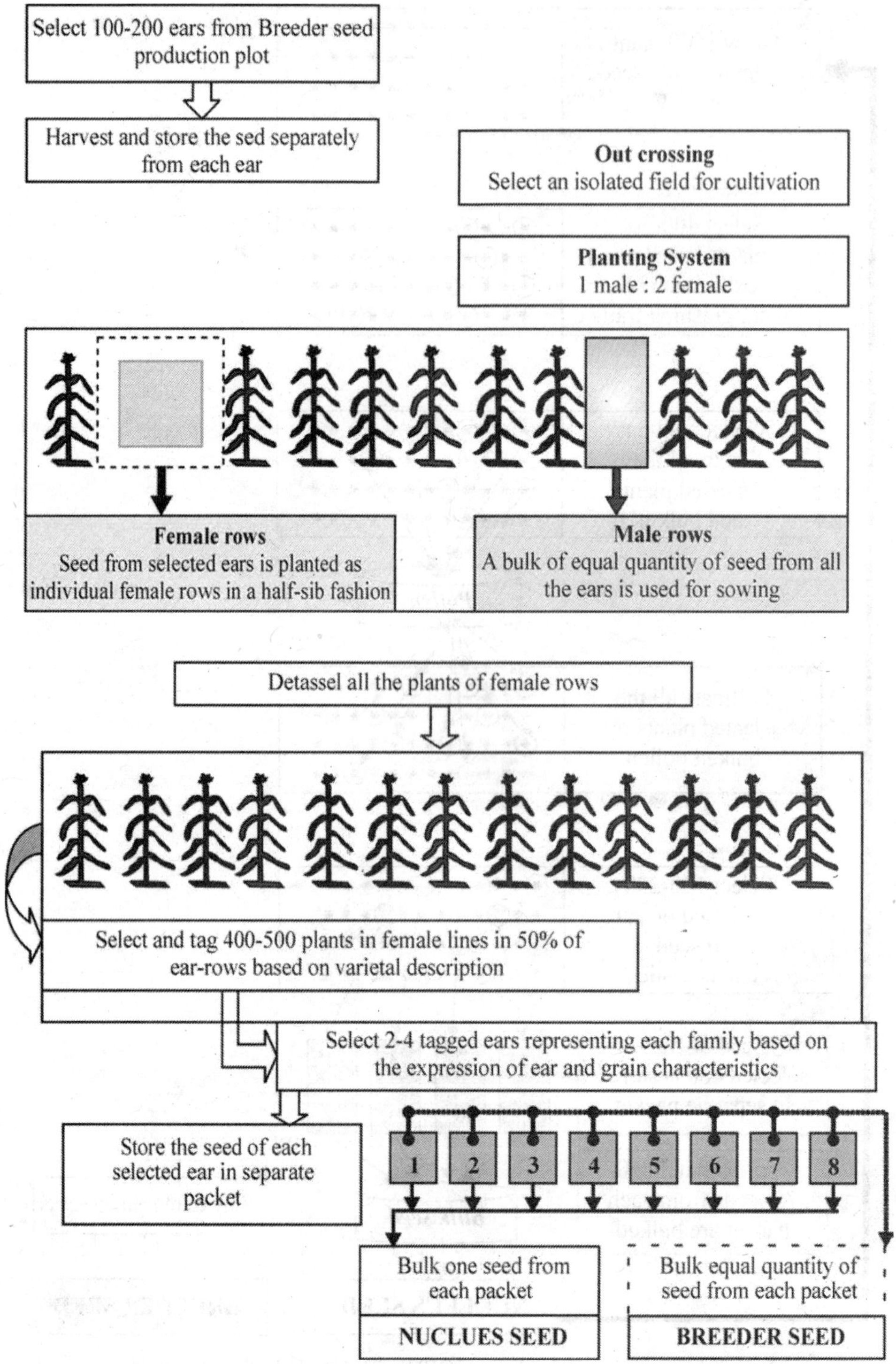

Fig. 4. Maintenance breeding of OPV of maize by half-sib ear-to-row crossing

Maintenance breeding of parental lines of rice

- Select and tag 200 true to the type male sterile (A) and maintainer (B) line plants from Breeder seed production plot of A line (AXB).
- Similarly select and tag true to the type 200 plants of Restorer (R) line from Breeder seed production plot.
- Individual tagged plants of all the three lines have to be visited several times to observe the expression of diagnostic traits and health.
- Tagged plant with any deviation in expression of distinguishing traits or health is rejected.
- Seed of single plant with the claimed expression of distinguishing traits is harvested separately.
- Harvested plant is threshed by hand to avoid mechanical damage to seed.
- Produce of one plant is stored in one packet i.e., number of packets will be equal to the number of plants.
- Seed of individual plant is examined for expression of distinguishing traits of the variety at seed level, seed health and number of seed in the laboratory.
- The whole produce of single plant with any variation in the expression of one or more than one seed from the claimed expression is rejected.
- The seed of one selected plant is stored in two packets with same serial number on both the packets.
- Grow A X B and A X R lines side by side in two fields as paired row i.e., one progeny row of A line followed by progeny row of B line; similarly in second block progeny row of one A line followed by R line. These lines are labeled as

  A1 XB1, A2 XB2, A3 XB3

  A1 XR1, A2 XR2, A3 XR3
- In both the cases A1 represents seed from same packet, same for A2, A3 etc.
- In this way paired crosses are prepared and seed from individual row are harvested as seed of A1, B1, R1 and hybrid.

Identification nursery

- 10-20 seeds from each of the AXB (seed of A) and AXR (hybrid seed) are planted to grow the identification nursery. Remaining seed of A (Produce of AXB), B line (selfed B) and R (selfed R) are sown 21 days after identification nursery as progeny row.
- The progeny row of AXB and AXR are planted according to serial number.
- The progeny of AXB is observed for expression of distinguishing traits, uniformity and stable expression of male sterility. The serial number of A line with off-type and fertile or partial sterile expression is recorded.
- In AXR identification nursery the serial number of R lines with poor restoration is recorded.

Multiplication nursery

Seed production of A and B line

- An isolated field with 20m space isolation is selected with provision of barrier isolation upto 2m height.
- Staggered sowing of B line i.e., 50% 3 days and 50% 5 days after sowing of A line is necessary for proper nicking.
- Spacing

A and B lines	30cm
Between A lines	15cm
Between B lines	30cm
Plant to plant	15cm

- Paired progeny row of A and B lines are transplanted according to serial number same as in identification nursery for AXB.
- Recommended package of practice for hybrid seed production is adopted.
- The progeny rows of corresponding A lines of the AXB cross expressed deviation for distinguishing traits, male fertility or partial sterility in identification nursery are removed from multiplication nursery before flowering.
- Remaining AXB lines are supplemented for cross-pollination.
- At the time of harvesting B lines are harvested first to use as NUCLEUS SEED.
- Then, A lines are harvested and threshed separately to use as NUCLEUS SEED.

Seed production of R line

- The seed of R lines are planted as progeny row with the same serial number as of AXR line in identification nursery.
- All the plants of the progeny rows of R line not showing proper restoration in identification nursery are removed before flowering from multiplication nursery.
- After assessing spikelet fertility of A X R line in identification nursery the seed of corresponding progeny row of R lines are bulked to use as NUCLEUS SEED.

Purification of A, B and R lines in pigoenpea

- 100 plants from A and 100 from B lines are selected.
- Crosses between these plants are prepared manually with serial number of plants on price level.
- The pods formed by hybridization are harvested plant wise.
- All the pods formed on each of the 100 plants of B line by self-pollination are harvested separately and stored according to serial number.
- In the next season, few seeds from each of 100 plants of A line formed by hybridization are planted in plant to progeny row according to serial number of plant.

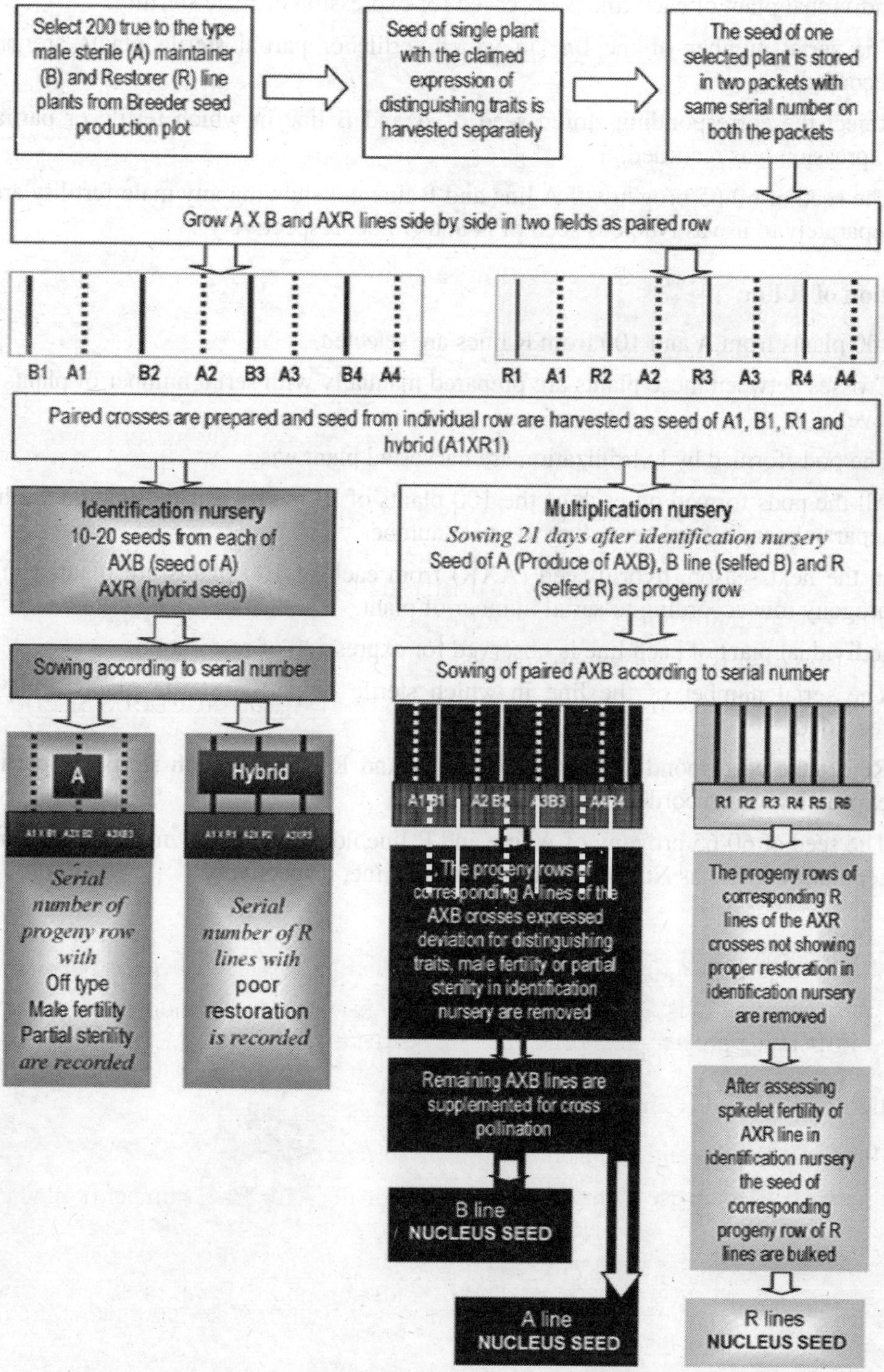

Fig. 5. Parental lines maintenance of hybrid rice

- Individual plant of each line is observed for expression of male sterility.
- The serial number of the line in which fertile or partial sterile plants are present is recorded.
- Reject the corresponding stored seed of A and B line in which fertile or partial fertile expression was recorded.
- The seed of 60-65 progeny of A line and B line not showing any male fertility are bulked separately to use as Nucleus seed of A and B line, respectively.

Purification of R line

- 100 plants from A and 100 from R lines are selected.
- Crosses between these plants are prepared manually with serial number of plants on price level.
- The pods formed by hybridization are harvested plant wise.
- All the pods formed on each of the 100 plants of R line by self-pollination are harvested separately and stored according to serial number.
- In the next season, hybrid seed (AXR) from each of 100 plants are planted in plant to progeny row according to serial number of plant.
- Individual plant of each line is observed for expression of restoration.
- The serial number of the line in which sterile or partial sterile plants are present is recorded.
- Reject the corresponding stored seed of A and R line in which sterile or partial fertile expression was recorded.
- The seed of 60-65 progeny of A line and R line not showing any male sterility are bulked separately to use as Nucleus seed of A and R line, respectively.

■■■■

Chapter 7

BREEDER SEED PRODUCTION

Fulfilment of the requirement of Breeder seed is one of the mandates of ICAR. The responsibility is shouldered by State Agricultural Universities and ICAR Research Institutes. The indent is placed by the State Governments to the Department of Agriculture and Cooperation, Government of India. It is compiled by the Department of Agriculture and Cooperation, Government of India and forwarded to ICAR for execution.

The indent for production of Breeder seed is allocated to different institutions in the annual group meeting of the respective crop. Allocation of Breeder seed production to different institutes, name and address of the concerned breeder, variety and quantity to be produced, name of indenters and quantity required, and name of the members of monitoring team are finalized by the Project Coordinator of the respective crop. The information is forwarded by the Project Coordinator of the respective crop to Assistant Director General (Seeds) ICAR, Directorate of Seed ICAR, Division of Seed, Ministry of Agriculture, Government of India, NSC, producing breeder, actual indenters and members of monitoring team as per **Proforma BSP I.** The members of the monitoring team are finalized by the Council. The representative of NSC acts as convener of monitoring team.

BSP I. Allocation of breeder seed production (national varieties)

The responsibility of Breeder seed production programme is given to the breeder who knows breeding behaviour of the crop, diagnostic characters of the variety and specific requirements for seed production.

Procedure

i. Nucleus seed is sown with reduced seed rate in line on well prepared and fertile land fulfilling the preceding crop requirement and isolation distance with prescribed package of practice for water, nutrient, weed and pest management. The main aim to follow the package of practice is to achieve high seed multiplication ratio i.e., to harvest high quantity of quality processed seed.

ii. After every 20-30m, a gap of 1.0-1.5m is provided for proper monitoring and operations of the seed production programme.

iii. After sowing, the proforma **BSP II** is submitted by the seed producing breeder/ agency to ADG (Seeds) ICAR, Division of Seed DAC, Ministry of Agriculture, GOI, NSC, Director Seed Certification of the State, Project Director/Coordinator of respective crops, Directorate of Seed Research ICAR and member of monitoring team.

BSP II. Time Table of Breeder sed production and its availability

iv. Rouging is utmost important and should be completed before flowering.

v. The field is to be monitored meticulously to avoid genetic impurity, presence of objectionable diseases, insect pests, parasites and weeds.

vi. The seed production programme is monitored by a team consisting of

- crop breeder / representatives of crop coordinator (ICAR)
- Breeder responsible for seed production
- Representative of State / National Seed Certification Agency
- Representative of State/National Seed Corporation

vii. The report of the monitoring team containing information about genetic purity alongwith other recommendations is submitted by the convener of monitoring team in proforma **BSP III** to ADG (Seeds) ICAR, Division of Seed DAC, Ministry of Agriculture, GOI, NSC, Director Seed Certification of the State, Project Director/ Coordinator of respective crops, Directorate of Seed Research ICAR with signature of all the members of monitoring team. Proper care is taken during harvesting, threshing and processing to avoid any type of ad-mixture and damage to seed.

BSP III. Inspection report of the monitoring team

viii. The final quantity of processed seed is submitted through proforma **BSP IV** to ADG (Seeds) ICAR, Division of Seed DAC, Ministry of Agriculture, GOI, NSC, Director Seed Certification of the State, Project Director/Coordinator of respective crop and Directorate of Seed Research ICAR by the producing breeder/agency. Seed is tested for sowing seed quality and a yellow coloured certificate is attached with tag containing all the required information, seal of producing agency and signature of the breeder.

BSP IV. Breeder seed actually produced

ix. The Grow Out Test of the sample is performed in off-season to test the genetic purity as per procedure given in Hand Book of Seed Testing. The final report is submitted through proforma **BSP V** to ADG (Seeds) ICAR, Division of Seed DAC, Ministry of Agriculture, GOI, NSC, Project Director/Coordinator of respective crops and Directorate of Seed Research ICAR by the producing breeder/agency.

BSP V. Report of Grow Out Test

x. Proforma **BSP VI** containing information on quantity of Breeder seed allotted, lifted and the Foundation seed produced from the lifted one is submitted by DAC to ICAR.

BSP VI. Utilization of Breeder seed

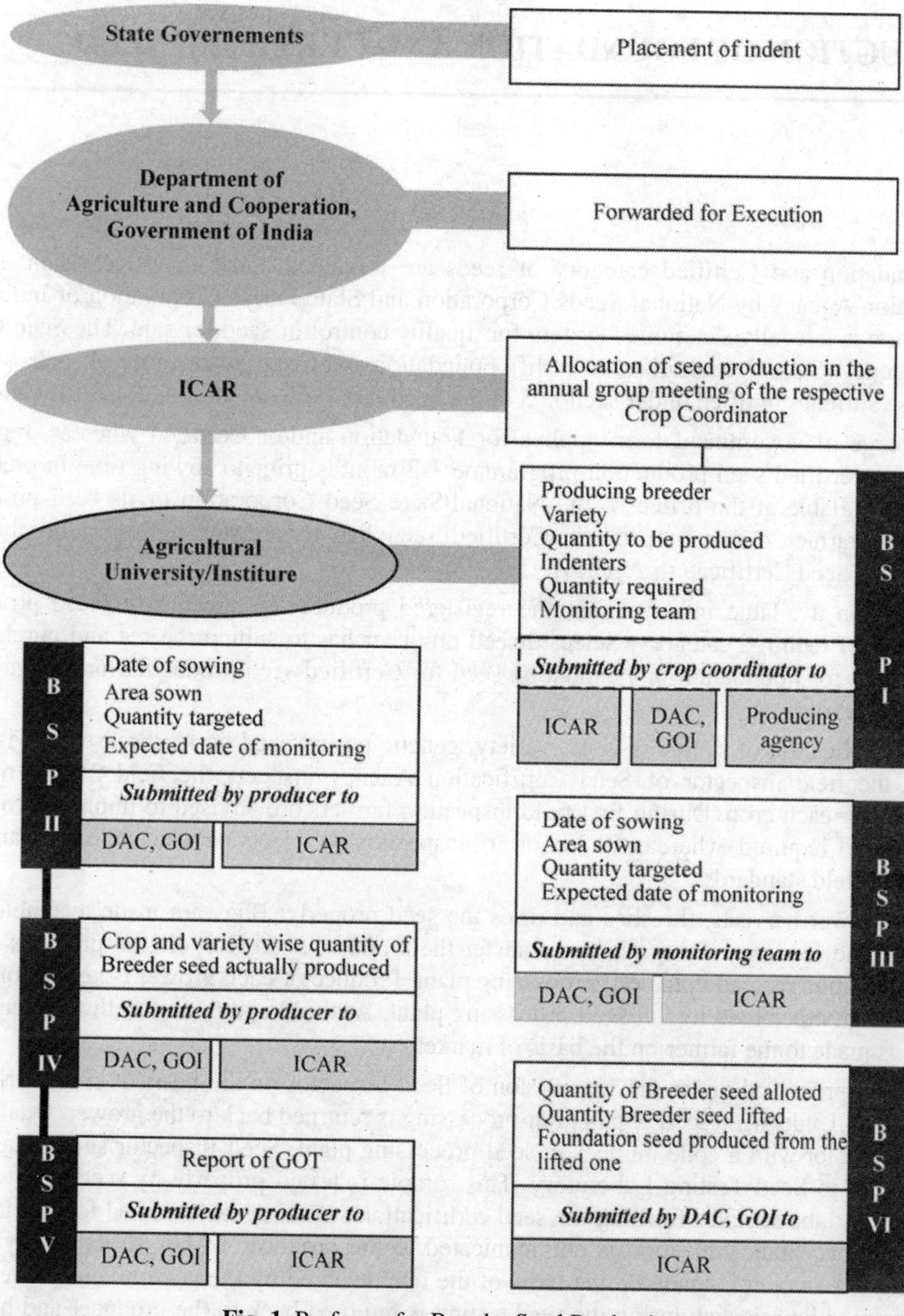

Fig. 1. Proforma for Breeder Seed Production

■■■■

Chapter 8

PRODUCTION OF FOUNDATION AND CERTIFIED SEED

Foundation and Certified category of seeds are produced under the supervision of Seed Certification Agency by National Seeds Corporation and State Farms Corporation of India. Seed certification is a legally sanctioned system for quality control of seed for sale. The Seed Certification Agency (SCA) is authorized to certify Foundation and Certified category of seed. Seeds of only those varieties notified under section 5 of the Seed Act 1966 are eligible for certification.

Incharge of government farm applies for Foundation and/or Certified whereas, registered farmers for certified seed production programme 1-2 months prior to sowing time in prescribed proforma available at the office of the National/State Seed Corporation or its seed processing plant. Any farmer desire to produce Certified seed has to register himself with the Seed Corporation/ Seed Certification Agency.

Based on available infrastructure the registered producer is selected for seed production programme of required category. Selected seed producer has to submit the fee and purchase the Breeder seed for Foundation or Foundation seed for Certified seed production programme from Seed Corporation.

For verification of source of seed, variety, genetic purity, seed borne diseases and isolation distance the field inspector of Seed Certification Agency inspects the field 2 to 3 times as prescribed for each crop. During first field inspection farmers are advised to improve production programme, if required, whereas, in 2nd or 3rd inspection he rejects or selects the programme on the basis of field standards.

The grower harvests, threshes and dries the seed properly. Bags are made available to the growers at the seed processing plant to transfer the produce. Seed Corporation provides charges for transportation of seed upto seed processing plant. Produce of each grower is kept as one lot at isolation from other seed lots in seed processing plant. At this stage three fourth payment of the produce is made to the farmer on the basis of market rate.

Seed is processed under the supervision of Seed Inspector on the basis of crop, variety and class of seed. Undersized seed obtain after processing is returned back to the grower. Quality seed is stored as a lot with a code number in seed processing plant. Seed Inspector submits a sample from this lot to Seed Testing Laboratory. This sample is tested properly by seed analyst in the laboratory and the samples fulfiling the seed certification standards are certified for its class. The decision on acceptance/rejection is communicated to the producer within 10 days. Payment of remaining 1/4 amount is made on the basis of the rate declared by the government in the current year. The seed lot rejected during the seed testing is returned back to the producer and he has to return the amount paid to him earlier.

■■■■

Chapter 9

AGRONOMICAL PRINCIPLES AND CONSIDERATIONS OF SEED PRODUCTION

Objective

Production of high quantity of quality seed production of healthy and genetic pure seed is the most important components to act as catalyst for realising the potential of all other inputs in agriculture.

Package of practice to obtain maximum yield is adopted for harvesting maximum yield as grain. But, there is some difference between cultural practices for grain and seed production as the end product has to be used for sowing purpose. For seed production programme, at quantity level the aim is high productivity of processed seed whereas at quality level it is fitness for sowing purpose. Cultural practices for seed production are standard recommended package of practice adopted for crop production with some alteration to improve the quality of seed and maintain characteristics of the variety.

Agronomical aspects of seed crop management involves selection of field, use of seed certified by the authorized agency, sowing method, sowing time, treatment of seed before sowing for assured field emergence, water and nutrient management to improve seed multiplication ratio, practices to maintain physical and genetic purity of the produce, practices to ensure germination and storability of seed, control of biotic stresses with special reference to objectionable seed borne diseases and objectionable weeds, precaution during harvesting and threshing to minimize mechanical damage, loss and admixture, post harvest management of seed to avoid deterioration of seed during storage and proper management in the store to minimize deterioration of seed.

Province effect

Seed production programme should be organized at the agro-climatic zone/area suitable for crop or variety to harvest high quantity of quality seed with low cost of production. The major consideration for selecting area to produce seed is environmental conditions viz., temperature, photoperiod and amount and distribution of rain that affect flowering, seed set, seed development and maturation. In India, hybrid seed production programme of rice is concentrated in Warangal and Karimnagr districts of Andhra Pradesh due to high hybrid seed producibility; sorghum in Warangal due to minimum infestation of diseases of seed borne nature; seed of hybrid maize in Eluru district of Andhra Pradesh during *kharif* for timely availability of seed for cultivation of rabi maize in Bihar. Seed of a very popular variety of Egyptian clover 'Mescavi' and seed of onion are still imported in India from Egypt due to high quality and significantly lower cost of cultivation. Seed production of vegetable crops is preferred in the area characterized by low humidity and limited rainfall during seed maturation and crop harvest. For hybrid seed production of tomato, brinjal, chilli, cucurbits (gourds, melons, cucumber, squash and pumpkin), okra and capsicum many locations in north Karnataka (Ranebannur, Haveri, Dahrwad, Koppal, Belgaum etc); Andhra Pradesh (Guntur, Karimnagar, Annatpur), Maharashtra (Jalna, Pune and Aurangabad) and Gujarat (Rajkot and Jamnagar); carrot and cabbage Haryana; radish Taran Taran, Punjab; cowpea Mirjapur UP; variety Arkel of vegetable pea Kanpur UP; cauliflower Hazaribag Bihar; Fennel Banswara Rajasthan; hybrids of temperate vegetable North hill zone (valleys of Sproon, Kullu, Kalpa, and Kashmir) are best suited, therefore commercially exploited. Quality seed of jute, a crop of West Bengal is produced in dry weather of Akola in Maharashtra. Gadwal, Karnool, Mahaboobnagar and Nandyal districts of Andhra Pradesh; Akola and Parbhani of Maharashtra and Dharwad of Karnataka have been identified as cost effective quality hybrid seed production area for cotton. Jamnagar and Hyderabad are most suitable for hybrid seed production of castor.

Production of seed in the area where climatic conditions are different than the adaptation, the problem of genetic shift can arise that may lead to deterioration of variety. It is gradual but critical especially for cross pollinated crops. Region of moderate rainfall and humidity are much more suited for seed production than regions of high rainfall and humidity. In general, regions with extreme hot summer and very cold winters should also be avoided for seed production unless particular crop is especially adapted to grow and product under these conditions.

Season

Season of seed production is important for harvesting of quality seed in high quantity. Seed of mung bean produced in summer has better quality than that of Kharif in Central zone of India. A comparison of seed of groundnut produced in Kharif and summer for sowing seed quality shows better performance of summer produce. Flowering, pollination and seed setting in temperate crops require low humidity while dry weather at seed maturity.

Selection of field

Seed production should be performed on well drained and leveled field with fairly deep and fertile soil that helps in uniform germination, growth and development of plants. The selection of field is the first and foremost task for maintaining the prescribed seed standards. It includes cropping history of the field to ensure same crop/ variety not grown on the same field. This precautionary measure prevents the appearance of volunteer plants of the previous variety to ensure genetic purity. It also helps to check the spread of pathogen and pest, which might have been built in the soil by the previous crop. Therefore, to overcome the problem of volunteer plants, objectionable weeds and pathogens and its adverse effects on seed quality recommended gap in crop/variety is recommended. The requirements of the field may be modified by pre-irrigation, deep polughing and other operations to destroy the sources of contamination. Seed infected/ infested with seed borne pathogen causing diseases viz., Karnal bunt of wheat Downy mildew of sunflower, nematode of wheat or weeds, viz., Mexican prickly poppy, wild morning glory etc., have perennial impact on the quality of seed.

Field preparation

It is desirable to prepare the field of *kharif* crops several weeks prior to planting to encourage pre germination of weeds and volunteer plants. Timely preparation of pulverized clean seedbed is essential for better and uniform germination that ensure establishment of desired plant stand with uniform growth. The soil should be made perfectly well drained and leveled. Deep ploughing during summer in alternate year is advocated to avoid biotic stresses. In light textured soils, well-decomposed farmyard manure @ 10 tones/ha is applied before last ploughing and incorporated in the soil for maintaining the fertility.

Population dynamics

Seed of required category should be purchased from authentic source. Before sowing, the seed has to be properly treated with recommended fungicide/chemical (Thiram, carboxin, carbandazim etc.) to check the infection/infestation of seed/soil borne diseases or/and insect pest and microorganism (Rhizobium, Mycorhiza etc) to make nutrient available to the crop plant. Pre-sowing irrigation is recommended for uniform germination. Sowing of seed crop should be done in rows at uniform depth on time with the help of seed drill in sufficient residual moisture for getting uniform germination. The seed drill should be properly checked and cleaned before use. A track of about one metre or more, depending upon the row spacing is left after every ten meter i.e., after sowing of 4-5 rounds of seed drill for proper monitoring, rouging, spray and certification at field level.

Time of sowing plays an important role in quality seed production of various field crops. Timely sowing of the seed crops is of paramount importance for proper growth and development of the plants and expression of the characteristics of the variety. Resowing in some part of the field should be avoided.

In order to facilitate proper monitoring and roguing of the off-type plants from the seed plots and to obtain good quality of seed a reduced population density is useful. Therefore, the seed rate should be around 10 per cent lower than the recommended seed rate for grain production. High seed rate / population reduce the seed size resulting in increased percentage of undersize seed in seed lot. Observations and monitoring of the seed crop become difficult under dense plant population. Hence, the plants should be properly spaced among as well as within the rows.

Water management

There must be assured supply of irrigation for growing of any seed crop. The dry region with assured irrigation is considered as the best seed production area. The crops should not encounter deficiency of water, particularly at critical growth stages. No moisture stress at reproductive and seed filling phase is essential for harvest of quality sound seed in all the crops. Irrigation should be given uniformly to the seed crop. Water stagnation or moisture stress or uneven application of water may cause variations in crop growth, which create problem during inspection of seed plots. The number of irrigations to various crops depends on the soil type, climate and crops/varieties. The quality of irrigation water has significant influence on quality of seed, as seed of rice varieties grown with saline water have poorer storability as comparison to seed produced with non saline water. Sowing of crops like soybean on raised bed or with furrow in between rows is recommended for production of high quantity of quality seed by avoiding excessive moisture stress. Application of water with sprinkler at the time of flowering should be avoided to enhance fruit setting as it is disadvantageous for pollination and enhanced foliar and seed borne diseases.

Integrated nutrient management

To obtain high yield of quality seed from the different field crops, balanced quantity of nutrients should be supplemented with recommended doses of the manures and fertilizers. The level of nutrients would however vary depending upon the variety, crop, season, soil fertility status, previous crop and soil moisture. One must try to apply recommended levels of nutrients based on soil test values for better results. A crop plant requires sufficient quantity of nitrogen for healthy growth, phosphorous for fruiting and seed development and potash for flowering and seed development. Phosphorus and potash are comparatively more important for seed crop in comparison to food crop to harvest quality seed with special reference to pulses.

Application of high dose of nitrogen to field crops leads to lodging due to excessive growth, encourage foliar diseases and reduction in yield due to poor source and sink relationship and partitioning of energy. It is recommended that in sugarcane, additional does of 20-30 kg nitrogen 20 days prior to harvest of seed can improve the germinability.

Several micronutrients are essential for improvement of sowing seed quality viz., magnesium and calcium to improve germination; boron for flower development and pollen germination; zinc for seed formation; and molybdenum for endosperm development. Minor elements are also important for development of healthy and sound seed. Calcium has positive effect on sowing seed quality of groundnut; sulphur, zinc and boron on oil seed crops. Deficiency of soil boron causes 'hollow heart' defect in seed of vegetable and field pea. Deficiency of copper in rice leads to loss of viability. Deficiency of calcium in groundnut forms darkened plumule. Deficiency of Manganese and Molybdenum forms 'marshy spots' on the seed of field pea that enhance hardseededness.

Weed management

The seed plots should be kept free from all weeds, specially the objectionable ones throughout the crop growth period. Weed competes with crop plant for water, nutrients and sunlight. With its smother effect for sunlight the maturity of the crop is delayed. Field infested with weeds is difficult to prepare and the crop infested with weeds is difficult to harvest. Sap of green weed enhances the infection of microorganisms, delayed drying and reduces longevity of seed. Presence of weed seeds at the time of harvest leads to mixing of weed seeds with crop seeds which is difficult to separate during seed processing. Harvesting of crop infested with green weed by combine harvester enhances the temperature of the harvested seed in the tank resulting in seed deterioration. Sowing of the seed crop in well-prepared clean seedbed minimizes the problem of weeds to a greater extent. Infestation of parasitic weeds viz., *Cuscuta, Orobanche* etc., reduces the quantity and quality of seed drastically. Weeds harbor insect pests and pathogens. The seed plots, should be kept completely weed free till harvesting of the crop with the help of appropriate herbicide or by mechanical hand hoeing and manual hand weeding. Use of herbicide may lead to deformation in the expression of morphological traits of seed crop e.g., deformation in spike of wheat and leaves of cotton by delayed application of 2,4D.

Pest management

Seed crop has to be protected against infestation of insect pest and diseases to harvest quality seed. Special care is required against seed borne diseases. Prophylactic measures against pathogens of prevailing diseases in the area associated with the crop are essential. Presence of soil borne pathogens/diseases should be considered during selection of field for seed production. Removal of plant debris and rotation of crop are important to reduce the risk of biotic stresses. Seed treatment is an important agronomical practice to protect the crop against seed and soil borne diseases. The crop is protected against some air borne diseases by space isolation. Diseased plants should be rouged out from the field with first expression to avoid its transmission. Population of insect pest has to be managed below threshold level with the spray of appropriate dose of recommended insecticide at correct time.

Isolation requirements

Recommended isolation distance must be kept between the seed and commercial crops of the same species to avoid out crossing and disease infection by wind and insects from neighboring fields. Volunteer plants in and around the seed production field should also be removed to avoid contamination. The seed crop must be isolated from other nearby fields of the same crop or any other contaminating crop as per certification standards. Time isolation could be adopted to reduce space isolation in maize. The harvested seed should also be placed in isolation to avoid mechanical contamination.

Thinning and gap filling

Thinning or gap filling in seed production plot should be avoided. If required, should be done immediately after emergence of the seeds to maintain the required plant population of same age group and uniform expression for plant height, days to flowering and maturity.

Rouging

Removal of off-type and rogue plants (pollen shedders, diseased plants, etc.) at the earliest possible, preferably before flowering is essential to maintain the genetic purity of the seed crop. It is one of the effective means of maintaining the genetic purity of seed stock particularly in large-scale seed multiplication programme. It is thoroughly performed at different growth stages viz., vegetative, pre-flowering, and flowering based on expression of distinguishable morphological traits. Its number varies from crop to crop.

Supplementary pollination

In hybrid seed production programme supplementary pollination by honeybees, hand pollination, floral mechanism, rope pulling etc., is essential to increase the seed yield.

Harvesting

Harvesting of seed crop should be done immediately after its physiological maturity i.e., the seeds are fully developed and have attained their maximum weight to obtain good quality seed. Delayed harvesting may deteriorate the quality of seed due to weathering, mechanical damage during harvesting and threshing and loss of seed due to shattering which becomes the source of volunteer plants.

Post-harvest management

There are considerable losses of seed produce in quantity as well as its quality during the post harvest operations like threshing, winnowing, transportation, grading, processing and storage. Therefore, best possible efforts should be made to minimize these losses. All these operations must be strictly supervised.

Threshing

Threshing floor must be cleaned thoroughly to avoid mechanical mixture. The thresher should be calibrated perfectly to minimize mechanical damage to seed. After threshing, the seed should be cleaned with the help of winnowers. Harvesting and threshing of seed production plots through combine harvester should be avoided particularly for pulses and in smaller plots, because the chances of mechanical mixture and damages to the seed are more.

Seed processing

The harvested seed has many factors that affect its sowing seed quality and storability. These factors viz., inert matter, weed seed, other crop seed, immature seed, damaged seed, diseased, undersized seed, other variety seed etc. are to be reduced upto threshold level to improve sowing seed quality. The storability of the seed reduces drastically during storage as these factors invite insect and microorganism due to their hygroscopic nature. All the operations from harvesting to storage are performed carefully to improve the sowing seed quality and its longevity.

Drying of seed

It is necessary to dry the seed to safe moisture content just after threshing to maintain seed viability and vigour. It is again dried upto safe moisture level after processing and before storage to enhance seed longevity. Moisture content in seed varies from crop to crop for safe storage. Orthodox seed can be dried upto low moisture content, whereas recalcitrant has to be stored with high moisture content.

■■■■

Chapter 10

SEED DETERIORATION

Deterioration of seed starts with its formation on the mother plant till development of seedling by the seed in the next generation. The factors responsible for seed deterioration are

- **Genetic ability of the crop:** Seed of many crops (eg., *Paspalum scorbiculatum*) are long lived under ambient condition, whereas others are short lived (soybean, onion etc.).
- **Nature of seed:** Seed of the crops with recalcitrant nature deteriorate at faster rate in comparison to the crops with orthodox seed.
- **Weathering of seed:** The seed remain stored from physiological maturity till physical (harvestable) maturity on plant. At this stage quality of mature seed is affected by environmental fluctuations i.e., variation in day and night temperature, relative humidity, rains, sunlight, UV rays etc. The damage of mature seed on plant due to environmental fluctuation is considered as field weathering. Seed of soybean is highly susceptible to field weathering due to thin seed coat that results in poor seed longevity.
- **Season of crop cultivation:** Seed of rainy season crops deteriorate at faster rate than the winter season crops due to more environmental fluctuation at the time of maturity, while seed of winter season crops are deteriorated at faster rate in store during rainy season.
- **Agronomical practices adopted**: Timely sowing of recommended variety with water, nutrient, weed and pest management yielded high quantity of quality seed that can be stored for longer period of time under ambient condition.
- **Climatic condition at the time of seed formation**: High humidity and forced maturity due to high temperature at the time of maturity leads to weathering of seed.
- **Forced maturity due to abiotic stresses**: Shrivelled, undersized and undeveloped seeds formed due to forced maturity deteriorate at faster rate in comparison to healthy seeds.
- **Development of seed on unhealthy plant:** Seed formed on diseased, deficient or infested plants deteriorate at faster rate in comparison to seed formed on healthy plant.
- **Duration of storage of seed on harvested plant:** Seed remains on the harvested plants from 5-10 days for threshing depending on the crop. The harvested plants of *Kharif* crops are stored as stack before threshing to complete the sowing of *Rabi* season crops. This stage is prone to field weathering.
- **Stage of crop plant for harvesting:** To reduce the field weathering, mechanical damage and loss of seed, harvesting should be performed at appropriate stage. Less than 15% moisture content in seed is considered appropriate for harvesting depending on the crop.

Early harvesting leads to formation of shriveled seed with reduced seed size whereas, late harvesting leads to field weathering and seed loss.

- **Use of combine harvester:** Harvesting of seed with the help of harvester leads to mechanical damage. Harvesting with high humidity enhances loss in texture of seed whereas late harvesting leads to mechanical damage. Harvesting of some immature plants or green weeds alongwith matured plants of crop may reduce the lusture of seed. Formation of a layer of soil around the seed with juice of green part deteriorates the quality of seed. Damage in the tip of the rice seed as mechanical damage results in loss of seed germination.
- **Method of threshing:** It is one of the most important stages of seed deterioration as damage of seed enhances with improper calibration of thresher, method of threshing and moisture content of seed.
- **Method of drying:** Variation in day and night temperature during drying of seed leads to seed deterioration. Direct sun drying with high moisture content leads to seed coat bursting, whereas high relative humidity in the night and low in day is responsible for wrinkling of seed. Drying of seed under high temperature with blower also leads towards deterioration of seed. Seed stored in the container immediately after drying (under sunlight or hot air) in the evening with high temperature deteriorates the sowing seed quality.
- **Radiation:** Seed may be deteriorated by exposure to radiation. Its effect may differ from crop to crop and variety to variety. Seed exposed to radiation with high moisture content is deteriorated at faster rate.
- **Quality of seed used for storage:** Properly processed seed i.e., no inert matter, other crop seed, weed seed and low moisture content can be stored for longer period of time under ambient condition with minimum deterioration. Infested seed deteriorates at faster rate in comparison to healthy seed in store.
- **Preconditioning:** Seed of many crops viz., maize, groundnut etc. stored after preconditioning deteriorates at faster rate.
- **Container for seed storage:** Sowing quality of seed stored in polylined gunny bags has lesser influence of environmental fluctuation in comparison to seed stored in gunny bags.
- **Method of storage:** Seed stored in lowest bag of a stack containing more than six bags; stack not on wooden plank and stack adjacent to wall deteriorates at faster rate in many oilseed crops.
- **Management in storage**: Seed deteriorates at faster rate under poor management during storage.
- **Location of seed store (climatic condition):** As per Harrington's rule, high relative humidity and temperature at the place of seed store may enhance the seed deterioration.
- **Infestation of microorganism:** It may lead to fast deterioration by reducing vigour and germination capacity.
- **Infestation of insect pest**: Presence of insect pest viz., internal feeder, external feeder or secondary feeder in the stored seed leads to fast deterioration.

- **Age of seed:** With the enhancement of age, deterioration in rate of vigour and germinability accelerated. Old seed with germination capacity may show poor field emergence. During germination of old seed more nutrients are leached out that stimulate infection of fungus like *Pythium* and *Rhizoctonia.*
- **Depletion of food reserve:** Bond among nutrients viz., phosphorous, zinc, magnesium, calcium and other minerals in seed is known as phytate. During germination under high moisture, phytate is broken down by the action of enzymes for growth of seedling. Under high moisture regime phytate bond may be broken down in the store resulting in deterioration of seed.
- **Hydrolysis:** Breakdown of fat by the action of lipase enzyme produces glycerol and free fatty acid. Free fatty acids are further oxidized into simple compound by α or β oxidation. It leads to deterioration in seeds of oilseed crops.
- **Accumulation of anti-nutritional substances:** Accumulation of enzyme coumarin and thiourea inhibit the activity of lipase (fat deficiency); gallic acid and aflatoxin inhibit activity of amylase (starch deficiency); coumarin inhibits activity of enzyme protease (protein deficiency). Whereas, in many crops strong phytate bond prevents availability of nutrients during germination.
- **Accumulation of toxic substances:** Accumulation of abscisic acid, ethylene gas, maleic hydrazide and other toxic substances in seed leads to fast seed deterioration.
- **Hormones:** Auxin is an important enzyme that regulates photoperiodism and geotropism during germination. Any mismatch in hormone cytokinins (root), auxins (shoot) and gibberellins (leaf) results in seed deterioration that leads to formation of abnormal seedling.
- **Inability of ribosome to form protein:** Inability of polyribosome to dissociate before attachment of RNA results in no protein synthesis that leads to deterioration of seed. The ATP content, synthesis of protein, synthesis of RNA, processing of ribosomal RNA, synthesis of mRNA reduces with the age of seed.
- **Deterioration in the action of mitochondria:** Mitochondrion present in the cells of embryo become more active with the start of germination but the ageing of seed mitochondria starts swelling and finally distorted. The activity of cytochrome oxidase, succinic, glutamic, malic peroxidase, dehydrogenase and other gradually decreases with the enhancement of age.

Chapter 11

DETERIORATION OF VARIETY

Deterioration of variety refers loss of uniformity, change in agronomical performance and change in expression for interaction with biological stresses.

Variety may be deteriorated due to genetic (out-crossing, mutation, residual segregation etc) or non-genetic causes (mechanical mixture, adverse agro-climatic conditions etc) as explained for individual factors.

1. DETERIORATION IN UNIFORMITY

Uniformity means freedom from admixture of other varieties of the same crop and freedom from genetic contamination by out-crossing. As per Hardy Weinberg law frequency of genes (alleles) and genotypes in a population remains constant from generation to generation with the following considerations

- No out-crossing
- No migration
- No mutation
- No genetic drift
- Infinite large population
- No selection
- Normal segregation of genes at gamete formation
- Equal viability of gametes
- Same gene frequency in male and female
- Equal fertility of parents
- Equal fertilization capacity of gametes

A population (variety) with constant allelic and genotypic frequency is said to be in equilibrium. Any deviation in above mentioned point is responsible for deterioration in the uniformity of a variety.

Uniformity of the varieties of self-pollinated crops developed from inbred line, hybrid varieties and variety/hybrid multiplied by clonal propagation is set at very high level.

An important attribute of an open pollinated variety (OPV) is to maintain its uniformity. Composite and synthetic varieties are maintained by open-pollination that forms continuous variable population. Therefore, the uniformity level is reasonably low and practically undesirable.

Within variety diversity in open pollinated population is powerful in maize. Most of the OPV are produced by recombination of elite and relatively similar genotypes. Therefore, certification standards for OPVs should be flexible, realistic, and appropriate for the conditions prevailing in a given geographical situation.

1.1. Out-crossing/ self-pollination

Pollen of other varieties within or outside crop has been recognized as a source of off-types. In an open pollinated variety, it occurs mainly due to inadequate isolation distance during seed production. It has no effect on morphology of produced seed baring xenia effect however variations are visible only at plant level and on seed harvested from those plants. In a homozygous variety the appearance of off-types depends upon the dominance of characters. Heterozygous may not be detected when the expression of the trait for identification of off-type is recessive. Accidental self-pollination in female parents during hybrid seed production programme is also responsible for genetic impurity.

1.2. Mutation

Spontaneous mutation by sunlight, fluctuation in temperature, water, atomic rays, thunder etc. occurs in the seed plot without any human control with expression in the next generation of multiplication e.g., bolters and wildings in potato and *speltoids* and *fatuoids* in wheat and oat. Mutation leads to formation of new genes leading to genetic variability. A change in gene frequency due to mutation depends on the rate of mutation. The only effective way to maintain genetic purity in this case is vigilant periodical rouging. The change due to mutation is random, relatively rare and unusual. Some crop species are more susceptible than the others for spontaneous mutation. In clonally propagated crops maintenance of uniformity is not a problem but recurrent somatic mutation may cause variability.

1.3. Residual segregation

Variability/segregation for some minor characteristics are neither considered by the breeder during the development of variety nor noticed during seed multiplication. It remains unnoticed in the population in a range of state. The proportion of different states of unnoticed trait reaches at equilibrium by natural selection. In wheat, subtle colour developed in few plants of some varieties only under drought which is not expressed in normal optimum condition. It provides a new category of offtypes.

1.4. Minor genetic variation

Many genes are not able to express in particular environment. When the variety is cultivated in new environment than only the expression of these traits are noticed. These traits may influence the agronomical performance of the variety that leads to its deterioration. Minor genetic variation is a common feature in often cross-pollinated crops. In open pollinated population, many new genetic combinations appear in each generation. Some of these combinations fall outside the acceptable range of the variety, regarded as offtypes. Rouging of these off-types is complicated in advance generations. Therefore, utmost care should be taken in the early generation to reduce chance of recombinants in subsequent generations.

1.5. Irregularities during meiosis

Variation in chromosome number is of two types

(a) Somatic complements are exact multiples of the basic haploid chromosome number which is termed as *polyploidy*.

(b) Somatic complements are irregular multiples of the basic number which is termed as *Aneuploid.*

Tall off-types in semi dwarf wheat varieties may appear due to inherent problem of aneuploidy, arising from chromosome pairing failures at meiosis. Other irregularities during meiosis that lead to alteration in the structure of chromosome are deficiencies, duplications, inversions, translocations, ring chromosome, bridge formation, iso-chromosome etc that leads to reduction in uniformity of a variety.

1.6. Premature release of varieties

Off-type may appear in the variety when the variety was not converted into a homogeneous population of homozygous plants before release in case of self pollinated crops. Release of a variety in early generation of selfing i.e., premature release of a variety is one of the most important cause of variety deterioration. Sonalika and Kalyansona varieties of wheat introduced form Mexico were released in India. Several secondary selections were made by the breeders due to existing genetic variability in both the varieties and developed as variety. Varieties with heterogeneous population of homozygous plants, many time exhibit wider adaptability.

1.7. Techniques of the breeder

Varieties released or bulked at about F_6 or early generation always exhibit visible segregation for minor morphological traits in late multiplication generation of the seed.

1.8. Mechanical contamination

This is the major source of contamination of the variety during seed production. Mechanical mixtures may take place right from sowing to harvesting and processing in different ways such as the seed may be contaminated by admixture during

Sowing: Use of one seed drill for sowing of different varieties of the same crop.

Field: Cultivation of two varieties of the same crop adjacent to each other without providing the required isolation.

Harvesting: Two varieties of the same crop are harvested continuously with the help of combine harvester without proper cleaning.

Transportation: Transfer of harvested plant or threshed seed of different varieties of the same crop in one/same trolley.

Threshing: Two varieties of the same crop are threshed with one thresher in the same threshing floor.

Drying: The seed of two or more varieties of the same crop spread in one threshing floor for drying.

Processing: Two varieties of the same crop are processed continuously with the same processor without proper cleaning.

Packing: Use of old unclean bags for packing of seed.

Human error: At various stages.

1.9. Volunteer plants

Seed of a variety cultivated in a particular field may shatter and remain in the field under dormant stage. Next year when new variety of the same crop is cultivated in that field for seed production, the seed of last year variety also geminate and developed into the plants of the same crop but of different variety. It leads to deterioration in uniformity mainly in rice, lentil, groundnut, sorghum etc. Therefore, the minimum period requires to take the seed production programme of the same crop but different variety in the same field is one to two years. Plants present in the field due to shattered seeds of previous season variety along with the current season variety are considered as volunteer plants. Volunteer plants are the potential source of genetic contamination.

---

2. GENETIC SHIFT

Change in the claimed average performance of agronomical traits viz., plant height, days to flowering, days to maturity and seed yield of the variety is considered as genetic shift. This shift may arise due to following reasons

2.1. Developmental variation

Shift may arise as developmental variation when the seed production of the crop is taken in new climatic condition i.e., new area/season/package of practice for several consecutive generations as differential growth response. It may be due to

- Cultivation of the variety in new area/season/ crop husbandry for which it is not released.
- Shift in the mean performance of agronomical traits due to change in climatic condition for seed production of the variety for which the variety is not recommended for cultivation.
- Selection pressure in new environment i.e., magnitude of abiotic stresses such as different soil (pH and salinity) and fertility conditions, variation in photoperiod, temperature or different elevations etc. may lead to expression of some minor gene that may cause developmental variation.

To avoid or minimize developmental variations the Breeder seed production programme of the variety should always be planned in area of variety adaptation as recommended in the release proposal.

2.2. Selection

The frequency of gene may change due to selection in favour of one or two alleles of a gene during maintenance breeding. The mean performance for the trait under selection pressure is altered and in the offspring the performance of the variety will differ than the claimed performance. Selection pressure arise at following levels that may lead to shift in the performance of the variety

- Segregation of genes at gamete formation
- Unequal viability of gametes
- Different gene frequency in male and female
- Unequal fertility of parents
- Unequal fertilization capacity of gametes

2.3. Migration

The rate of change of gene frequency in a population subject to immigration depends on immigration rate and the difference of gene frequency between immigrant and native. The process of migration is driving the changes in gene frequencies in natural populations. Gene flow between plant of immigrant and native variety is an important process that has strong implications in change in gene frequency that creates diversity within the variety. Migration of seeds or gametes between varieties increases the level of genetic diversity within populations.

2.4. Acclimatization

The process of introducing varieties of crop from the place of their cultivation to a place with different climate is termed as introduction. The adjustment of introduced variety to the new climatic conditions is known as acclimatization. During the process of acclimatization expression of many traits may alter. The alteration in expressions is more in polygenic traits. It may lead to deterioration of the variety. It is more deleterious when the seed of the variety is produced in new/ different climatic condition for which it is recommended. The low yield of popular rice variety IR 8 at few locations across the globe was due to lack of adaptation to changed environmental conditions.

2.5. Genetic drift

Small size of population from which single plants are selected to start the seed production programme or only small quantities of seed is taken for multiplication of seed in large area and preserved for the next years sowing. Because of such sub-sampling, all the genotypes will not be represented in the next generation that led to change in genetic composition. This is known as genetic drift.

2.6. Maintenance breeding

The technical requirement to maintain the basic and nucleus seed should be exactly same as described during the release of variety at all the centres as any selection pressure may lead to deterioration by different expressions of the same variety. Selection of single plants for production of Nucleus seed from the Breeder seed production plot i.e., large population. Selection of single plant from rows of single plant progeny raised for nucleus seed production may lead to genetic drift.

2.7. Isolation

Divergence within population can be caused by genetic drift because of complete or partial isolation.

3. INTERACTION WITH BIOLOGICAL STRESSES

It is an important factor that may deteriorate a variety in following ways

- Biological stress may deteriorate the quality of seed that reflects the performance of the variety in the farmers' field.
- Seed production of the variety of sorghum and sunflower in the field infected with downy mildew or wheat in the field infected with Karnal bunt may produce infected seed. The performance of the seed of the variety infected by these diseases deviates from claimed one.
- Variety claimed as resistant for particular disease may become susceptible because new race/biotype may appear due to genetic change in pest e.g., resistance against rust in wheat and gall midge in rice.

4. SEED CERTIFICATION AND MULTIPLICATION PROCESS

The farmer may not prefer to grow the variety in the event deteriorated seed of the variety is made available continuously for 2-3 years for cultivation.

- Improper seed certification is a factor that makes the poor quality seed available to the farmers for cultivation.
- Limited generation for seed multiplication should be adopted as more number of generations involved in seed multiplication lead to variation between Nucleus seed and the seed made available to the farmer. It is not applicable on hybrids.
- Low seed certification standard for genetic purity may also deteriorate variety at a faster rate.

Deterioration of hybrid

Apart from factors mentioned for varieties few factors are also associated for deterioration of hybrids

- **Base of male sterility:** To achieve homogeneous population of heterozygous plants, hybrid variety produced commercially by hand emasculation or use of gametocide or genetic male sterility is very difficult because of tedious and poor physiologically interacted procedure.
- **Breakdown of male sterility:** For hybrid seed production of most of the field crops cytoplasmic genetic male sterility is used. In many crops expression of male sterility is influenced by the environment (pigeonpea) or segregates (pollen shedder) as sterile and fertile or express partial sterility on the same plant (rice). In such cases genetic purity of hybrid seed cannot be maintained.

- **Identification of pollen shedder:** Pollen shedders are always identified after opening of flower therefore, out-crossing during seed production is unavoidable, that leads to genetic impurity in seed provided to farmers.

Chapter 12

FIELD INSPECTION

1. OBJECTIVE

Verification of the factors adversely affecting the quality of produced seed under field condition.

2. FACTORS THAT MAY DETERIORATE QUALITY OF SEED IN FIELD

A. *Factors deteriorating genetic purity:* Genetic purity of the seed used for sowing purpose, isolation distance, volunteer plants, presence of offtypes and rogue.

B. *Factors deteriorating seed health:* Presence of seed borne pathogens in seed used for sowing purpose, soil infested with pathogens of seed borne nature and isolation distance.

C. *Factors deteriorating soil health:* Presence of objectionable weeds and pathogens of seed borne diseases

D. *Factors deteriorating physical purity:* Presence of other crop and weed plants, infestation of insects and lodging.

3. UNIT OF CERTIFICATION

The entire area planted under seed production programme by an individual constitutes an unit, provided

i. All the seed production programme is to produce seed of one category and one variety.

ii. Area does not exceed 10 hectares and scattered fields constituting one unit (10 ha) should not be separated by more than 50 meters distance .

iii. The crop of the entire area is more or less of the same growth stage. The seed crop is grown as sole crop and it is not heavily and uniformly lodged.

4. METHOD OF FIELD INSPECTION

About 60-80 per cent field is covered during field inspection (Table 1). Number of plants/ earheads that should be observed during field inspection as one unit is standardized on the basis of area (Table 1), crop, method of sowing and spacing (Table 1, 2, 3).

Table 1. Minimum number of field counts required on the basis of field size

Size of field (hectare)	Minimum number of counts
Up to 2	5
2.1- 4.0	6
4.1- 6.0	7
6.1- 8.0	8
8.1-10	9

Table 2. Number of plants/earheads to be observed in one count

Habit	Crop	Number of plants/ear
Wide spread non tillering	Cotton, groundnut, maize, pigeonpea	100 plants
Narrow spaced non tillering	Chickpea, field pea, moong, mustard, sesame, urid	500 plants
Mainly broadcasted	Egyptian clover, lucerne, soybean	1000 plants
Tillering	Wheat, rice	1000 ear

Table 3. Percent coverage of area for seed certification

Method of sowing and spacing	Crop	Area covered
Broad cast	Egyptian clover, lucerne	60%
Line sowing, medium spaced	Chickpea, field pea, mustard, soybean	60-70%
Non tillering cereals in line	Maize, pearl millet, sorghum	75%
Line sowing wide spaced	Castor, cotton, groundnut, pigeon pea	85%

Field count: Number of plants/ earheads of seed crop, which should be observed during field inspection as one unit depending upon the cropping pattern and crop is known as field count.

Steps: The number of steps, which should be taken in the seed plot to observe number of seed crop plants requires to complete one field count.

Example: If one field count requires 1000 plants/ earheads and in one step one can observe 25 plant/ ear heads of seed crop. Then, to complete one field count one has to take 40 steps i.e., 25 x 40 = 1000 plants/ earheads. Observation of one count indicates seed plants, rogue and other observations taken in 40 steps movement in the field.

$$\text{Number of steps require for one field count} = \frac{\text{Number of plants/ears to be observed in one field count}}{\text{Number of plants observed in one step}}$$

Counting procedure: For accurate estimation of seed crop purity the maize inbred should be inspected at randomly, thickly sown crop at clockwise and broadcasted crop at randomly in 1 sq m as one field count.

5. CROP STAGES FOR FIELD INSPECTION

i. *At the time of sowing:* For verification of source of seed, land requirement, cropping history of the seed field, isolation distance and other agronomical and seed production practices.

ii. *Pre flowering*: Offtypes, rogues, weed, objectionable weeds and other crop plants.

iii. *Flowering stage*: Offtypes, rogues and seed borne diseases.

iv. *Post flowering and pre harvest stage*: Rogue, offtype, objectionable seed borne diseases and weeds, general performance of the crop.

v. *At the time of harvesting*: Objectionable weeds, objectionable seed borne diseases and offtypes.

During the field inspection at early stage of the crop, field inspector may suggest the farmer to improve the performance by rouging and weeding. In later inspection, he verifies the suggestions that were to be followed by the farmer and the improvement in the field performance. At the time of harvesting field inspector may suggest the precautions to be taken to maintain the seed quality.

6. NUMBER OF FIELD INSPECTIONS

Minimum 2 and maximum 4 field inspections are standardized for certification of different seed production programmes. Atleast one inspection at flowering stage is recommended to observe offtype, rogue and objectionable disease (Table 5).

Table 5. Minimum number of field inspections and recommended crop stages for seed certification

Crop	Number	Stage
FIELD AND FODDER CROPS		
Maize (synthetic and composite)	2	1st before; 2nd during flowering
Chickpea, cluster bean, grass pea, horse gram, Indian bean, kidney bean, lentil, moong, pigeon pea, rajmash, rice bean, urid bean	2	1st before; 2nd during flowering and fruiting
Soybean	2	1st flowering; 2nd at shedding of leaves
Napier grass	2	45 and 100 days after planting
Castor, cotton, Egyptian clover, groundnut, Indian clover, linseed, lucerne, minor millets	2	1st flowering; 2nd before harvesting
Barley, oat, rice, triticale wheat	2	Between ear emergence and harvesting
Buffel, Daraf, Dinanath, guinea, jute, marvel, mustard, Niger, pea, pearl millet, safflower, sesame, *Setaria*, sorghum, sunflower	3	Before flowering, flowering to fruiting and at maturity/ edible pod stage for green pea
Sugar cane	3	1st at 45-60 days; 2nd at 120-130 days after planting and 3rd 15 days prior to harvesting
VEGETABLES		
Radish, turnip	1	During flowering
Cowpea, cluster bean	2	1st before and 2nd after flowering
Fenugreek, spinach,, amaranth	2	Before and at the time of flowering
Little and pointed gourd	2	1st after transplanting and 2nd before cutting of the planting stock

Crop	Number	Stage
Garlic	2	1st at well developed plants and 2nd leaves begin to fall and before lifting of bulbs
Brinjal, okra, tomato, chilli, capsicum, celery, parsley, cucurbits (baring pointed and little gourd)	3	1st before and 2nd during flowering; 3rd at maturity prior to harvesting
Lettuce	3	1st before and 2nd at the time of head formation and 3rd at flowering
Cole crops	3	1st before marketable stage, 2nd at marketable stage and 3rd at flowering
ROOT, BULB AND TUBER CROPS		
Potato	4	35-45 and 60-75 days after planting, immediate and 10 days after haulm cutting
Carrot, onion	4	1st before, 2nd and 3rd during flowering, 4th at maturity
Sugar beet, garden beet, radish	2	1st before flowering 2nd during flowering
Turnip	1	During flowering
Lesser yam	3	90,150 , 200 days after planting
Colocasia	3	60, 90 and 160 days after planting
Asparagus	3	1st at crowning, 2nd during flowering, 3rd prior to harvesting
Sweet potato	2	1st at nursery; 2nd shortly after transplanting

7. KEY FACTORS TO BE OBSERVED

Following key factors are to be observed during first field inspection

7.1. Cropping history of the field

Year/ season	Crop	Variety	Class
Last			
Present			

The result is based on

a.	Last year / season crop was different	
	i) Seed is separable	*permitted*
	ii) Seed is not separable	*not permitted*
b.	Last year / season crop was the same	
	i) Variety was different	*not permitted*
	ii) Same variety but class was lower	*not permitted*
c.	Last year / season crop was same with same variety and same or superior class	
	i) But, field /seed was rejected due to seed borne diseases	*not permitted*
	ii) But, field/seed was rejected due to objectionable weed	*not permitted*
	iii) Passed during seed certification	*permitted*

The minimum period require to take the seed production programme of the same crop but different variety in the same field is one to two years (Table 6).

Table 6. Minimum gap required for seed production programme of different varieties of the same crop in selected field

Gap	Crop
One season	Rice, wheat, pearl millet, sorghum, soybean, sesame, mustard
Two seasons	Groundnut, Egyptian clover

The field of potato infested by brown rot (*Ralstoria solanacearum*), wart (*Synchytrium eondobioticum*) and common scab (*Streptomyces scabies*) should not be selected for seed production of potato tubers. Similar precautions should be taken in area where Karnal bunt of wheat and downy mildew of sunflower is a problem.

Seed production programme of different varieties of the same crop in the selected field as of last season should be destroyed by irrigation at least three weeks before sowing and ploughing just after emergence of volunteer plants.

7.2. Source of seed

Source of seed used for sowing purpose should be verified on the basis of the factors presented in Table 7.

Table 7. Verification of the seed used for sowing purpose

Factors for verification	Class of seed used for sowing propose		
	Breeder's	**Foundation**	**Certified I**
Source	Seed Corporation	Seed Corporation	Seed Corporation
Certificate	Breeder	Foundation	Certified I
Colour of certificate	Yellow (No.356)	White	Azure blue (No.104)
Tag	Breeder	Foundation	Certified
Seal	University/institute	Certification Agency	Certification Agency
Information on container	Breeder	Foundation	Certified
Purchase receipt	Seed Corporation	Seed Corporation	Seed Corporation

7.3. Isolation distance

To avoid out-crossing a minimum isolation distance between different cultivars of the same crop and cross compatible species should be maintained as per seed certification standards (Table 3). To avoid transmission of loose smut in barley, wheat, oat and triticale required minimum isolation distance should be maintained (Table 8).

Table 8. Minimum isolation distance to avoid transmission of loose smut

Crop	Field of crops infected with loose smut	Isolation distance (m)	
		F	C
Wheat	Rye, triticale, wheat	150	150
Barley	Barley	150	150
Triticale	Rye, triticale, wheat	150	150
Oat	Oat	150	150

7.4. Presence of rogue and offtype

Rogue and offtypes should be removed from the seed production plot. The field should be certified on the basis of minimum seed certification standard (Table 9) at proper stage of field inspection (Table 10).

Table 9. Maximum permitted limits for offtypes (%) during seed production

Crop	Foundation	Certified
FIELD AND FODDER CROPS		
Sugar cane (sett)	None	None
Napier grass (*Pennisetum purpureum*)	0.01	0.3
Linseed, minor millets, niger, pearl millet, safflower, sorghum	0.05	0.1
Barley, oat. rice, triticale, wheat	0.05	0.2
Castor, chickpea, cluster bean, cotton, cowpea, fieldpea, grass pea, groundnut, horse gram, Indian bean, kidney bean, lentil, moong, pigeonpea, rajmash, rice bean, sesame, Sunflower*, urid bean	0.10	0.2
** Sterile plants of the variety under seed production programme are not considered as offtype.*		
Rape seed and mustard, soybean, taramira, teosinte	0.10	0.5
Buffel, dharaf, Dinanath grass, guinea grass, *setaria*, stylo	0.10	1.0
Egyptian clover, Lucerne, marvel grass, Indian clover (*Melilotus* sp.)	0.20	1.0
Jute, dhaincha (*Sesbania*), sun hemp(*Crotolaria*), mesta (*Hibiscus*)	0.50	1.0
Maize#(composite, synthetic and open pollinated varieties)	1.00	1.0
# MSCS for pollen shedder when 5% or more of the plants in seed field have receptive silk.		
VEGETABLES		
Sweet potato	None	None
Lesser yam, potato including tissue culture-raised mini-tuber (PTCM)	0.05	0.1
Amaranth, asparagus, brinjal, cabbage, capsicum, carrot, cauliflower, celery, chilli, Chinese cabbage, cluster bean, Cucurbits (all), fenugreek, garlic and knol khol, lettuce, okra, onion, parsley, radish, rat tail radish, spinach, sugar beet (mother bulb/ root production), tomato, turnip	0.10	0.2
SPICES		
Ajawain (*Trachyspermum ammi*), cumin, coriander, fennel, Indian dill (Suwa)	0.1	0.5
Ginger, turmeric	0.5	1.0

Crop	Foundation	Certified
ORNAMENTAL		
Aster, carnation, chrysanthemum, marigold, petunia, snapdragon	0.05	0.1
Sunflower	0.1	0.2

Table 10. Field inspection stages for verification of offtypes during seed production

Stage	Crop
	FIELD AND FODDER CROPS
During flowering	Maize (composite, synthetic and OPV)
At any one inspection at and after flowering	Buffel grass, castor, cotton, Egyptian clover, forage sorghum, guinea, Lucerne, marvel grass, Niger, pearl millet, pigeonpea, rape seed and mustard, safflower, sesame, setaria grass, sorghum, stylo, sunflower, taramira, teosinte
Final inspection before harvesting	Barley, chickpea, cluster bean, cowpea, dharaf grass, Dinnanath grass, field pea, grass pea, groundnut, horse gram, Indian bean, jute, kidney bean, lentil, linseed, *Melilotus* sp., Mesta (*Hibiscus* sp.). minor millets, moong bean, Napier grasses, oat, rajmash, *Sesbania*, sunhemp, rice, rice bean, soybean, triticale, urid bean, wheat
	VEGETABLES
At plant bed and after transplanting	Sweet potato
I and II Inspection (35 and 65 days after planting)	Potato including tissue culture-raised mini-tuber (PTCM)
II inspection	Carrot, onion (mother bulb/ root production), radish
At and after flowering	Amaranth, asparagus, brinjal, cabbage, capsicum, carrot, cauliflower, celery, chilli, Chinese cabbage, cucurbits (all), knol khol, okra, onion, parsley, radish, rat tail radish, spinach, sugar beet (mother bulb or root production), turnip
Final inspection	Lettuce, colocasia, lesser yam, tomato
	SPICES
45 days after planting at 50% flowering	Ajawain (*Trachyspermum ammi*), cumin, coriander, fennel, Indian dill (Suwa)
180-190 days after planting	Ginger
120-180 days after planting	Turmeric
Final inspection	Fenugreek, garlic
ORNAMENTAL	
At and after flowering	Aster, carnation, chrysanthemum, marigold, petunia, snapdragon, sunflower

7.5. Presence of other crop plants

Seed production plot should be under sole cropping. Some crop plants are designated as inseparable crop plants for specific crop (Table 11). Number of other crop plants should not be more than the maximum seed certification standard at final inspection.

Table 11. Inseparable other crop plants and their maximum seed certification standard (%)

Crop	Inseparable other crop plants	MSCS(%)	
		F	C
Barley	Wheat, oat , triticale, chickpea	0.01	0.05
Wheat	Barley, oat, triticale, chickpea	0.01	0.05
Oat	Wheat, barley, triticale, chickpea	0.01	0.05
Triticale	Wheat, barley, rye, oat	0.01	0.05

7.6. Presence of objectionable weeds

Weed plants designated as objectionable for associated crop should be removed from the seed plot before flowering. Objectionable weeds should be verified at specific stage of field inspection (Table 12). The maximum permitted limit of objectionable weed in seed plot is presented in Table 13.

Table 12. Stage of field inspection for verification of objectionable weeds

Stage of field inspection	Crop
FIELD AND FODDER CROPS	
At any one inspection at or after flowering	Egyptian clover, lucerne, safflower, sunflower
Final inspection	Napier grass, oat, rape seed and mustard, rice
VEGETABLES	
At and after flowering	Amaranthus, bitter gourd, musk melon, okra, snake gourd, Snap gourd, water melon
Final inspection	Fenugreek, lettuce

Table 13. Maximum permitted limit (%) of objectionable weeds in seed production at field level

Crop	Weed		Foundation	Certified
		Field crops		
Rice	Wild rice	*O. sativa* var. *fatua*	0.01	0.02
Mustard	Mexican Prickly poppy	*Argimone mexicana*	0.05	0.1
Safflower	Wild safflower	*Carthamus oxyacantha*	None	None
Sunflower	Wild sunflower	Wild *Helianthus*	None	None
		Fodder crops		
Egyptian clover	Chicory (Kasni)	*Cichorium intybus*	None	0.05
Napier Grass	Canada thistle	*Cirsium arvense*	None	None
	Johnson grass(Saru)	*Sorghum halepense*	None	None
	Quack or Thitch or couch grass	*Agropyron repens*	None	None
	Wild morning glory (Hirankhuri)	*Convolvulus arvensis*	None	None
Oat	Wild oat	*Avena fatua*	0.01	0.02

Crop	Weed		Foundation	Certified
		Horticultural crops		
Amaranthus	Wild amaranthus	*A. spinosus*	0.01	0.02
Bitter gourd	Balsam apple (*mokha*)	*Momordica balsamina*	None	None
Bitter gourd	*Jangli karela*	*M. diocia*	None	None
Bitter gourd	*Bhat kharela*	*M. cochinchinensis*	None	None
Coriander	Grass pea	*Lathyrus sativus*	None	None
Cucumber		*Cucumis hardwickii*	None	None
Fennel	Dodder	*Cuscuta* spp.	None	None
Fenugreek	*Senji*	*Melilotus* spp	0.01	0.02
Lettuce	Wild lettuce	*Lactuca scariola*	0.01	0.02
Long melon		*Cucumis prophetarum*	None	None
	Weed melon (Tekmek)	*C. melo* var. *argrestis*	None	None
	Other non desert forms of	*C. melo*	None	None
Musk melon		*Cucumis prophetarum*	None	None
	Weed melon(Tekmek)	*C. melo* var. *argrestis*	None	None
	Other non desert forms of	*C. melo*	None	None
Okra	Wild okra	*Abelmoschus ficulneus*	None	None
		A. manihot	None	None
		A. moschatus	None	None
Snap melon		*Cucumis prophetarum*	None	None
	Weed melon (Tekmek)	*C. melo* var. *argrestis*	None	None
	Other non desert forms of	*C. melo*	None	None
Snake Gourd		*Trichosanthes palmata*	None	None
	Jangli chachinda	*T. lobata*		
	Rambel	*T. cucumerina*		
Watermelon	Wild water melon (*Indrayan*)	*Citrullus colocynthis*	None	None

7.7. Seed borne diseases

Different crops have designated and specified seed borne diseases objectionable at field level (Table 14,15,16 and 17).

Table 14. Seed certification standard for objectionable diseases of crops

Crop	Disease	Maximum infection (%)	
		Foundation	Certified
Field			
Barley	Loose smut	0.10	0.50
Moong	Halo blight	0.10	0.20

Crop	Disease	Maximum infection (%)	
		Foundation	Certified
Pearl millet	Grain smut	0.05	0.10
	Downy mildew (Green ear)	0.05	0.10
	Ergot	0.02	0.04
Certification standard for grain smut is verified at final inspection prior to harvest. Seed fields with incidence of grain smut more than the maximum permissible level can be certified if seed is treated with an approved organo-mercurial fungicide a month prior to its sowing. *Seed production programme of pearlmillet should not have infection more than the certification standard for downy mildew/ green ear disease at any field inspection by considering complete stool as one infected unit.*			
Sesame	Cercospora leaf spot	0.50	01.0
Sorghum	Kernel (grain) smut and Head smut	0.05	0.10
Seed fields of sorghum infected by grain and head smut can be reinspected (only once). Diseased earheads are removed and burnt and the field shown on re-inspection should not have infection more than the maximum permissible limit.			
Sunflower	Downy mildew	0.05	0.50
Infection of downy mildew is verified at each inspection for both the category.			
Triticale	Loose smut	0.10	0.50
	Ergot	0.02	0.04
Wheat	Loose smut	0.10	0.50
Forage			
Forage sorghum	Kernel and head smut	0.05	0.10
Oat	Loose smut	0.10	0.50
Sudan Grass	Kernel smut	0.05	0.10
	Head smut	0.05	0.05
Other crops			
Jute	Jute chlorosis	1.00	2.00
Vegetables			
Brinjal	Blight and fruit rot (*Phomopsis vexans*)	0.10	0.50
Brocoli	Black leg, Bacterial black rot, *Erwinia* soft rot	0.10	0.50
Cabbage	Black leg, Bacterial black rot, *Erwinia* soft rot	0.10	0.50
Cauliflower	Black leg, Bacterial black rot, *Erwinia* soft rot	0.10	0.50
Celery	Phoma root rot, Septoria leaf blight	0.10	0.50
Chilli	*Alternaria* leaf blight, anthracnose	0.10	0.50
Chinese cabbage	Black leg, Bacterial black rot, *Erwinia* soft rot,	0.10	0.50
Cluster bean	*Anthracnose*, *Ascochyta* blight , *Xanthomonas* bacterial blight	0.10	0.20
Cowpea	*Macrophomina* ashy stem blight (MASB), *Anthracnose*, *Ascochyta* blight, mosaic	0.10	0.20

Crop	Disease	Maximum infection (%)	
		Foundation	Certified
French bean	*Anthracnose, Ascochyta* blight, *Xanthomonas* bacterial blight, bean common mosaic	0.10	0.20
Indian bean	*Anthracnose*, *Ascochyta* blight, bacterial blight	0.10	0.20
Knol khol	Black leg, bacterial black, *Erwinia* soft rot	0.10	0.50
Lettuce	Lettuce mosaic virus	0.10	0.50
Muskmelon	Cucumber mosaic	0.10	0.20
Parsley	Septoria leaf spot	0.10	0.50
Radish	Black leg, Bacterial black rot, *Erwinia* soft rot	0.10	0.50
Summer squash	Cucumber mosaic, Watermelon mosaic	0.10	0.50
Tomato	Alternaria early blight, Stemphylium leaf spot, Tobacco Mosaic virus	0.10	0.50
Turnip	Black leg, Bacterial black rot, *Erwinia* soft rot	0.10	0.50

Seed production programme of cowpea, French bean, cluster bean and Indian bean should pass the seed certification standard for infection of Ascochyta leaf blight only in hilly areas.

Table 15. Seed certification standard for objectionable diseases of ornamental crops

Crop	Disease	Inspection stage	Field standard (%)	
			Foundation	Certified
Annual carnation	Streak mosaic virus	at final prior to harvesting	0.10	0.20
Chrysanthemum	Grey mould, Blotch	at final prior to harvesting	0.10	0.20
Marigold	Leaf spot, Flower bud rot, Collar rot	At final (III) in variety and at final (IV) in hybrids	0.10	0.20
Petunia	*Cercospora* Leaf blight, *Ascochyta* Leaf spot, *Phyllosticta* Leaf spot, *Alternaria* Leaf blight, *Phytophthora* Crown rot, Tobacco mosaic virus, Cucumber mosaic virus	at final prior to harvesting	0.10	0.20
Snapdragon	Anthracnose , Blight	at final prior to harvesting	0.10	0.20
Sunflower	Downy mildew	In all three	0.05	0.50

Table 10.16 Seed Certification standard for objectionable diseases of fruit crops

Crop	Disease	Inspection stage	Field standard (%)	
			Foundation	Certified
Banana	Bunchy top Panama disease	In both the crops minimum one inspection at the time of fruit maturity for health of mother plant	Not specified	Not specified
Grape	anthracnose		Not specified	Not specified

In both the crops mother plant should be healthy (free from diseases) certificate be given to nurseries

Table 17. Standard for objectionable diseases of vegetatively propagated crops

Crop	**Disease**	**Inspection stage**	**Field standard (%)**		
			Foundation		**Certified**
Ginger	Rhizome rot	I	None		None
		II-IV	None		None
	Phyllosticta leaf spot	III	1.00		1.00
	Bacterial wilt	III	1.00		1.00
	Rhizome rot	I	None		None
Planting on soil infested with Pythium sp., Pseudomonas solanacearum and Meloidogyne incognita should be avoided. Inspection I at the time of planting, II 45-125, III 180-190 days after planting and IV before harvest or between 240-250 days after planting.					
Potato			F1	F2	
	Mild mosaic	I and II inspection	1.0	2.0	3.0
	Severe mosaic, leaf roll and yellows	I and II Inspection	0.50	0.75	1.0
	*Total virus	-	1.0	2.0	3.0
	**Bacterial wilt	I and II Inspection	0.0	0.0	3 plants/ ha
True Potato Seed					
	Mild mosaic				3.0
	Severe mosaic, leaf roll and yellows				1.0
	Pseudomonas brown wilt				3 plants/ ha
Potato tissue culture raised mini tuber (PTCMT)					
	Plants showing symptoms				0.05%
	*Mild mosaic				0.05%
	*Severe mosaic, leaf roll, yellows and apical leaf curl				
	**Plant infected with *Ralstonia* brown rot Syn. Bacterial wilt				0.0
** Maximum permitted before dehalming*	***To maintain the crop hygiene, the disease must be watched*				
Sugarcane	Red rot	I,II, III	None		None
	Smut	I	0.02		0.10
		II	0.01		0.10
		III	None		None
	Grassy shoot	II	0.05		0.50
		III	None		None
	Cephalosporium Wilt	III	0.01		0.01
	Xanthomonas Leaf scald	II	0.01		0.05
		III	None		None
**Subject to immediate rouging of the whole clamp along with roots and destroyed. Inspection: Stage I at 45-60 and II at 120-130 days after planting whereas, III at 15 days prior to harvest.*					
Sweet potato					
Plant bed	Black rot		None		None
	Fusarium wilt		None		None

	Monilichates scurf		None	None
Seed field	*Fusarium* Wilt		None	None
	Mosaic		0.05	0.10
Taro	Phytophthora rot		0.0	0.00
	Dasheen mosaic		0.50	0.10
Tapioca	Mosaic		0.10	0.50
Turmeric	Rhizome rot	I	0.0	0.0

The seed turmeric is not eligible for certification on the land infested with Pythium sp., Pseudomonas solanacearum and Meloidogyne incognita. Inspection I at the time of planting, II at 45-50, III at 120-180 days after planting and IV before harvest or between 240-250 days.

7.8. Presence of objectionable insect pests

Some crop plants have designated objectionable insect pests at field stage which, should be verified during the field inspection at any stage (Table 18).

Table 18. Objectionable insect pest of designated crops and their maximum seed certification standard (%)

Crop	**Insect**	**Stage of inspection**	**Foundation**		**Certified**
Colocasia	Scale insect	Final	0.0		0.0
	Mealy bug	Final	0.0		0.0
Lesser yam	Scale	Final	0.0		0.0
Sweet potato	Wireworm	Final	1.0		1.0
	Weevil	Final	0.0		0.0
Tapioca	Scale insect	Final	0.0		0.0
Sugarcane			I	II	Certified
	Gurdaspur borer	III	5.0	5.0	5.0
			None	None	None
	Plassey borer	III	5.0	5.0	5.0
			None	None	None
	Mealy bug	III	5.0	5.0	5.0
			None	None	None
	Scale insect	III	5.0	5.0	5.0
			None	None	None
	Internode borer	III	10.0*	10.0*	10.0*
			None	None	None
	Stalk borer	III	20.0	20.0	20.0
			None	None	None
	Top borer	II and III	5.0	5.0	5.0

None: the infestation should be none in the area where presence of the pest has not been recorded
III field inspection is made 15 days prior to the harvesting of seed canes.

7.9. Presence of objectionable parasites

Some crop plants have designated objectionable parasites at field stage which, should be verified during the field inspection at any stage (Table 19).

Table 19. Seed certification standard for infestation of objectionable parasitic weeds

Crop	Parasitic weeds and nematodes	Inspection stage	Standard (%)	
			F	C
Sunflower	*Orobanche* spp.		0	0
Lucerne	Dodder		None	0.05
Napier Grass	Dodder		0	0
Ornamental sunflower	*Orobanche spp.*	I at 6-7 pair of leaves, II during flowering and III prior to harvesting	0	0

7.10. Presence of objectionable nematodes

Some orchard crop plants have designated objectionable nematodes at field stage which, should be verified in both the crops at the time of fruit maturity for health of mother plant during the field inspection at any stage (Table 20). In both the crops health (free from diseases) certificate is given to nurseries.

Table 20. Seed certification standard for infestation of objectionable nematodes

Crop	Disease	Field standard (%)	
		Foundation	Certified
Banana	Nematode	Not specified	Not specified
Grape	Nematode	Not specified	Not specified

8. RE-INSPECTION

In the event of unsatisfactory performance of the seed production plot the grower can apply for reinspection of the field by submitting half of the fee required for field inspection. Grower should improve the performance of the seed crop prior to reinspection with main emphasis to the factors, which were responsible for the rejection. Field inspector should consider that the contaminators removed by the grower have no adverse affect on the quality of seed.

■■■■

Chapter 13

MAINTENANCE OF GENETIC PURITY

Objective

To control the factors affecting genetic purity of a variety at field level.

Genetic purity can be maintained by

- Control of source of seed
- Preceding crop requirement
- Isolation
- Rouging
- Seed certification
- Testing of seed for genetic purity
- Change of seed after every three years

Out of these factors isolation and rouging are discussed in this chapter.

Isolation

The main objective of isolation is to keep the seed crop away from the sources of genetic contamination. Maintenance of genetic purity of a seed crop by keeping the variety in isolation from varieties of the same species and some cross compatible other species to avoid out-crossing is known as isolation. Isolation is of following types:

Space isolation: Seed plot is located at a particular distance from the sources of genetic contamination so that the foreign pollens are unable to reach in viable state.

Factors determining distance for space isolation

Mode of pollination: Self-pollinated crops (bisexual, homogamous and cleistogamous), vegetatively propagated and apomictic crops require minimum isolation distance. Cross-pollinated crops (unisexual, dichogamous, chasmogamous, herkogamous, heterostyly, self-incompatible and male sterile) require more isolation distance. Often cross-pollinated crops (chasmogamous, attractive for insects and stigma remains receptive and pollen viable for longer period of time) also require more distance to keep the crop in isolation (Fig. 1, Table 1).

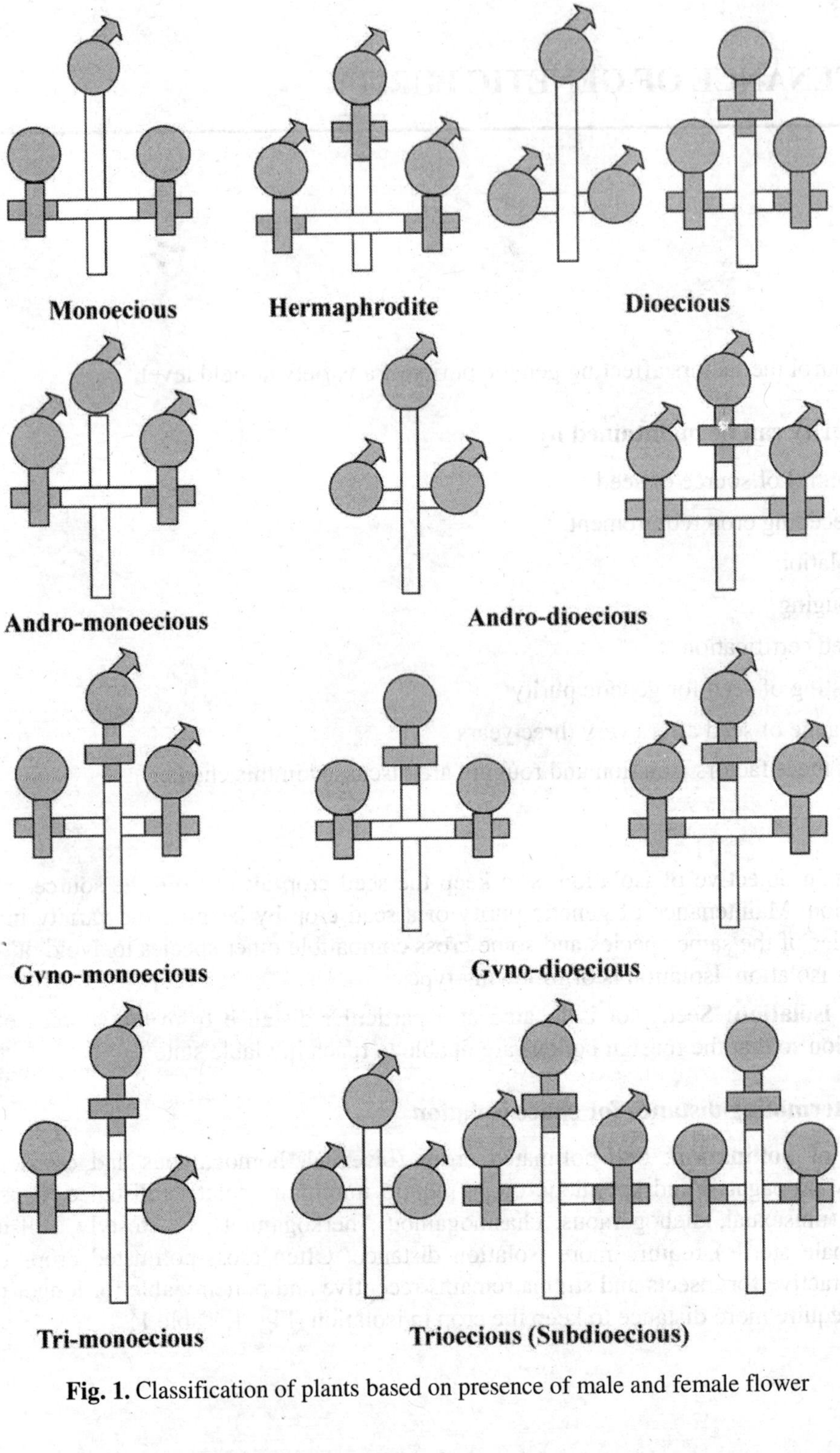

Fig. 1. Classification of plants based on presence of male and female flower

Table 1. Classification of crops on the basis of mode of pollination

Pollination		
Self	**Often cross**	**Cross**
Chickpea, field pea, French bean, ground-nut, lentil, moong, rice, soybean, urid, wheat	*Brassica sp.* (self compatible), cotton, linseed pigeonpea sesame, sorghum	*Brassica sp.* (self incompatible), castor, lucerne, maize, niger, pearl millet, safflower, sunflower, sugarcane
Chilli, lettuce, potato, tomato	Egg plant, Okra	Cabbage, cauliflower, coriander, cucumber, melon (musk and water), onion, pumpkin, radish, squash (Indian, summer and winter), gourd (ash, ridge, bottle, bitter, snake) and turnip

Pollinating agent

High isolation distance is required for cross pollinated crops with wind driven pollen because wind velocity is always additive. Greater isolation distance is also required for cross and often cross-pollinated crops with insect driven pollen, as high density and movement of insect responsible for pollination are always additive.

Any chemical that secretes by an organism and influences the behaviour of another species in such a way that the change is beneficial only to the producer host is known as *Allomone*. Its effect is beneficial for producing plant (host) by causing a reaction in pest that is favourable for plant but not for insect. It prevents or reduces contact between insect and plant as repellent and leads to self-pollination. The adaptive benefit of a trans-specific chemical messenger to the recipient rather than the emitter is known as *Kairomone*. It is normally non adaptive to the transmitter. It is a substance produced by the plant that causes a favourable reaction in insect as attractant/ arrestant/ excitant/ feeding stimulant/ oviposition inducer. When the action of semio-chemical is as *kariomone* and the action of the insect is beneficial for the plant and insect both then the chemical is known as *Synamone* (Honeybee for pollination). It leads to cross-pollination.

Some insects are involved in transmission of pathogen from one plant to another known as vector and the disease escape accomplished by the avoidance of insect vector is known as *Klenducity*.

Tripping: Exposure of male and female parts of a flower due to pressure of insect during nectar or pollen collection is known as tripping. Lucerne a self-incompatible crop of family Leguminoceae has sexual organs concealed by keel (petal). On visit, the pollen grains are deposited on the lower surface of the bee. The bee carries the pollens from one flower to other in search of nectar. The tripping exposes the stigma. During the process of tripping stigma strikes the bee on the lower side (where pollen grains are already deposited) with considerable force that results in pollination.

Determination of isolation distance

STEP I : Examination of floral structure : The floral structure is studied to verify the mode of pollination of the crop.

Stigma receptivity: To ascertain the duration of stigma receptivity by <u>fruit set method</u> controlled pollination is attempted at different stages of stigma development, corresponding to the period mentioned below :

a) two days before anthesis b) one day before anthesis

c) on the day of anthesis d) one day after anthesis
e) 2 days after anthesis f) 3 days after anthesis
g) 4 days after anthesis h) 5 days after anthesis

For each selected day minimum 20 fully developed but unopened buds are emasculated and bagged about 3 days prior to anthesis. On the selected day the emasculated buds are pollinated with freshly collected pollens. On the basis of percentage of fruit setting the duration of stigma receptivity is calculated.

Pollen viability: The pollen grains collected at the time of anthesis are tested at different time interval for viability using 2% aceto-carmine solution. The pollens are transferred in a cavity slide containing aceto-carmine solution with the help of a fine brush. It is covered with a cover glass and pollen grains are examined under the low power microscopic field to test the viability. Deeply stained pollen grains are considered as normal.

$$\text{Pollen viability}(\%) = \frac{\text{Number of viable pollen grains}}{\text{Total number of pollen grains in the microscopic field}} \times 100$$

Category of seed: Isolation distance is also influenced by the category of seed under multiplication. High category of seed requires more isolation distance than the lower category seed because of its importance and value.

STEP II: Identification of marker characters: Marker characters of the variety are identified so that it can easily be distinguished from other varieties of the same species.

STEP III: Genetics of marker character: Inheritance of the selected character is worked out by making the reciprocal crosses with genotypes having different expression of the marker character.

STEP IV: Homogeneous population of homozygous plants for marker character at dominant and recessive levels should be prepared by continuous selfing.

STEP V: Wind direction of the area (if specific): Wind direction of the area is determined prior to start of isolation studies.

STEP VI: Wind velocity of the area in the season (if specific): If wind velocity remains high in cropping season then the isolation distance should be determined in that area in cropping season.

STEP VII: Insect: Behaviour, density and movement of insects responsible for cross-pollination should be worked out.

STEP VIII: Nicking: Anthesis period of contaminator (dominant gene for marker character) and stigma receptivity of tester (recessive gene for marker character) should be similar, otherwise required staggering period should be provided.

Designs for determination of space isolation

In all, three designs (Table 2) are used for determination of isolation. In design 1 when the crop is cross-pollinated, the contaminator is grown at East and the tester plots towards West at the interval of 50m (Fig. 2). The size of the contaminator should be three times of tester plot to provide sufficient pollen load. Use of insecticide in the experiment should be avoided. In design 2 when the crop is often cross-pollinated the contaminator is grown in the centre and tester at 25m intervals as a circle in all the directions. Whereas, in case of self- pollinated crops contaminator is grown in the centre and tester plot at 5,10,15 and 20 meters distance, each located in any one direction (Fig. 3).

Table 2. Specifications for determination of space isolation

Factors	Design 1	Design 2	Design 3
Mode of pollination	Cross	Often cross	Self
Marker	Violet vs. green stem	Green vs. yellow cotyledon	Violet vs. white flower
Genetics	Violet is dominant over green	Green is dominant over yellow	Violet is dominant over white
Wind direction	East to West	Not specific	Not specific
Wind velocity	High	High	High
Insect responsible for pollination	Honey- bee Low density and slow movement at low and high temperature	High density and movement for pollination between 9-12am	No

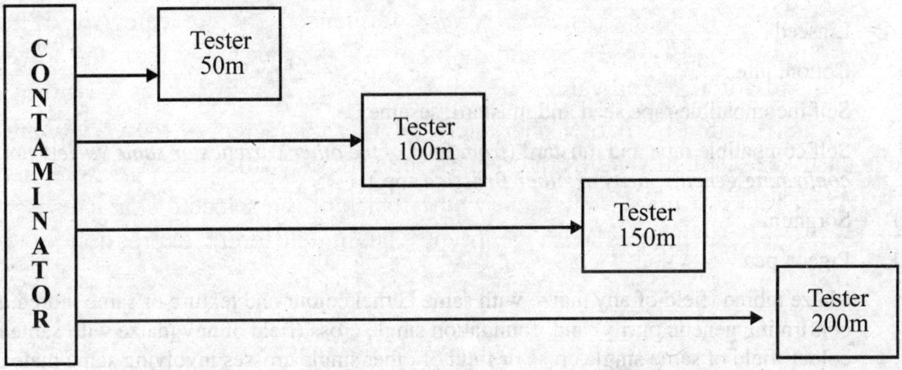

Fig. 2. Determination of isolation distance design 1

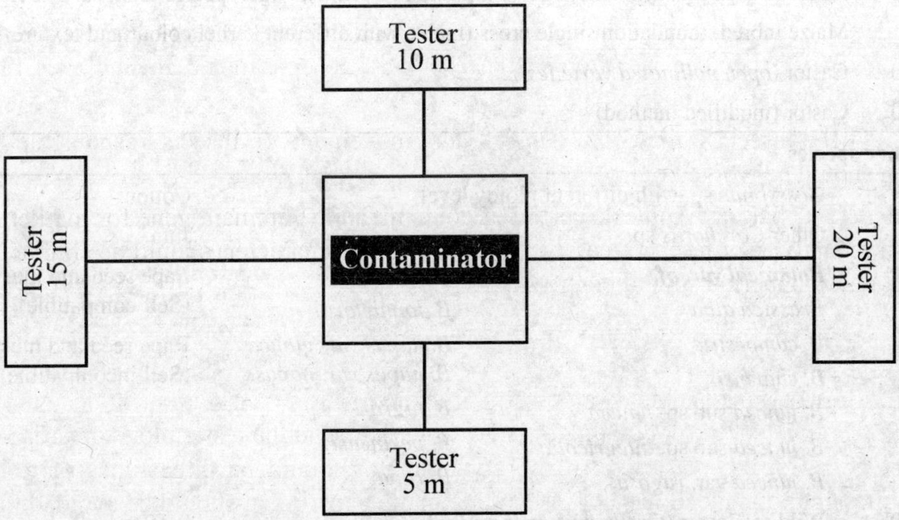

Fig. 3. Determination of isolation distance design 3

Each tester plot is harvested separately and the produce of each plot is raised separately. Plants are evaluated at the time of expression of marker characters. The percentage of contaminator in the produce of each tester plot is worked out to know the safe isolation distance. The minimum isolation distances recommended for seed production of different crops from other varieties and cross compatible species are presented in Table 3, 4, 5 and 6.

Table 3. Minimum distance (m) recommended to keep the seed production of field crops under isolation

Distance		Crops		
F	C	**From other varieties of same species**		
3	3	Barley, groundnut, minor millets, oat, rice, soybean, triticale, wheat		
5	-	Single cross maize field from other single crosses with genetic pure common male parent		
10	5	Chickpea cluster bean, cowpea, fieldpea, grass pea, horse gram, Indian bean, kidney bean, lentil, moong bean, rajmash, urid bean		
50	20	Rice bean		
50	25	Linseed		
50	30	Cotton, jute		
100	50	Self incompatible rape seed and mustard, sesame		
200	50	Self compatible rape and mustard (*from field of the other varieties or same variety not confirming genetic purity or other Brassica* spp.)		
200	100	Sorghum		
250	100	Pigeon pea		
400	-	Maize inbred (field of any maize with same kernel colour and texture or same inbred not confirming genetic purity) and foundation single cross (field of any maize with same kernel colour, field of same single cross or field of other single crosses involving same male parent but not confirming the genetic purity)		
400	200	Maize (*composite, synthetic, open-pollinated varieties*), Niger, pearl millet, safflower, sunflower		
600	-	Maize inbred, foundation single cross (maize with different kernel colour and texture)		
600	300	Castor (*open pollinated varieties*)		
1000	300	Castor (modified method)		
		Other species		
5	5	*Gossypium* sp. with different ploidy level		Cotton
5	5	Other *Corchorus* sp.		Jute
50	25	***Botanical var. of*** *Brassica alba*	*B. lonniefortii*	Rape seed and mustard (Self compatible)
100	50	*B. campestris* *B. chinensis* *B. juncea* sub sp. *juncea* *B. juncea* sub sp. *integricolia* *B. juncea* var. *rugosa*	*B. napus var. glauca* *B. napus var. napus* *B. nigra* *B. pekinensis* *B. rapa*	Rape seed and mustard (Self incompatible)
400	200	Wild safflower (*Carthamus oxyacantha*)		Safflower

Distance		Crops	
400	200	Wild sunflower	Sunflower
400	400	Teosinte	Maize (*open-pollinated composite, synthetic*)
600	-	Teosinte	Maize inbred line, foundation single cross

Table 4. Minimum isolation distance (m) recommended for seed production of horticultural crops in India

Distance		Crops	
F	C	**From other varieties of same species**	
3	3	Ginger, turmeric	
5	5	Colocasia, garlic, lesser yam (*Dioscorea esculenta*), sweet potato	
5	5	Potato	
20	20	Little and pointed gourd	
50	25	Fenugreek (field of other /same variety not confirming the varietal purity), tomato	
100	50	Lettuce (field of other /same variety not confirming the varietal purity)	
200	100	Ajwain (*Trachyspermum ammi.*), coriander, fennel, Indian dill (*A. suwa*)	
300	150	Brinjal (field of other varieties/same variety not confirming the varietal purity)	
500	250	Capsicum and chilli (field of other / same variety not confirming the varietal purity / cross compatible species), Okra (field of other /same variety not confirming the varietal purity)	
500	300	Asparagus, celery (A*pium graveolens* var.*dulce*), parsley (*Petroselinum crispum)*	
800	400	Cumin	
1000	500	All cucurbits baring pointed and little gourd	
1600	1000	Cabbage, cauliflower, Chinese cabbage, konl khol, spinach	
		Ornamental crops	
400	200	Sunflower (*variety*)	
600	300	Aster, annual carnation (*Margurite or Chaubad type*), snapdragon (*variety and hybrid*), chrysanthemum, petunia (*variety*), marigold (*variety and hybrid*)	
600	400	Petunia (*hybrid*) sunflower (*hybrid form field of other varieties and commercial hybrids*)	
		From other species	
400	200	Wild *Amaranth (spinosus)*	Amaranth
1600	1000	Botanical varieties of *Brassica oleracea viz., oleracea, ramosa, millecapitata, gemmifera, acephala, viridis, gongylodes,costata, subauda, italica, botrytis*	Cabbage, cauliflower, knol-khol
1000	500	Balsam apple, bhat karela, jangli karela	Bitter gourd
500	300	*Apium graveolens* var. *rapaceum*	Celery
500	250	*Capsicum*	Chilli
1600	1000	Botanical var. of *Brassica juncea, compestris, chinensis, napus, tournefortii, nigra, alba, rapa*	*Chinease cabbage*

Distance		Crops	
1000	500	*Cucumis hardwickii*	Cucumber
100	50	Wild lettuce (*Lactuca scariola*)	Lettuce
1000	500	Snap melon (var. *momordica*), oriental muskmelon, pickling melon (var. *conomon*), weed melon (var. *agrestis*), and other non desert form of *Cucumis melo* and *C. prophetarum*	Long melon
500	250	*Abelmoschus ficulneus, A. manihot, A. moschatus*	Okra
1000	500	Winter and summer squash, cushaw (*Cucubita mixta)*	Pumpkin
1600	1000	*Raphinus caudatus*	*Raphinus sativus*
1000	500	Sponge gourd	Ridge gourd
1000	500	*Trichosanthes palmata, T. lobata, T. cucumerina*	Snake gourd
1600	1000	*B. vulgaris* var. *cicla* and *rubra*	Spinach
1000	500	Ridge gourd	Sponge gourd
1000	500	Pumpkin, winter squash, cushaw	Summer squash
1000	500	Summer squash, pumpkin	Winter squash

Table 5. Minimum isolation distance (m) recommended for seed production of crops propagated by bulb and tubers

Distance				BULB AND TUBER CROPS	
Mother bulb		Seed production stage		Objectionable weed	Crop
F	C	F	C		
5	5	1000	1000	Other botanical varieties	Onion, carrot
5	5	1600	1000	Red tail radish	Radish
5	5	1600	1000	Other botanical varieties	Sugar beet
5	5	1600	1000	Other botanical varieties	Turnip

Onion hybrid	Field of varieties				Field of hybrid			
	of same skin colour		of different skin colour		same but not confirming certification standard		Other with common male parent	
Production	F	C	F	C	F	C	F	C
Mother bulb	5	5	-	-	5	5	5	5
Seed	1200	600	1500	750	1200	600	-	-

Table 6. Minimum isolation distance (m) recommended for seed production of fodder crops in India

Distance		Crops
F	C	**From other varieties of same species**
10	5	Dhaincha (*Sesbania aculeata*)
10	10	Napier grass (*Pennisetum purpureum*)

Distance		Crops	
20	10	Buffel grass, Dharaf grass (*Chrysopogan fulvus*), Dinnanath grass (*Penisetum pedicellatum,)* Guinea (*Panicum maximum*), marvel (*Dichathium annulatum)*	
50	25	Indian clover (*Melilotus* sp.), stylo	
50	30	Mesta (*Hibiscus* spp. *from field of other varieties and from field of the same variety not confirming to varietal purity requirements for certification)*	
200	100	Sorghum, teosinte	
250	100	Sun hemp	
400	100	Egyptian clover, lucerne	
400	200	*Setaria* grass	
		From other species	
200	100	Other *Chrysopogon* sp.	Dharaf grass
5	5	Other spp. of *Hibiscus*	*Hibiscus (Mesta)*
200	100	Other *Dichanthium* sp.	Marvel grass
400	400	Johnson grass (*S. halepense*) and forage sorghum with high tillering and grassy panicle	Sorghum

Time isolation

Cross pollination may be checked by adopting different time of sowing for different varieties, so that their blooming period will not coincide with each other. Time isolation is permitted for hybrid seed production of maize with modification in isolation distance (not less than half of the recommended distance), provided 5.0% or more plants in seed production plot do not possess receptive silks when, more than 0.1% of plants in the fields within the prescribed isolation distance start pollen shedding. Cutting can be done to alter flowering in seed production programme of grasses.

Barrier isolation

Isolation can be maintained by providing physical barrier of optimum height around the seed production plot. It reduces the movement of windblown pollen by controlling the velocity of air. The stems and leaves of the barrier crop act as an air filter removing pollen from the air entering in the seed production plot. Physical barrier of polythene sheet of 2 meter height or planting of 5 m wide strip of pearl millet / Dhaincha (*Sesbania aculata*) 15 days before transplanting of rice around the plots maintaining the A, B, and R lines of rice is recommended to maintain the genetic purity.

Border row

Hybrid seed production plots of maize are surrounded by the male parent used for seed production programme at the same time of sowing and with same seed rate as of seed crop. These plants check the entry of foreign pollens as physical barrier and provide pollens of male parent for the seed crop present in the centre. Isolation distance of plot producing hybrid seed of maize can be reduced upto half of the recommended distance by planting of border rows. Number of border rows to be planted is determined by the size of the seed production plot and its distance from the contaminator.

Discarding the border rows

Foreign pollens mainly pollinate the plants present on the outer periphery of the seed plot. Therefore, by discarding the produce of 5-6 lines present on the outer periphery the isolation distance can be reduced.

ROUGING TO MAINTAIN GENETIC PURITY

Objective

To maintain the genetic purity of the seed crop by selective removal of the undesirable plants of the same species which deviate from the normal expression of the variety from the seed production plot on the basis of visual inspection.

The act of removing the plants of the same species, which is deviating from the normal expression of the variety in a seed production plot is designated as rouging. It may also be defined as a selective removal of undesirable plants of the seed crop on the basis of visual inspection in the field to improve quality of seed. Rouging may be defined as careful and systematic evaluation of a seed production field and the removal of all undesirable plants of the same crop.

The goal of this operation is to assure the desired genetic purity in the seed production programme at field level. It is the most effective and important method of contaminant removal as it prevents contamination by out-crossing and eliminates existing contaminants. At the same time, it reduces the incidence of seed transmitted diseases. It is a major field operation of seed production that differs from commercial cultivation to maintain genetic purity and health of seed in large-scale multiplication. Plants of different categories are removed form the seed production plot by the process of rouging to improve quality of seed (Table 7).

Table 7. Categories of plants of same species removed during rouging

Rogue	Purpose
Diseased plant	To produce healthy seed
Insect infested plant	To reduce infestation
Plant of other varieties	To improve genetic purity by • Discarding produce of other variety (avoiding admixture) • Avoiding out-crossing (change at genetic level)
Plant of same variety with different expression mainly for phenological traits	To reduce genetic shift

Classification of genetic impurity

Two major types of contamination that occur predominantly in seed fields are genetic and mechanical contamination. Variant of the following two categories are rouged out to maintain genetic purity and uniformity:

Offtype: A plant in the seed crop, which deviates from the norm for the cultivar, but does not obviously belong to another cultivar, is considered as offtype. Plant is similar for distinguishing character(s) but deviates in the agronomical trait(s) like, days to flowering, plant height, disease reaction and days to maturity.

Rogue: Plant showing variation in expression of distinguishing characters or abnormal performance of the plant in comparison to normal expression of the cultivar i.e., plant of other cultivar, other species or diseased plant of the same cultivar. A plant is classified as a "rogue" if it is atypical which does not conform to the varietal description.

Requirement for proper rouging

Knowledge of diagnostic and phenological traits of the variety in seed production is the pre-requisite for the persons engaged in rouging. In the event the variety is new then the details of distinguishing and agronomical characters of the variety are pre-requisite.

Procedure

A minimum standard for seed certification is set for each class of contaminants that is permitted in a seed crop. Rouging ensures standard for genetic purity at field levels. Before start of rouging a plant breeder/ experienced person to take decision of rouging should monitor the field.

Rouging is carried out at the time when rogues can be most easily identified and before any contamination occurs to the seed crop by observing each and every plant by systematic movement and critical observation.

To maintain genetic purity by rouging, plants entirely other than the selected variety are identified based on the expression of diagnostic morphological traits whereas, off-types based on phenological traits. Field is scanned at different stage of plant growth to perform rouging. Meticulous rouging is essential before flowering, mainly in cross-and often cross pollinated crops to avoid genetic contamination through out-crossing.

The procedure requires systematic organization of human resource for complete removal of undesirable plants. The effectiveness of rouging in a crop well below eye level is of high standard in comparison to tall above eye level. Provision of good growing conditions *i.e.,* avoidance of biotic and abiotic stresses for proper expression of the characters are the base of effective rouging.

Rouging at seedling stage

Seedling has few distinguishing traits to identify genetic impurity. Presence of anthocynin on hypo or epicotyl is one of the most important trait at seedling level. Limited growth with complete and uniform exposure to the sunlight *i.e.,* proper GXE interaction makes expression of colour very effective *e.g.,* purple and green seedling pigmentation. Each and every seedling may be observed critically due to limited growth of plant. The expression of the diagnostic trait and limited growth made the procedure efficient and effective. Difference in height of seedlings because of resowing, may pose problem. Effective rouging at this stage reduces the chance of elimination of rogue during later stage. Uprooting of unwanted seedling without any damage to other seedlings is easier than plant at before and after flowering stages. Limited diagnostic traits at seedling stage with limited expressions are the major draw back, *e.g.*, all the varieties may be categorized in only two group *viz.*, purple and green, therefore seedlings of other varieties with same expression can not be identified. Differential in sensitivity to herbicide is also an important method to eliminate genetic impurity at seedling stage.

Rouging at pre-flowering stage

In fact rouging should be completed before flowering to avoid out-crossing. Limited traits with limited expressions make it difficult. Important traits to be observed before flowering are presented in Table 6 along with their expression for different crops. However, growth of plant, within plant variation for expression of distinguishing traits and very high environmental sensitivity make identification of rogues difficult and doubtful. The stage is perfect to sort out off-types based on phenological traits *viz.*, plant height. The level of many crops at this stage is upto or below eye level that make identification of off-types with tall height very effective. But, identification of dwarf (short) off-type in a large population is not feasible.

Rouging at flowering stage

Flowering is the perfect stage to identify genetic impurity in most of the crops. Variation in flower colour and other traits on flower for identification of plants of other variety and variation in time of flowering/ear head emergence of off-types are the most distinguishable and stable traits. To avoid contamination by out-crossing, rouging is to be performed with the initiation of flowering. Uprooting of genetic impurity from the crops with tillering (wheat and rice), creeping habit (field pea) and high plant density (soybean, lentil, Egyptian clover etc.) is difficult. Growth habit is another important trait, but in leguminous crops its expression is highly influenced by environment. It is feasible to identify a genetic impurity with early flowering but difficult to identify impurity with late flowering.

Rouging at green fruit stage

Many diagnostic traits are visible on fruits, therefore as a corrective measure to remove the left over genetic impunity, vigorous rouging is recommended in most of the crops at this stage. In any case rouging should not be performed at only this stage as it cannot avoid out-crossing.

Rouging before harvesting

Theoretically, rouging has to be performed before flowering but at many times it is not feasible because of large area and/or scarcity of trained human resource. It shifted rouging at maturity stage *i.e.*, before harvesting. It should never be advocated. Few crops have marked difference at maturity especially for plant colour that becomes an effective tool for marking of genetic purity.

Rouging at post harvest level

Important post harvest operations for genetic purity maintenance as per certification programme is for the crops where seed is stored as fruit, inflorescence or other appendages. In these crops, the rouging may be performed before pre-conditioning.

To eliminate off-type plants a visual observation is made systematically and uniformly over all the plants in the field by the rouging team. Undesirable or doubtful plants are removed manually. For efficient rouging, the following recommendations should be considered

- For rouging a team of 10-12 members with experience of rouging and proper training on the identification of the variety is constituted less than one supervisor.
- The team starts rouging from one side of the field walking parallel and in the same direction down the rows.
- It should be started from a definite place and end at a definite place with continuity.
- The place upto which rouging has been performed should be marked to start in the next day.
- The lines covered by a member during rouging should be narrow *i.e.,* not more than five for normal crops (wheat, rice, soybean, chickpea etc.) and two for tall crops (maize, sorghum, pigeon pea etc.) depending on the growth stage of the crop.
- It is easy to identify off-type plants under low angle sunlight from backside of roguer, therefore rouging during the early morning or late afternoon hours are advocated. Sunny hours may be avoided for rouging.
- Sometimes due to luxurious vegetative growth movement and observation become difficult at flowering and before flowering stages *e.g.,* soybean, mung bean, urid bean, field pea etc. In such cases use of long stick for safety and to turn the plants for observation may be helpful.
- Rogues should be removed from the field and disposed off.
- In Breeder seed production programme plant breeder should provide a space of 1-2m after every 20-30m for observation/monitoring.
- Rouging should be performed at minimum possible intervals with importance to critical stages. This should be repeated at different critical stages.
- The supervisor should cover the area by walking diagonally concentrating on overseeing and inspecting the activities of rouging team rather than on actually doing the rouging.

The certified and Foundation seed production programmes are certified based on genetic purity at field level by seed certification officer and Breeder seed production by the monitoring team. In the event, the observed number of genetic impurity is more than the prescribed Indian Seed Certification Standard, then seed certification officer may suggest rouging during first visit at early crop growth stage however during the visit at advanced stage he may pass or fail the programme based on the performance.

Authentic source of seed, careful selection of land and use of properly cleaned implements may prevent many problems of genetic impurity. The field in which different varieties of the same crop were grown in the previous year should not be selected for seed production of same crop/other variety. The seed production programme of the crops with seed or plant parts capable to remain viable in the field after shattering and grow as volunteer plant should strictly be avoided in the same field. Seed production programme should be avoided in area close to seed store or threshing floor.

Genetic aspects of rouging

The operation of rouging maintains the original genetic base of a variety. Plant breeder selects a small homogeneous population of homozygous desirable plants of a self-pollinated crop as a variety with the aim of maximizing the frequency of desirable alleles in the population by positive selection pressure. The breeder with responsibility to maintain the frequency of the same allelic combination and frequency generation after generation has to apply negative selection *i.e.*, removal of undesired plants present in relatively small proportion.

The variation in the variety is restricted upto acceptable limit every year based on indirect judgement of genotype on the basis of phenotype, which is the result of genotype x environment (GXE) interaction. It limits the effectiveness of rouging. Plants with similar genetic constitution may express differently and plants with different genetic constitution appear as of plants of same variety because of interaction between GxE. It shows that plant identified as offtype may produce normal progeny and normal plant may produce impure progeny at genetic level.

The Potential for uniformity

Potential for phenotypic uniformity in the populations (clones, inbred lines, hybrid and open pollinated population) differ according to their genetic structure. Environment plays main role for variation in vegetatively propagated crops. Theoretically inbred lines are uniform at genetic level as repeated self-fertilization during selection process made it single homozygous line. Variability arising in these varieties is discontinuous and relatively easy to detect. Hybrid developed from controlled crossing of inbred lines is completely uniform and any variability reflects the non-uniform parental lines or problem of production of F_1 due to self-fertilization.

Wide range of genetic base and greater phenotypic variability in open pollinated population, composite and synthetic make rouging difficult as variation for these populations tends to continuous and individual plant at each extreme may be considered as off type. Composite and synthetic varieties by their nature cannot be uniform and forced uniformity will lead to inbreeding that is lethal for their performance. Complex nature of synthetic and multiline populations needs specific maintenance techniques. However these lines should not show any variation beyond component lines. In general, wider the range of genetic base and environment lower the potential for retaining original gene frequency by rouging.

Rouging is a very effective tool to maintain genetic impurity of homogeneous population *viz.,* varieties of self-pollinated crops and hybrids. Whereas, for heterogeneous population *viz.,* cross-pollinated crops (synthetic/composite) or varieties of crops like Niger (sporophytic self-incompatible) seed production in isolation is more effective than rouging. Rouging in open pollinated varieties, composite and synthetic population should be reasonable without any intension of forced uniformity.

Rouging and population dynamics

Off-types may occur in various ways and rouging provides a major opportunity to remove them. No selection should be carried out during rouging to keep the genetic make-up of the variety same. The proportion of off-types in any particular population (seed lot) depends on

- The rate of addition of off-type plants in each generation.

- The number of generation of multiplication.
- Proportion of progeny of off-types.
- Stability of off-types in subsequent generations.
- The relative rate of multiplication of off-type plants.
- The effectiveness of removal of off-types by rouging in each generation.

■■■■

Chapter 14

MANAGEMENT OF OBJECTIONABLE WEEDS

As per agronomical definition weed plants are unwanted and undesirable plants present in a seed production plot, which interfere with the development of crop plants and reduce the quantity and quality of the produce.

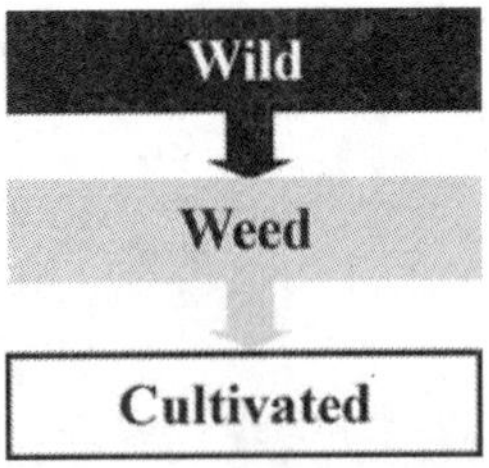

Seed science and technology considers that all the plant species were present as wild and during the process of evolution they were converted into weed and then into cultivated forms. Therefore, all the plant species not yet domesticated are considered as weed.

Noxious weeds

are legally considered as non-native plants that, once established, are highly destructive, competitive and difficult to control. They have economic and ecological impacts that go beyond agricultural lands. Some are toxic and pose public health threat to humans and animals. They may destroy native and beneficial plant communities. A noxious weed is in the country's noxious weed list, and their control is required by law. Weeds that may be undesirable, troublesome and difficult to control are termed as noxious weeds. They have immense capacity of reproduction and dispersal, and adopt tricky ways to challenge men's efforts to get rid of them.

Obnoxious weed (a weed of concern)

A weed of concern is in the state noxious weed list, but it is a plant that escaped from intentional plantings and is now widespread in the country. There are no regulations concerning these plants in the list. These plants often degrade native plant and animal habitat.

Objectionable weeds

Among various noxious weeds, some weed plants with following characteristics are considered as objectionable weed in seed production

- *Seed mimicry:* Seed is difficult to separate once mixed with crop seed viz., *Argemone maxicana* in mustard, *Cichorium intybus* in Egyptian clover. This type of objectionable weed is also known as satellite weed.
- High seed multiplication ratio eg., 200-400 seeds/plant in *Convolvulus arvensis,*16000 in *Cuscuta* and 20000 in *Argemone*.
- *Plant mimicry* Differentiation between crop and weed plant at early stage is difficult viz., wild rice in rice, wild oats in oat, etc.

- *Chronological mimicry :* These weeds have height and ripening time similar to the associated crop. For example, days to maturity of *Phalaris minor is* similar to wheat therefore harvested along with the crop and contaminates wheat grains.
- *Biochemical mimicry* **:** Herbicide recommended for any crop has no ill effect on the development of crop plant. A weed representing the same family or genus associated with the crop governs tolerance against the recommended herbicide. As a result, that weed may remain associated with the crop even with the application of recommended herbicide or herbicide with similar mode of action. This is considered as biochemical mimicry.
- Seed/plant is poisonous or injurious to human being.
- Smothering effect of weed on the main crop.
- Eradication of weed is difficult once established.
- Better dispersal mechanism.
- Dormancy and high seed longevity e.g., 22 years in *Convolvulus arvensis.*
- Low seed volume e.g., 1.0g of *Cuscuta* seeds is enough to infest one hectare area.
- Serves as an alternate host for crop disease and pests.

Sources of objectionable weed seed

- Infested field
- Infested seed
- FYM
- Movement of livestock and farm machinery
- Presence of weed in non-cropped area

MANAGEMENT STRATEGIES TO CONTROL WEED INFESTATION

Prevention

- Use of weed free seed is the foremost method to check the infestation.
- **Seed treatment:** to remove light weight weed seed from the crop seed, the seed is dipped in 5-10% solution of table salt for five minutes. The light weight seed of weed floats on the surface which is removed. Before sowing the crop seed is washed thoroughly with fresh water to remove traces of salt.
- ***Production of seed in weed free field.***
- Deep ploughing in summer.

Field preparation

- **Field preparation:** Proper tillage suppresses growth of weeds by desiccation and smothering by uprooting, cutting, or tearing. Repeated tillage also exhausts the food reserves of perennial weeds.

- **Preparation of field in dark:** Seeds of many weed require light to get rid of dormancy. Thus, germination of weed species whose seed require light may be impeded if the field is prepared in the night. For many buried seeds, it is the exposure to light during soil cultivation that stimulates germination. The exclusion of light during seedbed preparation has been shown to reduce weed emergence. Preparation of field in the night reduces weed emergence by up to 70%. However, not all weed species have light-sensitive seeds.
- **Light tillage:** For germination seed has to reach in the upper surface of the soil. To prevent the movement of seed present deep in the soil to the surface, light tillage is recommended e.g. *Phalaris minor.* It prevents germination and growth of weeds seed due to non-availability of nutrients from well placed fertilizer near the crop seed.
- **Stale seed-bed technique:** The field is irrigated and then prepared few days or weeks before planting or transplanting a crop to flush out germinable weed seeds before sowing. This depletes the weed present in the soil and reduces subsequent weed emergence. The emerged weed seedlings are killed by light cultivation or application of pre-sowing contact non-residual herbicides like paraquat.
- **Cultivation of crop on raised bed:** Weed seed present on top of the raised beds fails to germinate as the top of bed dry out quickly eg., *Phalais minor.* This method also facilitates mechanical weeding.

Crop rotation

- **Change of crop:** It helps in adopting control measures to eradicate the objectionable weeds. Identification of weed and application of selective herbicide is often difficult for control of objectionable weed. Therefore, change in crop permits weed control measures that were not possible.
- **Crop rotation:** Adoption of crop rotation may discourage development of weed populations that are well adapted to the growing conditions of any particular crop.
- **Keep field fallow:** In the event of heavy infestation field may be left fallow as during the fallow period, use of a different form of tillage implements or control method may be possible to reduce weed abundance.
- **Cultivation of crop with allelopathic effect:** The term Allelopathy is coined by Molisch (1937) that derived from two Greek words, *allelo* means mutual and *patho* means to suffer. It includes all stimulatory and inhibitory reciprocal biochemical interactions among plants. Allelopathy has potential in integrated weed management. It is a harmful, direct or indirect effect of one plant on another through the production of chemicals that enter the environment. Crop plants have capacity to produce and exude allele-chemicals into their surroundings to suppress the growth of weeds in their vicinity. Sorghum has allelopathic effect on *Amaranthus* spp.
- **Cultivation of crop with fast growth:** Cereal crops are more competitive against short-stature weeds e.g., wheat against *Melilotus alba* because of tall stature.

Inter-culture operation

- **Weeding:** Hand weeding and harrowing are the best method for control of any type of weed effectively.
- **Mulching:** It prevents germination of weed seeds and suppresses emergence and growth of seedling by allelopathic effect, toxic microbial products and change in soil pH. Mulching is very effective against annual weeds and some perennial weeds such as *Sorghum halepense*.
- **Soil solarization:** The basic principle of soil solarization is to trap heat in the soil received from the sun through light by covering the field with transparent polyethylene sheet. High temperature created by solarization kills the seed of wild oat and *Phalaris* present in the soil effectively.
- **Companion crop:** *Azolla*, is used in low land and irrigated rice as a companion crop. It reduces growth of weed by forming a blanket over the surface of water apart from supplementation of nitrogen in rice crop.
- **Farm-yard manure (FYM):** It is the main source for introduction of weed seeds to the field as the seeds of weed remain viable in the manure pits. 65-90 ^{0}C temperature of compost pit should be maintained for 4-5 months to kill the weed seeds. Treatment of FYM with chemicals like acrocyanamide, SMDC, DMTT, ammonium thiocyanate or uramon (synthetic urea) may kill weed seeds in 6-8 weeks.
- Weed seeds and vegetative parts of plants move with farm equipment and livestock. Therefore their movement from weed-infested areas to seed production area should be avoided.
- Non-cropped area viz., irrigation and drainage ditches, fence lines, farm boundaries, bunds, etc. should be kept free of weed plants.
- Prevention of seed formation on weed plants is an important control method as the seed multiplication ratio of weed plants is very high alongwith shattering habit and high seed longevity. These seeds remain in soil and cause problems for producers. Therefore weeding should be performed before flowering in weed plants.
- **Use of herbicide:** Use of selective herbicide is very effective and economical to control associated weeds. Control of *Phalaris minor* with the help of isoproturon is an important example. Most of the objectionable weeds belong to the same family as of crop therefore application of herbicide is not possible.

Seed processing

Weed seed carrying capacity depends upon the seed texture. Bigger size crop seeds carry less weeds seeds. Whereas small seeded crops are more prone to contamination particularly with weed seed of similar size and shape. Weed seed can be removed

- With the help of various mesh sizes vibrating sieves based on seed size.
- Dipping the seed in 5-10% common salt solution for settling the heavy seeds to the bottom and light weight are floated based on density.

- By specific gravity pneumatic and aspirator separators based on difference in specific gravity irrespective of shapes and sizes.
- By disc or cylinder separator grader based on difference in shape.
- By passing the mixture between two rollers covered with velvet cloth and rotating in opposite directions based on difference in seed surfaces and texture.
- By conveyer belts made of velvet cloth based on difference in texture.
- By magnetic separators that can remove rough, cracked, and sticky weed seeds from the smooth crop seeds.
- By disk separators based on difference in ability to slide or roll when they are subjected to centrifugal force in the separating drums.

Chapter 15

SEED TREATMENT AND MANAGEMENT OF SEEDBORNE PATHOGENS

OBJECTIVE

Improvement in shelf life of the seed, protection of seed during germination, seedling establishment and improvement in the vigour of seedling and plant with chemical or physical treatment.

A fairly large number of pathogens are carried **externally** or **internally** in the seed. The pathogen associated with the seed also become active when the seed germination takes place resulting in seed rot, seedling decay/ deformities or produce disease at a later crop growth stage. Use of healthy seed free from seed-borne pathogen is ideal for management of seed borne diseases. However, it is very difficult to obtain seed free from pathogen particularly in the tropical climate of India. The next choice remains to treat the seed with chemicals, use of physical factors or bio-control agents. Recently several novel methods have been developed to manage seed borne diseases by seed treatment.

Seed protection treatments

Seed is exposed to a wide range of biotic (seed and soil borne pests) and abiotic stresses during germination i.e., sowing to seedling establishment that has significant influence on plant population. Application of biological, physical or chemical agents and techniques on seed for protection during germination and seedling establishment to ensure optimum population of healthy plants with the use of required seed rate is known as seed protection treatment.

Merits of seed treatment

- Wide spectrum it controls both systemic and non-systemic pathogens
- Prevent spread of both systemic and non-systemic pathogens
- Protects seed from seedborne and soilborne pathogens
- It improves germination through control of surface moulds.
- It improves field emergence by enhancing vigour
- Enhances shelf life of seed
- Facilitate infected seed to form a healthy seedling
- Easy to apply

- Cost effective
- Optimum plant population with reduced seed rate
- Assured plant population in naturally infested soil

Biotic stresses at the time of seedling establishment: Following diseases and pathogen associated with seed results in stresses during seedling establishment

Disease: Symptoms

Seed rot: rotting of seed before germination.

Damping-off and seedling blight: soft rot of stem tissues near ground level and water soaking of seedling tissues.

Seedling wilt: Gray coloration initiated at the leaf tips and extending rapidly to the whole leaf, causing complete collapse of seedlings in 24 to 28 hr.

Root rot

Soilborne pathogen : (a) *Pythium* spp. (b) *Fusarium* spp. (c) *Sclerotium* spp. (d) *Rhizoctonia* spp.

In situations seed treatment is essential when

Damaged seeds: Damaged seed are used for sowing purpose. Injury on seed during post harvest operation or physiological process offers opportunity for the pathogen and moisture to deteriorate the seed by easy entrance. High electrical conductivity at the time of germination is the reason of seed damage that enhances the attack of soil borne pathogens.

Diseased seeds: Seed infected or infested by seedborne pathogen must be treated by appropriate chemical for assured germination and emergence.

Undesirable soil conditions: Cultivation of crop on infested soil (sick soil) warrants seed treatment for assured seedling establishment.

Susceptible variety: To establish optimum pant population of the variety susceptible to seed or soilborne pathogen require seed treatment.

Disease free seed: For production of disease free certified seed, the treatment of the seed to be used for sowing purpose is essential.

Underdeveloped seed: Use of underdeveloped seed for sowing purpose requires seed treatment to ensure formation of healthy seedling.

TYPES OF SEED TREATMENT

Seed disinfectant: It eradicates pathogen present deep in the seed. For effective control disinfectant should penetrate into the different parts of the seed.

Seed disinfectant: It is effective on the pathogen present on the surface of the seed i.e., contaminated the seed surface but not infected the seed. Chemical is applied on the surface of the seed as dust, slurry or liquids for disinfestation.

Seed protectant: It is effective to protect the germinating seed and young seedling from soil borne pathogen that may decay the seed before germination or seedling before or during emergence.

Agents for seed treatment

Fungicides: For dressing, seeds dried to low moisture content (8-10%) are thoroughly mixed with required quantities of fungicide e.g., Thiram or Carboxin @ 2 g/kg of seeds as powder or slurry. For dry application fungicide is moistened with water at the rate of 5 ml/kg of seed, mixed with seed and used for sowing purpose or shade dried before storage.

Elimination of a pathogen that has penetrated into living cells of the seed, infected it and become established is known as *seed disinfection.* Control of spores and other forms of pathogenic organisms present on the surface of the seed is known as *seed disinfestations.* Whereas, treatment to protect the seed and young seedling from pathogenic organisms of the soil is generally known as *seed protection.*

Fungicide is usually applied on seed as dust or slurry (a mixture of wettable powder in water) or liquid. Based on composition a fungicide may be organic or inorganic; Metallic or non-metallic. Presently mercurial fungicides are banned due to their lethal impact on human / animal chromosomes.

Mode of action: Fungicide that remain on the site of application i.e., surface of the seed and controls or retards the growth of certain fungi on contact without affecting the host's metabolic system of host is known as *contact fungicide*. Whereas the fungicide that moves inside the seed and control or retards the growth of certain fungi without affecting the host's metabolic system is considered as systemic *fungicide*. Fungicides that restrict the further growth of the organism is often called *fungistatic.*

Seed treatments recommend under Indian Seed Certification Standard: The seed of crops with designated objectionable seedborne pathogen may be treated with the designated fungicides (Table 1) before testing of seed for certification. The information related to seed treatment should be displayed on the container of both seed lot and submitted sample.

Table 1. Seed treatment recommend under Indian Seed Certification Standard

Crop	Fungicide	Rate g/100kg seed	Nature of treatment	Quantity of water to prepare slurry (l/100 kg seed)
Brinjal	Thiram 75%WDP or Captan dust	25	Dry	-
Chickpea	Thiram 75%WDP or Mancozeb	250	Dry	-
Chilli	Thiram 75%WDP or Captan dust	250	Dry	-
Cotton	Thiram 75%WDP or	250	Slurry	0.50
	Captan dust	110	Dry	-
Cowpea	Thiram 75%WDP or Mancozeb	70	Slurry	0.50
Finger millet	Thiram 75%WDP or Captan dust	75	Slurry	0.50
Groundnut	Thiram 75%WDP or	250	Slurry	0.50
	Captan dust	125	Dry	-
Jute	Captan dust	80	Dry	-
Maize	Thiram or Captan 75%WDP	70	Slurry	0.50
Mungbean	Activated clay	1000	Dry	-
Okra	Thiram 75%WDP or Captan dust	100	Dry	-

Crop	Fungicide	Rate g/100kg seed	Nature of treatment	Quantity of water to prepare slurry (l/100 kg seed)
Pigeonpea	Activated clay	1000	Dry	-
Rice	OMSD 1%	250	Dry	-
Sorghum	Thiram or Captan 75%WDP	85	Slurry	0.50
Soybean	Thiram 75%WDP or Mancozeb	70	Slurry	0.50
Sunflower	Thiram 75%WDP	250	Slurry	0.50
Tomato	Thiram 75%WDP or Captan dust	335	Dry	-
Urid bean	Thiram 75%WDP or Mancozeb	250	Dry	-

Insecticides: It is often applied to seed to control damage of insects such as wire worms and seed corn maggots at the time of seedling establishment and to check dispersal of seed by ant e.g., sesame.

Bio-agents: Seeds are treated with pre identified beneficial bio-agents to protect the seed and supplement the nutrients. Use of bio agent for seed treatment is ecologically sound, biologically sustainable. It is an essential component of organic farming.

Seeds are mixed with adhering agents and specific biological strains in appropriate proportions followed by surface drying.

Fungi: *Trichoderma* spp., *Aspergillus flavus, Curvularia pallescens* and *Chaetomium indicum*

Bacteria : *Pseudomonas, Bacillus subtilis Clavibacter* sp.

Botanicals: Leaf powder of *Azadirachta indica, Vitex negund, Prosopis,Acasia, Pungam etc.*

Nutrient supplementation: *Rhizobium* and *Azospirillum*

- The success of protection by bio-agents invariably depends on
- The potential of biocontrol agent
- The number of propagules applied to the seed
- The application technology

The concurrent inhibition of other microbes in the application process

TREATMENT WITH CHEMICALS

The chemical may be a fungicide (to kill fungus) or antibiotic (to kill bacteria) or nematocide (to kill nematodes) insecticides (to kill virus vectors) with different formulations including, water dispersible powder (WP), dust or slurries are used to control specific pathogen (Table 1). Based on the mode of action the chemical may be a **protectant** i.e., the chemical that becomes effective when applied prior to infection e.g., captan, thiram, dithan M 45, ziram, furbam etc., and **therapeutant** i.e., the chemical that is capable of eradicating the pathogen after it has caused infection and there by cure the plant e.g., Carbendazim (bavistin, derosol), Carboxin (vitavax), benzidmiazole (benomyl) etc. The chemicals those act by entering into the system of the plant are known as **systemic** e.g., therapeutant chemicals, whereas, the other chemicals, which do not act systemically are called **non-systemic** e.g., protectant chemicals.

Considering the ill effects and chromosomal abrasions effect mercury (the heavy metal) is not advocated and is banned for plant disease management, however, in some cases, the judicious use is followed. Chemicals use for seed treatment is mercurial or non mercurial compounds. In the present era mercury compounds are legally banned.

Non Mercurial compound may be organic or inorganic in nature. Organic Non Mercurial compound such as thiram, captan, Carbendazim etc. are widely used because of its effective control, compatibility with the symbiotic microorganism, less dangerous during handling wide spectrum, perform the role of both disinfectant and protectant, its over dosage is not harmful on germination of seed, long shelf life etc. The major inorganic Non Mercuric compounds used as fungicides are copper carbonate, copper sulphate and cuprous oxide. Copper carbonate and copper sulphate are used to control bunt of wheat and cuprous oxide to prevent seed decay and damping off of seedling of vegetable crops.

METHODS OF SEED TREATMENT

Seed dressing: Treatment of seed with agrochemicals / pesticides (fungicides, insecticides or a combination of both) before sowing to protect the seed during germination and seedling establishment from soil borne pathogens/insect pest is known as seed dressing.

Slurry treater: Large quantity of the seed in seed processing plant is treated by **slurry** treater. Slurry is prepared by mixing the chemical with water. The required quantity of the chemical is automatically mixed with the specified quantity of the seed before bagging. In a seed processing plant it is a continuous process and is more efficient than the rotary seed dresser.

Seed seeping in botanicals: Soaking of loose smut infected wheat seeds in 3% onion or 4 per cent garlic extract at 30ºC for 29 hrs bring down the infection, however, soaked seeds cannot be stored for more than 10 months. With the development of systemic fungicides the rest of the methods have only academic significance.

Rotary seed dresser*:* Small quantity of the seed is treated by rotary seed dresser. A simple mixer can be made by running a pipe through a drum at an angle. The drum is then mounted onto a stand. The seed and treatment are placed in the drum and it is rotated slowly until all the seeds are corned with the chemical.

Vacuum infiltration: Soaking of seed in captafol or benomyl (1g a.i./ 1 water + 0.1 ml Tween-20) solution for 7 days with change of solution every day. Keeping the seed in a Vacuum pump/ chamber has been reported most effective for controlling deep seated *Fusarium oxysporum* fsp. *elacidies* in oil palm.

Osmotic priming: Aqueous solution of polyethylene glycol (PEG) provides an osmotic environment. Immersion of seed in PEG solution followed by drying and storage does not impair germination and the method is called osmotic priming. Immersing rice seed in 20%PEG containing 0.1% Carboxin reduce seedborne infection of *Drechslera oryzae.*

Solid matrix priming: Pre sown hydration in a solid based media is called solid matrix priming. It increases the efficiency of the fungicide in control of seed borne infection. Seed treatment with standard fungicide (Thiram, Carboxin, Captan, Metalaxyl etc.) plus priming agent effectively control the seed and soil borne pathogens and increases the stand of maize in field whereas, seed treatment only with fungicide is ineffective.

Seed dip method: It involves the dipping of the seed or propagating material in a chemical (fungicide, antibiotic etc.) solution for a specified period prior to sowing. Seed soaking with organo- mercurial (for potato scab), dithan M 45 or ridomil (for potato late blight), streptomycin sulphate (for cabbage, cauliflower or mustard bacterial and fungal diseases) have been found effective. Hot water treatment followed by seed dip in streptomycin sulphate has given very effective results against bacterial diseases of cabbage and cauliflower. Seed soaking of cluster and Indian bean in antibiotics like streptocycline and aureofungin control *Xanthomonas* spp. Seed dip and treatment with 20 percent trisolium phosphate solution dilute the tobacco mosaic virus (TMV) concentration.

Hopper box treatment: Requisite quantity of the chemical (dust formulation) is mixed in the hopper box of a seed drill before sowing.

Table 1. Some commonly used fungicides for seed borne pathogens

Common name	Chemical name	Some Commercial formulation
Thiram	Tetramethylthiuram disulphide	Thiride
Captan	N- trrichloromethylthio-4 - cyclohexene-1, 2- dicaroximide	Orthocide
Carbendazim	Methyl-1H-benzimidazole-2-yl carbamate	Bavistin, Derosal
Carboxin	5,6- dihydro-2-methyl-1,4-oxathin-3-carboxanilide	Vitavax
Benomyl	Methyl 1 - (butylcarbamoyl)-2- benzimidazole carbamate	Benlate
Mancozeb	Manganese ethylenebis- dithio carbamate + zinc	Dithane M -45
Oxycarboxin	5,6- dihydro-2-methyl-1, 4-oxathin-3- carboxanilide-4, 4-dioxide	Plantvax
Metalaxyl	N- (2,6- Dimethyl phenyl)-N- (methoxyacetyl)- alanine methyl ester	Apron, Ridomil, Subdue
Propiconazole	1-[2-(2,4-dichlorophenyl)-4 -propyl - 1, 3- dooxlan - 2yl]	Tilt
Ziram	Zinc dimethyl dithio carbamate	Cuman, ziram etc.

Table 2. Fungicides with major target seed borne pathogens

Fungicide	Disease
Metalaxyl	Downy mildew of sunflower and pearl millet, late blight of potato
Benomyl	Dry rot of potato, wilt of sweet potato, *Ascochyta* blight of French bean, cluster bean and Indian bean, *Phomopsis* blight of brinjal, ashy stem of cowpea, A*nthracnose* of cow pea, Indian bean, French bean, *Cercospora* leaf spot of sesame
Carboxin	Loose smut of wheat, barley etc.
Thiram plus carbendazim	A*nthracnose* of French bean, cluster bean, cowpea, fruit rot of brinjal, tomato leaf spot, ashy stem of cowpea and potato *Phoma* rot
Mancozeb , Ziram (as foliar application)	Protective fungicide for major diseases (bean anthracnose, Alternaria leaf spot, Phomopsis fruit rot), Fusarial wilt, chilli ripe fruit rot, and ergot of sorghum and pearl millet early and late blight of tomato, potato, chilli, and anthracnose.
Propiconazole (as foliar application)	Bunt of rice, wheat and triticale

2. SEED TREATMENT INVOLVING PHYSICAL FACTORS

Hot water treatment: The seeds are dipped for specific period of time in hot water to a particular temperature safe to the seed embryo and deleterious to the associated pathogen. This method is useful for low volume high value seed. The method is effective for internally associated fungi and bacterial pathogens. Hot water (52ºC) treatment for 30 minutes ensure effective control of black leg of cabbage and black rot of cabbage and cauliflower.

Seed dip in water: Soaking of seed in plain water at 20ºC for 41 hours reduce the loose smut infection without damage to wheat seed. Soaking in water at 25ºC for 41 hours or 30ºC 28 hours eliminate the fungus from wheat seed effectively.

Solar energy treatment: The method is used for control of loose smut of wheat. The seeds are soaked in water for 6-8 hours thereby activating the dormant mycelium present in the embryo and subsequently exposing the seed lot to solar heat. As a result of the solar heat the activated mycelium is destroyed without damaging the wheat seeds. Soaking of wheat seed for 4 hours in cold water on a bright summer day and then spread over a hard cemented floor or on galvanized iron sheet exposing the seeds to direct sun light for 4-6 hours controls the disease.

Hot air treatment: The seed material is exposed to hot air stream for a specific period, which is safe for viability of seed. The method is more effective for the pathogens associated on to the seed surface. Treatment of set with hot air currents ensures effective control of smut pathogen associate with sugarcane set.

Seed seeping: Seed seeping in 20 per cent commercial salt (common salt) efficiently eliminate the lighter, chaffy, Karnal bunt infected seeds and nematode galls from wheat seed and seed of *Cichorium intybus* from Egyptian clover. This method can partially used to separate the ergot of pearl millet, sorghum and triticale. Floated material may contain the ergots. The seeds must be dried up before storage.

Grain auger: Seed is treated by spraying of chemical of required concentration on seed in an auger or straw conveyer. Individual seed is coated properly with a film of the chemical.

Spade: Seed is spread on a clean dry tarpaulin surface as a layer of 10 to15 cm. The required quantity of chemical diluted in required quantity of water is sprinkled over the seed. It is properly mixed with the help of spade for uniform seed treatment. Coloring agent is also mixed with the chemical to confirm the treatment. The most importantly the operation should not be done with seeds having thin fragile seed coat e.g. soybean.

3. TREATMENT WITH BIOLOGICAL AGENTS

Reports indicate that micro-organisms have also been successfully used as seed treatment. Treatment of Brassica seed with *Gilcoladium roseum* and *Trichoderma harzianum* reduces infection of *Alternaria brassicola* and increases the seedling emergence. Several formulations of biocontrol agents are available in the market. Solid matrix priming along with addition of a biological control agent is called bio-priming. Bio-priming of chickpea seed with *Pseudomonas fluorescens* effectively controls various seed and soilborne pathogens and also improve the emergence.

4. POLYMER COATING AND SPECIALIZED METHODS OF APPLICATION

Coating of seed surface with specific polymer such as *Ongard* has increased the efficacy of the fungicide to control the seedborne pathogen of maize. Treatment of seed with Carboxin fungicide by dripper/ augur equipment performed better than the seed treated with ordinary mist equipment.

Precautions required in seed treatment:

- Use appropriate recommended chemical for crop, pathogen and disease.
- Use precise dose of chemical and formulation for large scale .
- Never use chemical of expiry date for seed treatment.
- Seed should be properly dried before seed treatment.
- Seed should be properly mixed with chemical without any mechanical damage.
- Never use treated seed as food or feed.
- First treat the seed with fungicide, thereafter insecticide and finally with *Rhizobium* sp. (FIR), this sequence be followed wherever recommendation is made.

Management of objectionable diseases at field level

Loose Smut: Use disease free seeds from disease free locations. Maintain isolation distance of 150 m from the fields of wheat, triticale and rye infected with loose smut. Treatment of seeds with carboxin (Vitavax) 2.5 g/kg or carbendazim (bavistin, derosal, carbendazim etc.) 2.0 g/kg offers promising control the disease.

Grain smut of pearl millet: Removal of smutted earheads, use of clean seeds, deep summer ploughing and seed treatment with approved organo- mercurial fungicide controls the disease.

Downy mildew (Green ear) of pearl millet: Use seeds from disease free plants and location. Adjust dates of sowing to avoid the rains during early stage for commercial seed crop. Seed treatment and foliar application of metalaxyl (Ridomil MZ 0.2%) are effective and economical.

Ergot of sorghum and pearl millet: Use clean seeds. Seeping of seeds in 20-30per cent commercial NaCl solution removes lightweight sclerotia. Long crop rotation and deep ploughing in summer are beneficial. Destruction of weeds (especially belonging to Oryzae, Phalarideae, Agrostideae, Avenae, Festuceae and Hordeae tribes) which may harbour the pathogen is necessary to control the disease spread.

Grain smut of sorghum: Use of seed from disease free location and plants is of prime importance. Seed treatment with approved mercurial fungicide is useful as the spores are on the seed surface as contaminant.

Head smut of sorghum: The disease is managed by combination of practices such as sanitation, crop rotation and seed treatment. Monitoring and uprooting of the individual plants is economical since few plants are attacked in a field.

Downy mildew of sunflower: Healthy and clean seeds produced in disease free regions should be used for sowing purpose. Late planting in rainy season (July-August), excess irrigation and water stagnation especially during seedling stage should be avoided. Removal of affected plants from the field reduces the disease spread. Seed treatment with metalaxyl (e.g. Ridomil MZ 25 WP @ 0.3% or Apron 35 SP @ 0.6%) compounds followed by foliar application of the same fungicides controls the disease.

Early blight of tomato and chilli: The disease is seed and soil-borne, hence seed treatment with common seed dresser and foliar application of mancozeb (e.g. Dithane M 45 @ 0.25%) provide protection against secondary infection. Continuous cropping must be avoided.

Tomato mosaic: Seed treatment with 20 per cent trisodium phosphate solution and use of disease free seedlings are recommended. Field workers should avoid the use of tobacco products.

Anthracnose of chilli: Use seeds obtained from disease free plants. Seed treatment with Thiram (0.25%) is the best. Seedlings, plants and fruits should be protected by spraying mancozeb (e.g. Dithane M 45, 0.25%). Field sanitation is essential.

Fruit rot and blight of brinjal: Use seeds from disease free plants. Remove plant debris from the field. Dip fruits in mercuric chloride (25g /35 l water) for seed extraction. Plant protection with foliar application of mancozeb (e.g. Dithane M 45, 0.25%) or zineb (Dithane Z 78, 0.25%) should be adopted. Continuous cropping of brinjal should be avoided.

Anthracnose of cowpea, French bean, cluster bean and Indian bean: Use seeds from disease free plants. Seed treatment with Thiram (0.25%) or Carbendazim (e.g. bavistin 0.15%) followed by foliar application of mancozeb (0.25%), or Carbendazim (0.15%) at 15 day interval is recommended. Deep summer ploughing and removal of plant debris are also beneficial.

Ascochyta leaf blight of cowpea, French bean, cluster bean and Indian bean: Use clean healthy and disease free seeds. Field sanitation should be done to avoid secondary spread. Seed treatment with Thiram (0.25%) or Carbendazim (0.15%) reduces the seedborne inoculums.

Charcoal rot of potato and Ashy stem blight of cowpea: Disease free seeds and tubers obtained from disease free locations should be used. Deep summer ploughing and removal of plant debris reduces the inoculums. Furrow application of penta-chloro-nitro-benzene (PCNB) or Thiram at planting time is effective. Seed treatment of beans with Thiram (0.3%) reduces the inoculum. Development of stress condition for plants should be avoided to reduce infection in late crop growth stage.

Cowpea common mosaic for cowpea and French bean: Use disease free seed.

Black rot of cabbage and cauliflower: Use of disease free seed. Hot water treatment at 52°C for 30 min followed by dip in streptocycline 100 ppm solution for 30 min offer good control. Crop rotation for two years also reduces the soil borne inoculum. Soil application of bleaching powder @ 10 kg/ha is also recommended.

Black leg of cabbage, cauliflower, Chinese cabbage and knol khol: Hot water treatment at 50°C for 30 minutes is effective for fresh seeds, only. Use of disease free seeds, removal and destruction of plant debris is recommended.

Soft rot of multiplier onion: Precautions during harvesting, handling, transportation and storage of bulbs are quite necessary to avoid injury.

Brown rot of potato: Practically no control measure is economical, however, proper rotation with non-Solanaceous crops can reduce the disease. Use of disease free tuber is important to check the spread of the disease.

Mild and severe mosaic of potato: Use tubers produced by disease free plants grown in disease free locations. The tuber selected for sowing purpose should not be more than 4-5 cm diameter. Diseased plants should be destroyed. Remove the top foliage during critical active period of aphid infestation. Use of Thimet 10G prior to planting and spray of systemic insecticide (eg. metasystox 0.1%) reduce the development of aphid population.

Management of objectionable diseases at seed level

Karnal bunt of wheat: Use healthy and disease free seed. Foliar application of Mancozeb (e.g. Dithane M 45, @ 0.2%) or Carbendazim (e.g. bavistin, derosal, Carbendazim @ 0.2%) or propiconazole (e.g. Tilt 0.1%) at early heading stage gives effective control followed by second application at interval of 15 days.

Bunt of rice (Kernel smut): Use clean and healthy seeds for sowing. Avoid heavy application of nitrogenous fertilizers. Foliar application of propiconazole (Tilt 0.1%) applied at 35-45 days after transplanting (early boot) and second spray at interval of 10-12 days is effective.

Ergot of sorghum and pearl millet: Use clean ergot free seeds. Immersing the seeds in 5-20% commercial salt solution allows the separation of ergot.

Late blight of potato: Collect seeds from the disease-free plants and locations. Treatment of tuber and subsequent foliar application of Mancozeb (e.g. Dithane M45, 0.25%) or metalaxyl compound (Ridomil MZ, 0.2%) is beneficial. Haulm cutting in diseased area is recommended to check the entry of late blight pathogen to the tuber and viral pathogen.

Dry rot of potato: Use disease free tubers. Avoid injury on tubers during harvest, storage and transport. Storage at low temperature (5 to 10°C) checks the disease development. Deep summer cultivation.

Common scab of potato: Use clean skinned tubers. The population of the bacterial pathogen may be kept under control by alteration of soil pH towards acidic side (by application of elemental sulphur) or by increasing the soil moisture. Deep summer cultivation.

Black scurf of potato: Use clean skinned tubers. Amending the soil with Neem cake (25 q/ha) followed by application of 120kg/ha nitrogen, help in reduction of the damage by the fungus. Storage of tubers at low temperature (5°C) checks the development of the disease. Deep summer cultivation.

■■■■

Chapter 16

SEED DRYING

OBJECTIVE

Reduction of seed moisture content to the acceptable levels for enhancement of storability and longevity is known as seed drying.

Moisture content of seed is negatively related with its storability. As the moisture content of seed increases, its storability decreases. The removal of moisture from seed for safe storage to preserve viability is termed as seed drying. Adequate drying of seed prolongs viability for reasonably long periods without cold storage. The process of drying begins with physiological maturity till storage/packing of seed to avoid deterioration.

The seed is hygroscopic in nature, therefore, it adjusts its moisture content in equilibrium to the existing environment. Equilibrium means at this stage seed will not absorb or lose any moisture, its moisture content and the moisture in the environment is same.

Seed drying is the process in which environment with low humidity is created to remove moisture from the seed upto the desirable level without any damage to its viability and physical appearance. This seed is stored at low moisture content under low relative humidity to enhance its longevity.

High seed moisture causes a number of problems that includes:

- Moisture increases the respiration rate of seeds, which in turn raise the seed temperature. In large-scale commercial seed storage, the seed not dried sufficiently generate enough heat due to fast respiration. High humidity and temperature accelerate the process of ageing that deteriorates the seeds quickly.
- The deterioration of the seed with high moisture content is fast as it encourages growth of mold and activity of microorganisms. It damages the seeds either slowly or quickly, depending on the moisture content of the seeds.
- Store seed pests breed fast on the seed stored with high moisture in store that cause rapid destruction of seeds in a short period e.g., weevils.

Several methods are available for drying of seeds. Drying of seeds is faster in the atmosphere with reduced relative humidity. A relative humidity of 10 - 15% and a temperature of 15°C are recommended by the International Bureau of Plant Genetic Resources Advisory Committee on Seed Storage.

Drying mechanisms

Individual seed: In a seed, water is present in bound form with amino or carboxyl group; adsorbed form loose binding with hydroxyl and amide group; and free form present in the tissues as capillary force. Moisture level of seed reaches 10-12% by removal of moisture present as free form in the seed. It can be achieved by reducing relative humidity under ambient temperature. Removal of free water does not damage the seed. The basic mechanisms involved in drying of seed are

- Seeds are hygroscopic in nature and its moisture reduces upto equilibrium in relation to the surrounding air. The equilibrium moisture content (EMC) depends on relative humidity and temperature of the air. Under no circumstances moisture content of seed can be lowered down lesser than the associated EMC.
- The free bound water in seed or in nearby atmosphere is always in exchangeable condition. The movement of water from seed to air is termed as evaporation, whereas the movement of water molecule from air to seed is termed as absorption. Both are the continuous processes in seed. When evaporation exceeds absorption seed dries out and when absorption exceeds evaporation the seed takes moisture from atmosphere. When evaporation and absorption are equal then it shows equilibrium for moisture content in seed and atmosphere.
- Moisture is evaporated from the surface of the seed to the atmosphere due to low relative humidity and high temperature in the environment.
- Flow of air carries away the evaporated moisture that makes the low RH in the atmosphere.
- The moisture from the inner part of the seed is moved to its surface.

Seed lot: When air is forced through the seed for drying, all the seeds do not dry uniformly at the same time. During the process of drying the seeds of seed lot can be divided into three zones:

Dried Zone : Seed present near the inlet gets dried first and dry below the desired level earlier than the seed present away from the inlet. The moisture of seed and atmosphere at this region is at equilibrium.

Drying Zone : The air passing through the dried zone reaches in the next region, to carry moisture from the seed, until the moisture reaches at equilibrium. This region is termed as drying zone. The location of drying zone is in between dry and wet zone. With the process of drying the depth of dried zone is enhanced whereas wet zone is reduced. The active process of drying of seed occurs in drying zone.

Wet Zone : Refers to the region above the drying zone i.e., the seed in-between top of drying zone and the top surface of the seed, which is wet with 16-20 % moisture. The top most layer of seed in wet zone contains maximum moisture whereas, layer near the drying zone is driest to reach at equilibrium.

The difference in moisture content of the air entering and leaving the seed is known as stratification. The amount of stratification and width of drying zone depends upon the volume of air flowing through the seed and its relative humidity.

Factors determining rate of seed drying

Seed: Moisture Content of seed is an important factor for drying process. Moisture content has an appreciable effect on bulk density of seed (weight per unit volume). The energy require to force air through a bed of seed is dependent on the air flow, the depth of seed and physical properties of the seed viz., surface and shape, size, moisture content, and the quantity and nature of contamination (stones, straw, weeds etc.). Energy in the form of heat is supplied to evaporate moisture from the seed. The latent heat of vaporization of a seed depends on its moisture content and temperature and is appreciably greater than the latent heat required for evaporation of water. The temperature of the seed, physical structure of the seed, chemical composition of the seed and the seed coat permeability influence the rate of moisture migration from the centre to the surface of the seed.

Air: The following property of the air flowing around the drying seed is a major factor in determining the rate of removal of moisture

Temperature of air: Hot air has capacity to carry more amount of water. As per Harrington's report one kg air may carries 14.8 g water at 20ºC temperature whereas, at 40ºC it enhances upto 41.4 g.

Temperature ºC	Capacity to carry water (g water/kg air)
0	3.9
10	7.6
20	14.8
30	26.4
40	41.4

- *Relative humidity in air:* Surface saturation and relative humidity of drying air influence removal of moisture from the surface.
- *Velocity of air:* The rate of drying increases with increase of air velocity as it helps in removal of moisture from surrounding of seed. The moisture content falls rapidly at first but as the seed looses moisture the rate of drying slows down.

Damage to seed during drying

Temperature of air: In general, more than 40ºC temperature of drying air may reduce the viability of seed. Drying at low temperature maintains the viability of seed.

Low RH of air : Rapid drying with low relative humidity in air also damages the seed by developing cracks on the seed coat. Soybean seed dried with the air of less than 40% relative humidity looses its viability even at low temperature.

Velocity of air: Drying under high velocity of air may develop stress cracks on seed because of unequal drying of individual seed. Rapid drying deteriorates the seed due to quick evaporation of water from seed. Whereas, in slow drying seed deteriorates due to high humidity and temperature for longer period of time.

Damage due to excessive drying of seed

After drying, the moisture content of the seed should be suitable for storage. The desired moisture content depends on the type of seed, duration of storage, type of packaging and condition of store. Damages caused by excessive drying of seed are

Cracking of seed coat: Drying with heated air or excessive exposure to sun can raise the internal kernel temperature to such a level that the endosperm cracks. The extent of stress cracking is related to the rate of drying. Rapid cooling of seed can also contribute to stress crack development.

Cracking of seed: Excessive drying may separate cotyledons from embryo.

Nutritive value: High temperature during drying may denature proteins, sugars, enzymes and other complex structure of nutrients stored in the seed. It affects vigour of seed drastically followed by formation of abnormal seedling.

Seed viability: Drying under high temperature reduces the viability of seed. Seed embryos are killed by temperatures greater than 40-42°C.

Appearance : The colour and texture of seed surface can be adversely affected by high temperature during drying of seed.

Handling of seed: Over dried seed become fragile and susceptible to mechanical damage.

METHODS OF SEED DRYING

Sun drying

The sun is an appreciable and inexhaustible source of heat to evaporate moisture from the seed, the process of drying is enhanced by velocity of wind that removes evaporated moisture. In India, the mean level of insulation on the ground measured as a mean over the hours of daylight is more than 0.5 kW/m^2 at the harvesting stage of the crops of both the season. The assuming available heat in a 12 hour day is 21.6 MJ/m^2. Theoretically, this quantity is sufficient to evaporate 9 kg of water. Sun drying is the most common drying method in India in both the season. Sun drying is cost effective, environmental friendly with no damage on seed viability.

Stages of sun drying of seed: The moisture of seed is generally reduced in the field before harvest and later by sun drying on the threshing floor. The system involves harvesting of crops when they are fully dried in the field, leaving the harvested produce in the field for a couple of days for sun drying and later spreading the threshed and winnowed produce in thin layer on threshing floors for sun drying.

From physiological maturity to harvestable maturity on plant: After physiological maturity the seed has no association with the mother plant. The crop is left on the plant to dry the vegetative part of mature plant and loss of moisture from seed by evaporation.

From physical maturity upto threshing on harvested plant: In Central and North Indian states *Kharif* crop is harvested in the month of September and harvested plants are stored as stack or in bundle upto sowing of *Rabi* crops. In doubled cropped area, just after sowing of *Rabi* crops threshing of *Kharif* crops starts. At this time intensity of light is very high that helps in fast sun drying of seed on harvested plant. Stacking of harvested plants helps in movement of air through seed. Threshing of *Rabi* crops is not a problematic step because of no relay cropping, air current with high temperature and low humidity.

After threshing for storage Drying of threshed seed on flat exposed surface is the most common way of seed drying. Seed should be spread on black plastic sheet to get clean and uniformly dried seed. For large quantity of seed a thin layer should be spread on elevated concrete floor. To facilitate uniform drying the seed is raked 7-8 times in a day. Drying of seed under shed in ambient condition is recommended for many crops viz., soybean, rice etc. as it reduces the seed/seed coat cracking and maintains the sowing seed quality. Covering of seed around midday is beneficial under hot and sunny conditions of summer. After drying, seed is cooled down under low relative humidity before storage.

Forced air drying

Principle: The movement of moisture content of seed depends on temperature and relative humidity of surrounding air because of its hygroscopic nature. When the relative humidity of air is lesser than the seed then moisture from seed will move out and the process is enhanced under high temperature and velocity of air.

Procedure: The seed is placed in drying bin upto the recommended depth. Air of recommended temperature is forced through damp seeds to remove the water. The heat necessary for evaporating the water comes from the temperature of air. The rate of seed drying depends on rate of moisture migration from the centre of the seeds to surface and by the speed at which surface moisture is evaporated in the surrounding air. After completion of seed drying, natural air without heat is forced in the seed lot for 30-60 minutes to bring the temperature of seed upto ambient level.

Types of forced air drying: There are three major drying methods for drying with forced air based on the temperature of air:

Natural air drying: Natural air of ambient temperature is used for drying of seed.

Drying with moderately heated air: The temperature of the air forced for drying is raised to 10 to 20° F to reduce relative humidity of air and enhance evaporation of moisture from seed.

Drying with high heated air: The temperature of the air forced for drying is raised to 110° F to reduce relative humidity of air and enhance evaporation rate from seed.

Heated air drying systems: This involves construction of bins/storage structures for drying and air distribution system. The building for seed drying system depends upon size of operation, number of crops to be dried and level of mechanization desirable. Weather tight storage bin of adequate strength, good air distribution system, convenient to inspect, fumigate and clean, easy to fill and empty and in which seed of any crop can be dried and stored is required for seed drying. Air distribution systems for seed drying are:

Main and lateral duct air distribution system: In this system the main duct is located in the centre or one side of the bin.

Single central perforated duct system: In this system the main duct made of perforated metal is located in the centre of the bin. The seed is filled around the duct upto equal thickness but not exceeding 6 feet. The air is forced upwards in the perforated main duct that flow laterally to dry the seed.

Perforated false floor air distribution system: The air is introduced in the bin through the perforated false metal sheet floor to dry the seed.

Multiple storage bins : To dry seed of many crops simultaneously the seeds are stored in differ bins and the flow of air from one source is controlled by sliding air gate of the respective bin.

TYPES OF FORCED AIR SEED DRYERS

Layer in bin dryer: The bin is filled to a specific depth based on seed moisture, capacity of the drying unit and bin size. The hot air is forced into the bin to dry the seed. After drying of seed upto required level of moisture, more seed is added in the next layer for drying. In this method the seed is uniformly dried.

Batch in bin dryer: The bin is filled to a specific depth based on seed moisture, capacity of the drying unit and bin size. The hot air is forced into the bin to dry the seed. After drying, the seed is placed in storage bin after cooling. After transfer next batch of seed is placed for drying.

Batch dryer: The seed is placed in bin with inner air chamber surrounded by two parallel perforated steel walls that contain seed upto desired thickness. The heated air for drying and natural air for cooling are forced through the seed. After drying the seed is removed and replaced by another batch.

Continuous dryer: In this method flow of seed in bin for drying is continuous through feeder whereas dried seed is collected at the outlet. Heated air is forced through the upper 2/3 seed column.

Wagon drying: It is a type of batch drying in which seed is loaded from combine into a wagon with perforated floor for drying. The wagon is connected with the air distribution duct to force the hot air through the perforations of the wagon floor for drying of seed. After completion of seed drying, natural air without heat is forced in the seed lot for 30-60 minutes to bring the temperature of seed upto ambient level. The cool seed is stored.

Bag drying: The threshed and cleaned seeds are stored in gunny bags of small size. These bags are placed in the drying bin directly for drying of seed by hot air. It is an advantageous method to dry the small seed lots of many crops/ varieties simultaneously.

Recommended temperature and depth for heated air drying of crops in bin

Crop	Depth in cm	Drying temperature (°C)	Safe moisture (%)
Maize	125	40-45	13
Wheat	125	40-45	12
Sorghum	125	40-45	12
Soybean	125	30-45	11
Rice	112.3	40-45	12
Groundnut	287.3	36	

Solar dryers It is an improved technology for drying seed by heating air with the help of solar energy in a solar collector and then passed through beds of seed by a fan. The components of solar dryer are solar collector, drying bin, solar chimney and fan. Solar chimney provides a column of warm air that increases the flow of warm air in the dryer.

Ultra-drying of seed for long-term storage: Drying of seed down to a moisture content of 1-3% extend the seed life span upto 4-16 times (Harrington, 1972). Ultra-dried seed does not require very low temperature for long term storage. It is an effective method for storage of orthodox seeds. It may be achieved by silica gel drying.

Silica gel (silicic anhydride, SiO_2) is the most effective moisture absorbing material for ultra drying of seed. Silica gel is an amorphous (non crystalline) granular in texture, white and highly porous form of silica is obtained through an industrial process. It is available as a powder or as beads in different sizes. The best size bead for drying seeds is approximately 1/16" to 1/8" in diameter. Dehydrated silica gel can absorb up to 20% of its own weight of water. It balances itself with the confined atmosphere until this reaches approximately 10-12% of relative humidity under controlled environment. Silica gel treated with a small amount of cobalt chloride ($CoCl_2$) acts as an indicator of moisture level. It is of deep blue color under completely dry condition and with the absorption of moisture it gradually changes in color from deep blue to light pink. Recently, the use of cobalt chloride is banned because of its carcinogenic nature. Now a day, some iron salts, methyl violet and other indicators are used. The moisture levels of 2-3%, lower than those obtained by other methods can be achieved by silica gel method. It maintains the level of moisture content indefinitely in tight containers. With colored indicator it indicates capacity of water absorption. Silica gel is reused after drying in a controlled fashion at 220ºC. It delays ageing by absorbing toxic gases produced during the process.

Management of drying operations

Seed is dried immediately after receipt. In case of delay it is aerated by fan to prevent heating of seed. The temperature and moisture content of seed at different drying zone should be monitored regularly. Air flow, time of drying, static pressure, relative humidity and temperature of air, depth of seed etc. should be monitored regularly.

The air forced in for drying of seed should never be too hot as it may deteriorate the seed present at the bottom. During the process of drying the hot air gets cool as it rises through seed lot and its capacity to hold water decreases due to saturation. This air deposits water on the surface of the seed present at the top layer. That result in transfer of water from bottom to top layer. Therefore in early stage of drying when seed holds maximum amount of water natural air should be forced into and the temperature should be raised with advancement of drying period.

Tempering period: For minimum loss in viability of seed of maize cooling aeration is provided after high temperature air drying. Less than 1% moisture is removed when cooling is carried out immediately after drying. Whereas, more than 2% moisture is removed when tempering period is provided after high temperature and slow cooling. Tempering phase enhances drying efficiency and reduces damage to the seed.

Two-stage seed drying: During peak period, two stage seed drying may be adopted when limited drying resources are available. In the first stage of drying the seed is immediately dried upto intermediate moisture content and stored. In the second phase seed is again dried after 20-30 days upto desirable level of moisture content. The method is adopted for rice and maize as rice seed can be stored up to 20 days at 18% and maize at 20% moisture content without significant losses either in quantity or quality of seed. After 20 days seed is dried properly for long term storage.

Pre-drying aeration: Freshly harvested seed of rice with high moisture content can be stored safely upto 3-7 days with aeration of ambient air for 6-8 hrs each day. The duration upto which rice seed can be stored safely depends on the moisture content of the seed and ambient air conditions.

Drying of seed : Seed should be dried at low temperature under high air flow rates than recommended for grain. Dryers and associated equipment should be designed for easy and proper cleaning to avoid mechanical mixture. For drying of seed continuous-flow dryers are not recommended similarly proper care should be taken to avoid mixing of different varieties during batch drying.

The drying period is predicated by following methods:

Loss of weight after drying: The time required to dry the seed until the equilibrium moisture content is predicted by loss of weight in seed before and after drying:

- Initial weight of seed sample before drying.
- The moisture content of the seed before drying is measured by oven dry method.
- The percentage moisture content required for safe storage of seed.

The process of the drying is started with weighing of seed sample at regular intervals until the weight of the seed reaches to the calculated value

$$\text{Weight of seed for safe storage} = \text{Initial weight of seed} \times \frac{(100 - \text{initial moisture (\%) of seed before drying})}{(100 - \text{required moisture (\%) of seed for safe storage})}$$

■■■■

Chapter 17

SEED PROCESSING PLANT

PLANNING

Capacity of machines: The machines for seed processing must be efficient with capacity to process in unit time as per requirement. The machine should be compatible with each other to improve the efficiency of processing. The overall processing capacity should be such that apart from required processing some reserve capacity can also be processed.

Sequence of machines: Efficient machines at different stages are arranged in a sequence so that continuous flow of seed is maintained from beginning to end, and flexible enough to bypass any machine as per requirement. The machines should be arranged in the following series

Preconditioning: Seed lots of many crops are first preconditioned to start processing. Equipment required for preconditioning are arranged in one hall as per requirement.

Seed drying: Seed is dried upto desirable moisture content before processing.

Weighing: After drying the seed is weighed before start of basic-cleaning.

Basic-cleaning: Seed is cleaned with two screen seed cleaner with aspirator.

Grading: Different types of seed grader are used to grade the clean seed.

Seed treatment: Graded seed is treated with the help of seed treater.

Bagging and weighing: Defined quantity of processed seed is bagged in jute bag.

Store

Store for unprocessed seed: Seed received form field or from seed producer contains high moisture level with inert matter. The received seed is weighed and its moisture content is determined before processing. The seed is stored as unprocessed seed lot in the store under isolation with mark of identification on each bag. The quantity and date of receipt is entered in the register.

Store for processed seed: The processed seed is stored in bag under isolation with tag and certificate to supply to the indenter.

Planning

Capacity and number of equipments are fixed based on the number and size of the seed lots to be processed. List of all the equipments is prepared. The equipments are arranged sequentially as per requirement. Requirement of elevator between machines and their capacity is worked out.

Capacity of the store for unprocessed and processed seed lot is fixed based on the predicted size of seed lot and number of crops, varieties and class to be handled.

Detailed drawing should be prepared to show exact location of equipment and distance between them. Electric wiring and electric point has to be mentioned on the design.

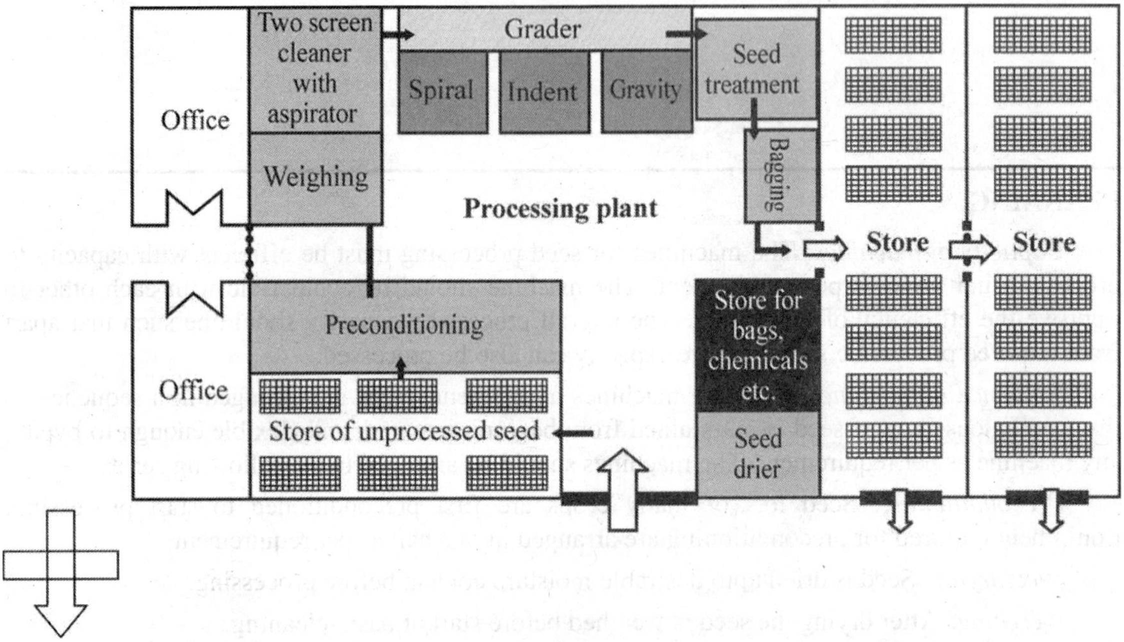

Factors and variables to be considered in planning and designing a seed processing plant:

- Number of crops, varieties and class to be handles
- Size of lot to be handled
- Requirement of the equipment as per crop, class and agronomical practices adopted for seed production
- The sequence of operations depends upon the kind of crop and the initial quality of seed lot, type of contaminants, moisture content of the seed lot, etc
- Location of the seed processing plant
- Source of electric power for running machinery
- System of seed delivery to processing plant
- Availability of labour/ man power

Chapter 18

SEED PROCESSING

The harvested seed has various factors that affect its sowing seed quality and storability. These factors (inert matter, weed seed, other crop seed, immature seed, damaged seed, diseased, undersized seed, other variety seed etc) are to be reduced upto threshold level to improve sowing seed quality. The storability of the unprocessed seed is reduced drastically during storage as these factors invite insect and microorganism due to their hygroscopic nature. All the operations from harvesting to storage (Fig. 1) are performed carefully to improve the sowing seed quality and longevity.

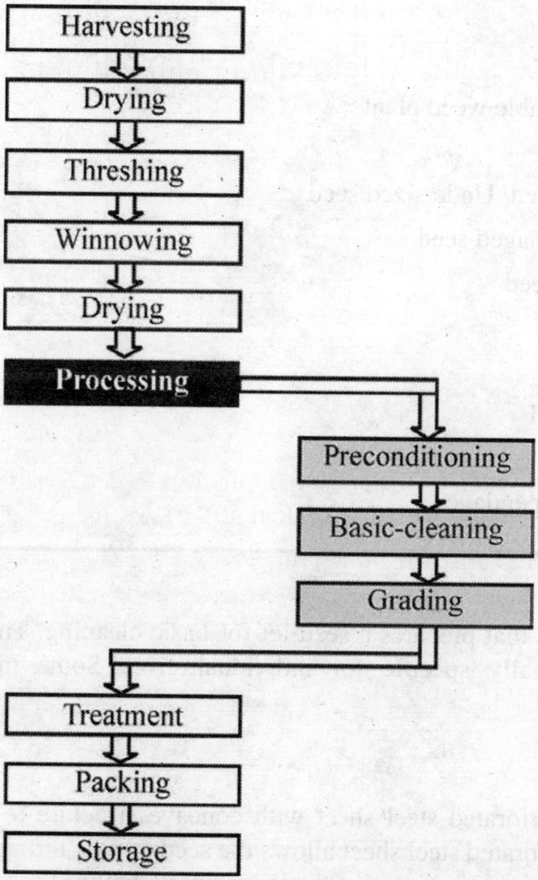

Fig. 1. Steps for post harvest handling and processing of seed

Seed processing reduces the heterogeneity of a seed lot arose due to variability in the factors viz., fertility, physical, chemical and biological properties of soil; irrigation, fertilizer, herbicide, and other management practices; management of biotic and abiotic stresses, position effect on development of seed on plant etc.

Improvement in physical quality of seed lot by removal of undesirable material and upgrading of seed quality through removal of damaged and undersized seed by mechanical devices with highest efficiency is defined as seed processing. It involves drying, pre-conditioning, basic-cleaning and grading as major steps.

Requirement of seed processing are

- Complete separation
- Minimum seed loss
- Up gradation upto required quality
- High efficiency
- Minimum requirement of labour

Types of impurities removed during seed processing

- Inert material
- Weed seed
- Seed of objectionable weed plant
- Diseased seed
- Immature/ shriveled/ Undersized seed
- Mechanically damaged seed
- Insect damaged seed
- Discolored seed
- Other crop seed
- Other variety seed
- Off size seed
- Crop and seed appendages

2. PRECONDITIONING

This is the operation that prepares a seed lot for basic cleaning. The equipment required for preconditioning is generally specific for individual crop. Some important preconditioning equipment is:

2.1. Sheller

It is made up of perforated steel sheet with concave structure to remove kernel (seed) of maize from cob. The perforated steel sheet allows the seed to pass through and retain the cobs and its parts. Shelled seeds are again cleaned with air pressure (shelling).

2.2. Huller

It removes tightly fixed husk from seeds of grasses to facilitate in the process of sowing and germination (hulling).

2.3. Debearder

It removes awn and other appendages from seed of oat and barley that hampers processing of seed lot (debearding).

2.4. Scarifier

It scratches the hard seed coat to facilitate process of germination by increasing exchange of water and oxygen in crops like lucerne, faba bean, rice bean etc. (scarification).

3. BASIC CLEANING

This step of seed processing removes the larger, smaller, lighter and thicker adulterants as compared to the crop seed, from the seed lot. Basic cleaning is done on the basis of weight, size and density using cleaner with air screen. This process involves following equipment

3.1. Grader

It separates the undersized seeds and coarse impurities viz., trash, soil pads etc. from the normal desirable seed on the basis of seed density and size with the help of screen of different mesh sizes and their vibration.

3.2. Scalper

High quantity of inert matter is present with seed after threshing, winnowing or pre-conditioning. Large size inert matters with high moisture content that may jam hopper or prevent movement of seed and reduce the efficiency of machine are removed with a scalper. It is the top most screen of a seed cleaner/ grader with larger holes than the desirable seed size to remove the inert matter of larger size than the seed (scalping).

3.3. Aspirator

It removes lighter inert matter and adulterant from the crop seed with the help of air pressure (aspiration).

4. SCREEN CLEANER

It is generally made up of 2-3 screens of different mesh sizes, which are agitated to provide proper speed and place for separation. The air operation removes light seed and inert matter.

4.1. Selection of screen

Most of the seed lots contain undesirable materials larger, smaller, heavier and lighter as compared to the desirable seed of the crop. Hence, one screen with larger mesh and second with smaller mesh than the crop seed are selected for proper cleaning. The shape of hole in the screen may vary from oblong, round or triangular in shape based on the shape of the seed of crop.

4.2. Selection of top screen

Screens (5-6) with the hole size slightly larger and smaller than crop seeds are selected. Stack of these screens is made keeping largest size on the top and smallest at the bottom in descending order. A quantity of 1-2 kg seed is placed on the top screen and the stack of screens shake vigorously until all the seeds have found their level. The screens starting from the top of the stack are removed and the material held in each screen is recorded by weight. The screen size that just allows the crop seeds to fall through but holds the material larger than the crop seed is used as scalping screen. The material remaining on this screen is weighed as X_1.

4.3. Selection of bottom screen

From the same stack of screens, the screen is selected which holds the crop seed but allows broken seed, weed seeds smaller than the crop seed and undersized crop seed to fall through. This is the mesh size best suited for bottom screen. The material which passes thorough bottom screen is weighed as X_2. The sum of X_1 and X_2 represent the clean out and is exhibited in percentage.

4.4. Selection of air pressure

The force of air is adjusted first to remove major light impurities and then to remove minor light impurities and seed components without any loss of the crop seed.

4.5. Slope of screen

The flow /speed of seed on the screen is maintained by the slope of screen. The speed of seed movement on the screen depends on desired separation and shape of seed. In the event of difficult separation the seed should remain on the screen for longer period of time to give every seed an opportunity to pass through an opening. When the separation is easy, screen in an acute slope can be used. Normally, scalping screen is set at acute slope for fast removal of large inert matter while, grading screen is set to flat slope for close separation. The pitch should be steeped for chaffy seed lot, while it should be flattened for round seed in order to prevent bouncing over the screen, which in turn affects the grading efficiency.

4.6. Vibration of screen

It is tuned to control stroke of seed on the screen. High speed vibration turns, tumbles and exposes the seed from all the sides to the openings of screen. Very high speed of vibration may damage the seed; reduce the processing efficiency in terms of quality and loss of seed due to bounce. Whereas, efficiency of processing may reduce with low vibration as seed may slide over the screen due to reduced bouncing action or may plug the hole of mesh. Therefore vibration is calibrated based on shape of seed and physical purity.

4.7. Feed rate

To receive optimum quality of processed seed with high efficiency the feed rate of the seed is properly adjusted. Feed rate can be increased or decreased with the help of knob provided at the base of the hopper.

4.8. Operation of seed cleaner

Screens of proper mesh size (Table 1, 2 and 3) are placed on top and bottom of the seed cleaner. The air vents and feed hoppers are kept closed. Empty bags are placed at exit. The seed lot is poured through the hopper after turning on the machine. The flow gate of hopper is opened gently to make the even flow of seed on the top screen in such a way that it covers only 1/2 to 2/3 portion of the upper screen. The air vents are adjusted until all material lighter than the crop is blown off. The operation is continued till all the seed in hopper have passed through the machine (Fig. 2).

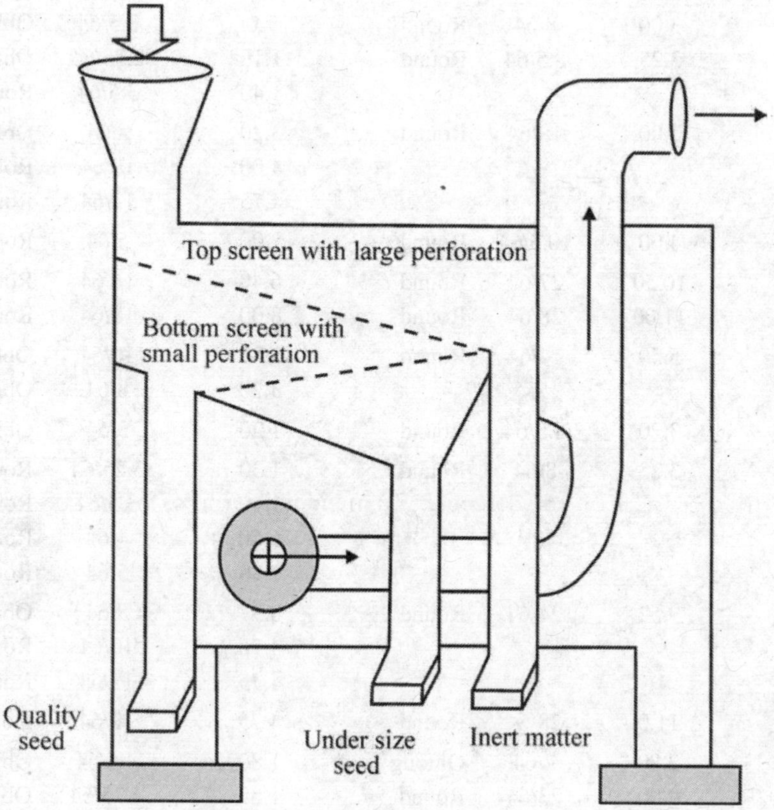

Fig. 2. Operation of seed cleaner

Table 1. Recommended screen based on mesh size and shape for basic cleaning of field crops

Crop	Top screen			Bottom screen		
	Size		Shape	Size		Shape
	mm	inch		mm	inch	
Castor	13.50	34/64	Round	4.40	11/64	Oblong
				6.00	15/64	Round
Chickpea	9.00	23/64	Round	5.00	13/64	Round
	10.00	26/64	Round	5.50	13.75/64	Round
				6.00	15/64	Round
Cluster bean	6.00	15/64	Round	1.80	4.5/64	Oblong
Cotton delinted	7.20	18/64	Round	3.90	10/64	Oblong

Crop	Top screen			Bottom screen		
	Size		Shape	Size		Shape
	mm	inch		mm	inch	
Cotton fuzzy	14.30	36/64	Round	5.20	13/64	Oblong
Indian rape seed and mustard	2.75	7/64	Round	0.90	2.25/64	Oblong
	3.00	8/64	Round	1.0	2.5/64	Oblong
	3.25	8.5/64	Round	1.10	2.75/64	Oblong
				1.40	3.5/64	Round
Lentil	7.00	18/64	Round	3.20	8/64	Oblong
				4.00	10/64	Round
				4.75	12/64	Round
Linseed	4.00	10.5/64	Round	2.00	5/64	Round
Maize	10.50	27/64	Round	6.40	16/64	Round
	11.00	28/64	Round	8.00	18/64	Round
Mung	5.50	14/64	Round	3.80	10/64	Oblong
				3.20	8/64	Oblong
Niger	3.20	8.5/64	Round	1.20	3/64	Oblong
Pearl millet	3.25	8/64	Round	1.30	3.25/64	Round/oblong
				1.40	3.5/64	Round/oblong
				1.60	4/64	Round
				1.90	5/64	Round
Pigeonpea	9.52	24/64	Round	3.97	8/64	Oblong
				4.76	10/64	Round
				4.75	12/64	Round
Rajmesh	11.0	28/64	Round	4.75	12/64	Oblong
Rice	2.80	7/64	Oblong	1.80	4.5/64	Oblong
	9.00	23/64	Round	1.85	4.75/64	Oblong
Safflower	7.25	18/64	Round	1.20	3/64	Oblong
Sesame	2.40	6/64	Round	1.60	4/64	Round
Sesame				1.90	4.5/64	Round
Sorghum (grain)	4.75	12/64	Round	2.10	5.25/64	Oblong
				3.50	9/64	Round
Soybean	8.00	20/64	Round	4.00	10/64	Oblong
Sunflower	9.00	23/64	Round	2.40	6/64	Oblong
Urid	5.00	13/64	Round	2.80	7/64	Oblong
Wheat	6.00	14/64	Round	1.80	4.5/64	Oblong
				2.10	5.25/64	Oblong
				2.30	5.75/64	Oblong

Table 2. Recommended mesh size and shape for basic cleaning of forage crops

Crop	Top screen			Bottom screen		
	Size		Shape	Size		Shape
	mm	inch		mm	inch	
Egyptian clover	2.00	5/64	Round	1.00	2.5/64	Oblong
Forage sorghum	4.00	10.5/64	Round	2.10	5.25/64	Oblong
	4.75	12/64	Round			
Guinea grass	2.10	5.25/64	Round	2.4 x0.65		Wire mesh
Indian clover	2.10	5.25/64	Round	2.4 x0.80		Wire mesh
Oat	7.50	19/64	Round	2.00	5/64	Oblong
Setaria grass	2.40	6/64	Round	1.90	4.5/64	Oblong
Sudan grass	4.00	10.5/64	Round	1.20	3/64	Oblong
				1.30	3.25/64	Oblong

Table 3. Recommended screen based on mesh size and shape for basic cleaning of vegetable crops

Crop	Top screen			Bottom screen		
	Size		Shape	Size		Shape
	mm	inch		mm	inch	
Asparagus	6.00	15/64	Round	2.40	6/64	Round
Bitter gourd	11.0	28/64	Round	6.50	16.5/64	Round
Bottle gourd	11.0	28/64	Round	6.50	16.5/64	Round
Brinjal	4.00	10/64	Round	0.80	2/64	Oblong
				2.10	5.25/64	Round
Cabbage	2.75	7/64	Round	0.90	2.25/64	Oblong
Carrot	2.30	5.75/64	Round	1.00	2.5/64	Round
Cauliflower	2.75	7/64	Round	1.10	2.75/64	Oblong
Capsicum	4.00	10/64	Round	0.80	2/64	Oblong
				2.10	5.25/64	Round
Celery	1.80	4.5/64	Round	0.40	1/64	Oblong
Chilli	4.00	10/64	Round	0.80	2/64	Oblong
				2.10	5.25/64	Round
Cucumber	8.00	20/64	Round	2.00	5/64	Round
				2.50	6.25/64	Round
Fenugreek	3.25	8.5/64	Round	1.20	3/64	Oblong
Indian squash	9.50	24/64	Round	6.50	16.5/64	Round
Lettuce	2.30	5.75/64	Round	0.80	2/64	Round
Muskmelon	5.00	13/64	Oblong	1.00	2.5/64	Round

Crop	Top screen			Bottom screen		
	Size		Shape	Size		Shape
	mm	inch		mm	inch	
Okra	6.00	15/64	Round	4.30	11/64	Round
Onion	3.80	9.75/64	Round	2.00	5/64	Round
Parsley	2.75	7/64	Round	0.75	1.75/64	Oblong
Pumpkin	11.0	28/64	Round	6.50	16.5/64	Round
Radish	4.50	11/64	Round	2.00	5/64	Round
Ridge gourd	9.50	24/64	Round	6.40	16.5/64	Round
Snake gourd	9.50	24/64	Round	6.40	16.5/64	Round
Sponge gourd	9.50	24/64	Round	6.40	16.5/64	Round
Sugar beet	9.00	23/64	Round	3.00	8/64	Round
Summer squash	8.00	20/64	Round	2.00	5/64	Round
Tomato	4.00	10/64	Round	0.80	2/64	Oblong
Turnip	1.80	4.5/64	Round	1.20	3/64	Round
Watermelon	6.00	15/64	Round	1.80	4.15/64	Oblong
				2.10	5.25/64	Round

5. GRADING

Classification of a seed lot on the basis of commercial usage *viz.,* size, shape, density and colour is known as grading. It is done for further improvement of seed lot as finishing operation. It improves the seed lot by removal of cracked, damaged, shrivelled and other defective seeds apart from inert matter. Such seeds may influence the establishment of plant due to poor germinability and reduced vigour. Grading requires different types of separators to remove weed and broken seed of cereals and materials longer or shorter than the desired crop seed.

5.1. Disc separator

Pieces of broken seeds, weed and other crop seed of round shape can be removed by the disk or cylinder separators. It has series of indented disks that are rotated inside a cylinder through which the seed moves. The size of indent can be selected to leave out shorter contaminating seed from crop seed of larger size or remove small crop seed from larger sized contaminator.

5.2. Indented cylindrical separator

It can separate the impurities (especially broken seeds, other crops seeds and weed seeds) that are either longer or shorter than the crop seed. It has a horizontal cylinder with inbuilt trough. The cylinder has many half round indents on the inner surface which rotates horizontally (Fig 3). A feeder is present on the upper side with discharge end at lower part. The revolving indent cylinder provides each seed a chance to fit into the indent by turning out the seed mass. The seed mass travels along the inside bottom of the cylinder from the feed end to the discharge spout.

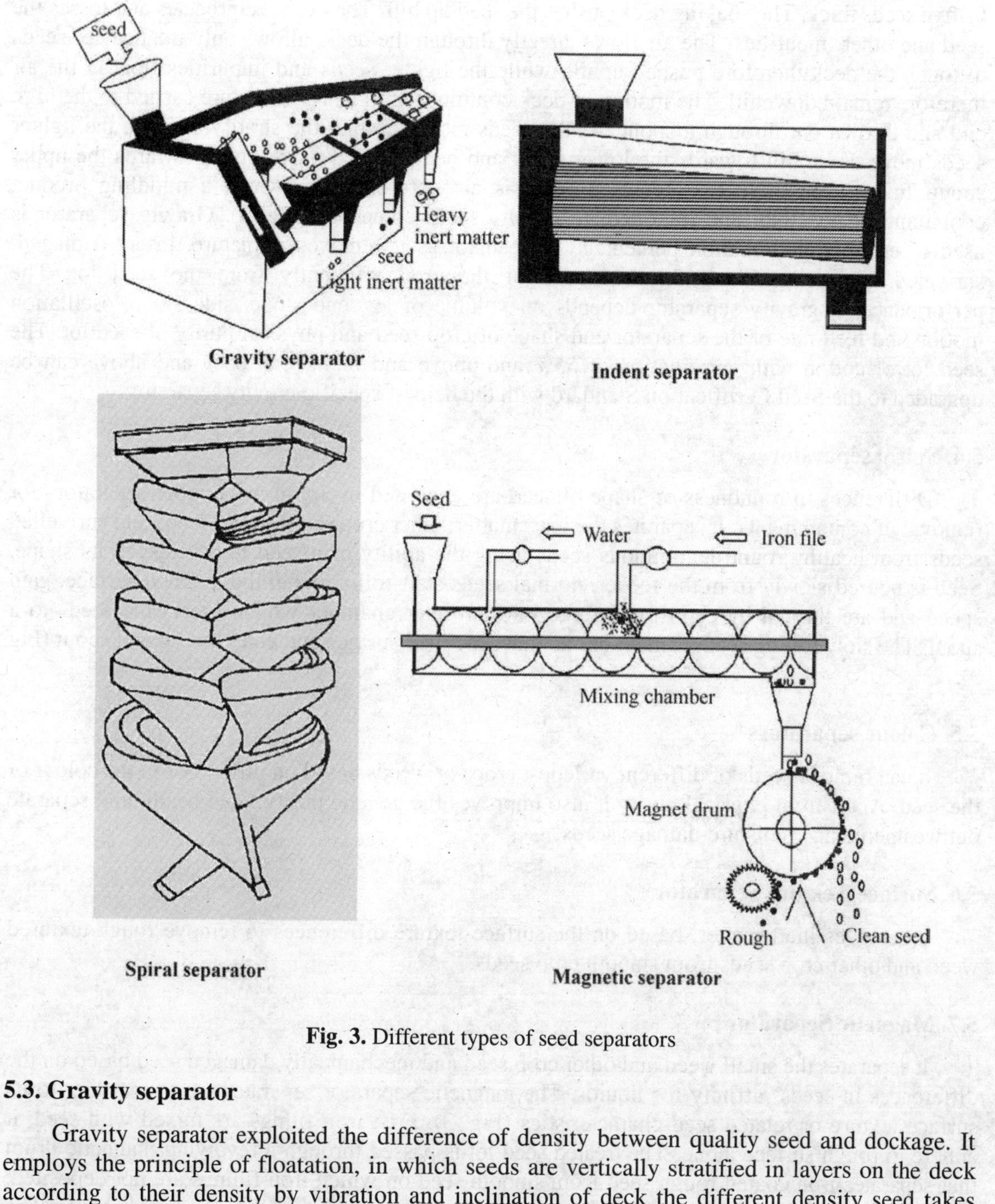

Fig. 3. Different types of seed separators

5.3. Gravity separator

Gravity separator exploited the difference of density between quality seed and dockage. It employs the principle of floatation, in which seeds are vertically stratified in layers on the deck according to their density by vibration and inclination of deck the different density seed takes separate path. Immature, shriveled, under sized and insect-damaged seeds have low whereas stones and soil lumps have more weight than the normal healthy seeds. These types of impurities may be removed successfully with the help of gravity separator. It consists of a stratifying deck surface mounted at a slight angle. Seeds are fed onto the specific gravity separator in layer three-to-five seeds thick. The shaking deck pushes the seed up hill. The desk reciprocates and tosses the seed and other impurities. The air flows directly through the deck, allows only the heavier seeds to touch the deck therefore pushed uphill, while the lighter seeds and impurities float in the air therefore remain downhill. The motion of deck continues until heavy seeds are carried to the high end and thrown out through a spout. As the seeds move through the stratifying area, the lighter seeds move down hill towards the lower spout and heavier seed move uphill towards the upper spout. In between these two spouts a spout is also present from which a middling product containing heavy, light and intermediate density seeds come out (Fig. 3). Gravity separator is used to enhance germination percentage of seed lot as it removes immature, insect damaged, shriveled, under sized and diseased seed of the crop efficiently from the seed lot. The performance of gravity separator depends on volume of air, end slope, side slope, oscillation motion and feed rate of the separator and shape of crop seed and physical purity of seed lot. The seed lot of cotton with germination of 55% and above and mungbean 65% and above can be upgraded to the Seed Certification Standard with the help of specific gravity separator.

5.4. Spiral separator

Differences in roundness or shape of seed are exploited in Spiral or Dropper separators for removal of contaminants. It separates the inert matter, other crop seeds, weed seeds and shrivelled seeds from healthy round leguminous seeds using the ability of a seed to roll due to its shape. Seed is poured slowly from the feeder, normal seeds start rolling from the inclined surface, gain speed and are thrown by centrifugal force into an outer chamber which directs the seeds to a spout. The slow rolling seeds remain on the inner inclined surface and enter the second spout (Fig 3).

5.5. Colour separators

It can remove seeds of different varieties, crops or weeds based on difference in the colour of the seed. Apart from physical purity it also improves the genetic purity, seed health and separate out weathered and moisture-damage seeds.

5.6. Surface Texture Separator

It removes inert matter, based on the surface texture differences to remove rough-textured weed and other crop seeds from smooth crop seeds.

5.7. Magnetic Separator

It separates the small weed and other crop seed and mechanically damaged seed based on the differences in seeds' affinity for liquids. The magnetic separator separates seed according to its surface texture or related seed characteristics (Fig. 3). First iron filings are mixed with seed, it adhere to rough surface alone. The treated seed lot is passed through a revolving magnetic drum that separates iron coated rough seed from smooth seed on which iron fillings are not deposited. Water is also mixed with iron filling for perfect deposition of iron filling on rough seeds. The greater the difference between surface textures of the seed lot's components, more effective will be the separation. It is used to remove seed of dodder and *Cichorium intybus* from Egyptian clovers and Lucerne.

5.8. Air Separators/Pneumatic Separators

In pneumatic separation light weight inert matter is separated with the help of air pressure. A fan is fixed to create air pressure greater then atmospheric in the separating column. From the unprocessed seed, the light weight seed and other contaminants with a terminal velocity less than the air velocity are lifted and rise through the column toward the top and removed through discharge spout. The seed with desired weight are deflected by the screen into the heavy seed discharge spout.

Precautions during seed processing

- All the bags of a seed lot should be marked perfectly and carefully.
- Seed lots representing one variety of one crop should be processed in the processing area at one time.
- To avoid admixture the processing machines and plant should be cleaned thoroughly before change of variety or crop.
- All the machines should be operated at the optimum speed to get high quantity of quality processed seed.
- Part of seed lot separated as undersize, impure and inert matter should be placed separately without any chance of mixture.
- Experience person should supervise the processing of seed.

■■■■

Chapter 19

SEED LOT

Lot is a specified quantity of processed seed of a variety and class produced by a grower.

Quantity of seed in a seed lot is specified for different crops (Table 1). The processed seed containing higher quantity than the prescribed ones in a seed lot is divided into two lots one containing maximum prescribed quantity and second with extra quantity. The seed is stored in bags or containers and marked for identification.

Table 1. Prescribed maximum lot size for various crops

Size (000 Kg)	Crops
Field and Fodder Crops	
10	Barnyard millet, common millet, finger millet, Italian millet, jute, kodo millet, lentil, linseed, little millet, pearl millet, rapeseed and mustard, safflower, sesame, sorghum, taramira, buffel, dharaf, Dinanath, Egyptian clover, guinea, Lucerne, marvel, rice bean, setaria grass, stylo
20	Barley, castor, chickpea, cotton, fieldpea, grass pea, groundnut, horsegram, kidney bean, moong bean, oat, pigeonpea, rajmash, rice, soybean, sunflower, teosinte, triticale, urid bean, wheat
40	Maize
Vegetable crops	
10	Amaranth, egg plant, cabbage, capsicum, carrot, cauliflower, celery, chilli, Chinese cabbage, cucumber, fenugreek, knol khol, lettuce, longmelon, muskmelon, onion, parsley, pumpkin, radish, rat tail radish, snap melon, spinach, tomato, True Potato Seed, turnip
20	Asparagus, bitter gourd, bottle gourd, Cluster bean, cowpea, French bean, Indian bean, okra, ridge gourd, snake gourd, sponge gourd, squashes (all), sugar beet, watermelon
40	Colocasia, lesser yam, potato tuber, sweet potato

3 Procedure for preparation of a lot: The name of crop and variety; month and year of harvesting; code of State, processing plant and lot are mentioned on the bag. Months are mentioned as Jan, Feb, March, April, May, June, July, Aug, Sept, Oct, Nov and Dec, whereas, year as 11 for 2011, 20 for 2020; a specific code number is allotted to each State, processing plant and each processing plant allots a number to a specific lot.

■■■■

Chapter 20

SAMPLING

1 OBJECTIVE

Samples are derived from different portions of a seed lot and mixed to obtain a sample of required quantity representing the seed lot in true sense. From this composite sample, small portion of required quantity is obtained in such a way that even after reduction, it represents the seed lot (Fig. 1).

2. EQUIPMENT AND MATERIALS

Trier, plastic tubs, bags, balance, seed divider, sticker and labels.

2.1. Trier

It is required to draw the primary sample from the seed lot stored in bags or containers. Two types of triers are required for sampling

Stick trier : It is made up of two hollow steel tubes with coinciding slots. Slots can be closed or opened by turning the inner tube within the outer tube with the help of a handle. Inner tube is divided into compartments by transverse partition, whereas outer tube has a solid end.

Nobbe trier : It is a long hollow metal tube with a sharp solid pointed end and an oval hole just near the pointed end.

2.2. Seed divider

It is an equipment used for getting desired quantity of true to the type sample for submission in laboratory or for individual test. Seed dividers are of following types

Boerner type divider (conical divider) : It consists of a funnel shaped hopper on the top with a valve at exit end. Just below the hopper a cone with its upward pointed end is present with numerous closely spaced channels at the circumference of the cone. Alternate port leads through duct to a common spout among the two. Two pans are provided to collect the seeds from spouts (Fig. 3).

Soil divider (multiple slot divider) : It consists of a hopper with rectangular ports on the lower end fixed in a straight row. Alternate port leads to left or right in receiving pan with the help of ducts (Fig. 3).

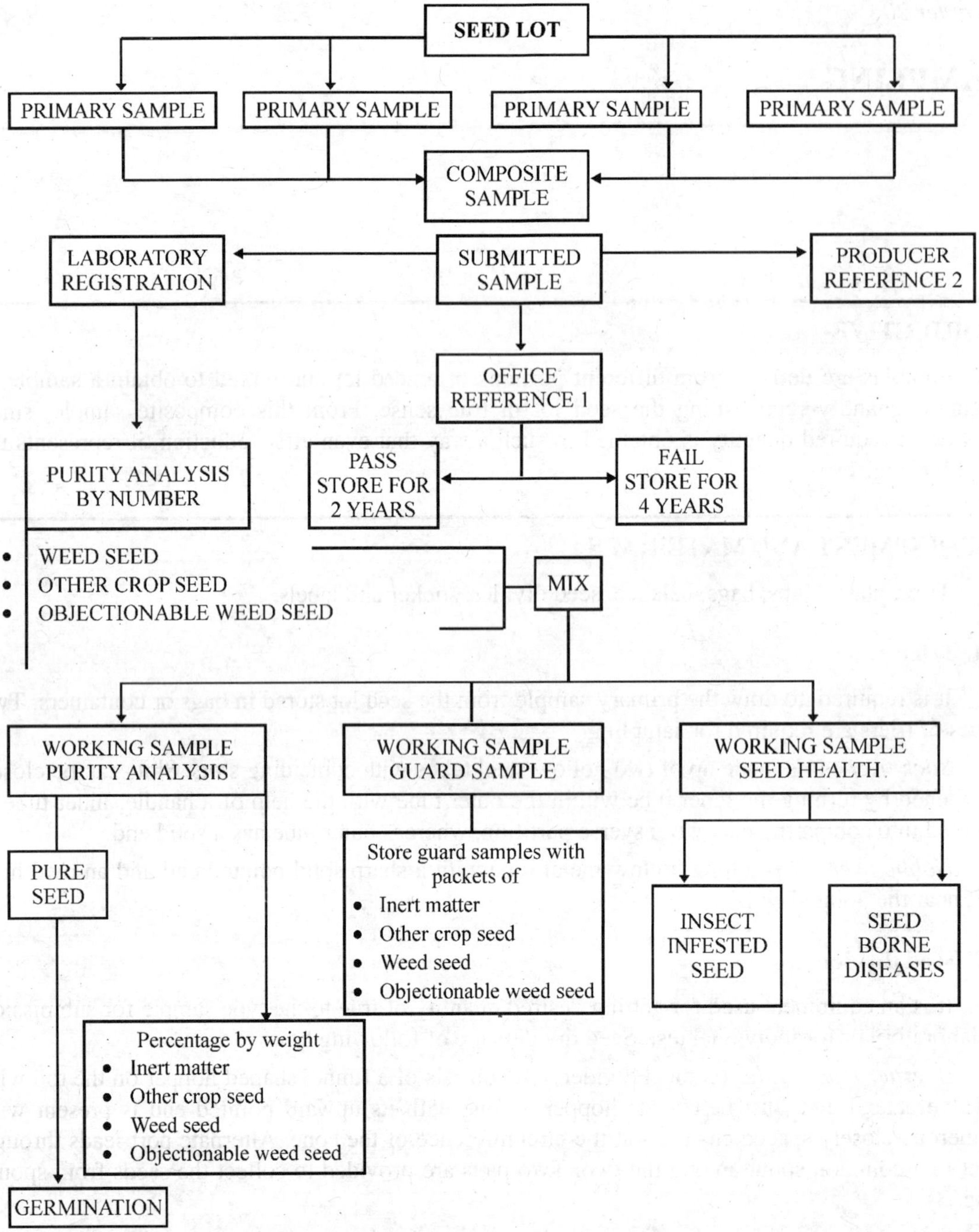

Fig. 1. Steps involved in sampling and seed testing

Gamet divider (centrifugal divider) : It consists of a hopper, spinner and electric motor. The periphery is divided into ports and half of the ports are connected with one spout and remaining with the second (Fig. 2).

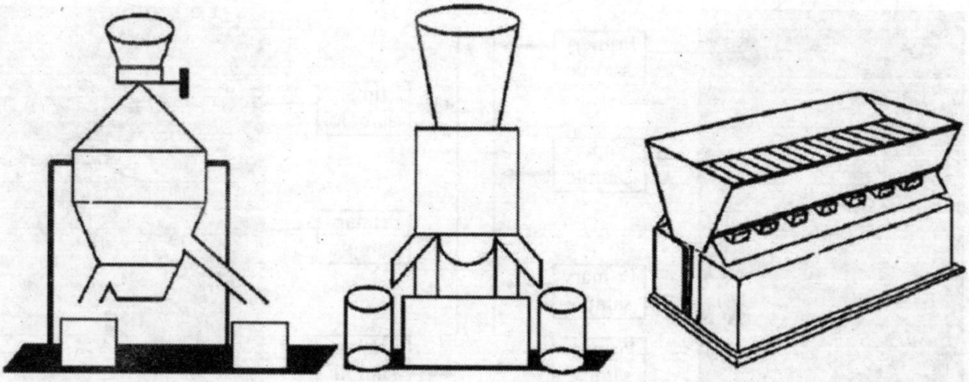

Fig. 2. Seed dividers used for sampling

3. PROCEDURE

3.1. Sampling in a processing plant

3.1.1 Primary sample: It is a small quantity of seed taken from one point of the processed lot. The seed lot is arranged to approach conveniently upto individual container. Primary samples are drawn from different portions and depth by inserting the stick trier with the closed slot diagonally in the seed bag or container upto desirable depth with minimum damage to seed. The flow of seed is facilitated in the tube by opening and closing of the slot. Finally, the trier is withdrawn with closed slot and collected sample is transferred to a container.

Spear trier is inserted into a bag upto a desirable depth at an angle of 30 degree with the hole present at the pointed end facing downwards. The spear is withdrawn gently, so that, equal quantity of seeds enter into the hole from centre to the side of the bag. The point of insertion is closed with the help of a sticker or by running across the trier on the hole a couple of times in opposite direction.

Minimum number of primary samples should be taken as per Table 1 and 2. The quantity of seed drawn in one primary sample depends on the sampling intensity, size of submitted sample and crop (Fig 3).

3.1.2 Composite sample : Primary samples drawn from different places of a lot are mixed and the mixture is known as composite sample. The size of composite sample should be 10 times more than the required submitted sample.

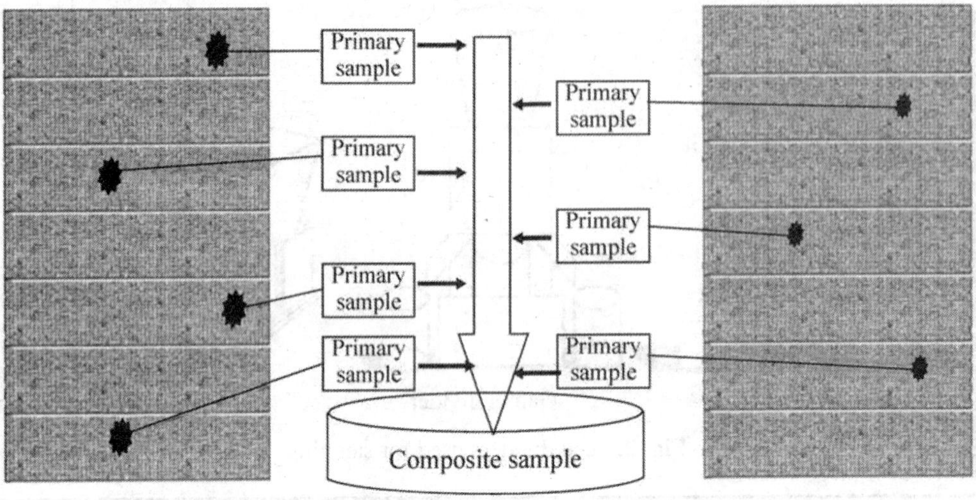

Fig. 3. Preparation of primary and submitted sample

3.1.3 Submitted sample : The required quantity of seed, which is sent to seed testing lab is known as submitted sample (Table 1 and 2). To prepare a submitted sample, the composite sample is mixed thoroughly and reduced upto required quantity with the help of seed divider or by repeated halving method.

Table 1. Sampling intensity for a seed lot stored in container

Number of container	Sampling intensity
1-5	Each container, from atleast 5 positions
6-14	Not less than 5 containers
15-30	At least 1 container in every 3
31-49	Not less than 10 containers
50-400	At least one container in every five
401-560	Not less than 80 containers
561 or more	At least one container in every 7

Table 2. Sampling intensity for seed stored as bulk

Lot size (Kg)	Sampling intensity
Upto 50	Not less than 3
51-1500	Not less than 5
1501-3000	At least 1 for each 300 kg
3001-5000	Not less than 10
5001-20000	At least 1 for each 500 kg
20001-28000	Not less than 40
28001-40000	At least 1 for each 700 kg

(a) **Use of seed divider** : The number of the sample is mentioned on the bag. The divider and its pans are cleaned properly. The valve of hopper in Boerner divider is closed. The pans placed at both the spout and the valve of Boerner divider is opened whereas, hopper of soil divider containing seed is put on ports and ducts. The seed is poured on rotated spinner of Gamet type divider. The divided samples from both the spouts are collected. The seed of one collecting pan is kept in a container and the seed of other pan is divided again and again to get the required quantity of submitted sample.

(b) **Repeated halving method :** The composite sample is spread uniformly on a clean and smooth surface. It is divided into two equal parts and one part is again thoroughly mixed and division is repeated to get the required quantity of submitted sample.

3.1.4 Sample for determination of moisture content : A quantity of 100 g seed of the species that has to be ground and 50 g of other species is submitted in moisture proof bag with submitted sample for determination of moisture content.

3.1.5 Dispatch of submitted sample : Sample should be dispatched to the seed testing lab as early as possible providing all the details like date of sampling, number of processing plant, crop, variety, class of seed, lot number, lot size etc.

Apart from this sample, two reference samples are also prepared by the same method. One reference sample is stored by the office and second by producer. Office sample of seed lot passed in seed testing is stored for two years.

---

4. SAMPLING IN SEED TESTING LAB

The submitted sample received in seed testing lab is registered and designated by a code number. Submitted sample is tested for determination of seeds of other crop, weed, objectionable weeds, objectionable diseases and other distinguishing varieties by number. Three working samples of the submitted sample, which passes the seed certification standard by number are prepared. Each working sample consists of at least 2500 seeds (Table 3).

4.1 Preparation of a working sample :

4.1.1 Mechanical divider : As described for preparation of submitted sample.

4.1.2 Repeated halving method : As described for preparation of submitted sample or the seed is poured on a clean smooth surface and shaped as a mound after thorough mixing. Mound is divided into two halves, each half is again halved, each portion is again halved giving total 8 portions. Alternate portions are combined i.e. 1st and 3rd of first row and 2nd and 4th of second row. The remaining portion is kept in a pan and the process is repeated to obtain required size of the working sample.

4.1.3 Random cup method : Six to eight small cups of equal size and shape are arranged at random on a tray. The seed is poured uniformly over the tray. The seeds, which fall into the cups, are collected as working sample. This method is useful for the crops with small seed size but not for chaff and round seeds.

4.1.4 Spoon method : The seeds are poured evenly in one direction over the tray. If required, seed can be poured second time in opposite direction. Shaking of the tray is avoided, small quantity of seeds are collected with the help of spatula from minimum 5 random places to make a working sample of required quantity.

The working sample is stored in paper bag marked with code number, name of the crop and purpose.

Table 3. Size of submitted and working samples required for different crops

Crop	Submitted sample (g)	Working sample (g)
FIELD AND FODDER CROPS		
Gueina grass, Setaria grass	0025	0002
Marvel grass	0030	0003
Brassica juncea, taramira	0040	0004
Lucerne, Indian clover	0050	0005
Egyptian clover, finger millet, buffel grass	0060	0006
B. campestris, sesame, stylo	0070	0007
Kodo millet	0080	0008
Italian millet	0090	0009
Linseed, jute, common millet	0150	0015
Lentil	0600	0060
Sorghum	0900	0090
Pearl millet	0950	0015
Safflower	0950	0090
Wheat, oat, triticale	1000	0120
Sunflower	1000	0200
Pigeonpea	1000	0300
Cotton	1000	0350
Horse gram, moong bean	1000	0400
Grass pea	1000	0450
Castor, soybean	1000	0500
Rice, rajmash, urid bean	1000	0700
Fieldpea, maize	1000	0900
Chickpea, groundnut	1000	1000
VEGETABLE CROPS		
Celery	0025	0001
Chinese cabbage, parsley	0040	0004
Carrot, lettuce	0030	0003
Tomato	0015	0007
Turnip	0070	0007
Onion	0080	0008
Brassica olerecea all varieties	0100	0010

Crop	Submitted sample (g)	Working sample (g)
Chilli, egg plantl	0150	0015
Cucumber, musk melon	0150	0070
Spinach	0250	0025
Radish	0300	0030
Pumpkin	0350	0180
Coriander	0400	0040
Fenugreek	0450	0045
Sugar beet	0500	0050
Cluster bean, asparagus	1000	0100
Okra	1000	0140
Water melon, sponge gourd	1000	0250
Ridge gourd	1000	0400
Bitter gourd	1000	0450
Bottle gourd	1000	0500
Indian bean	1000	0600
French bean and all squashes	1000	0700

■■■■

Chapter 21

PHYSICAL PURITY ANALYSIS BY NUMBER

1. OBJECT

To identify and determine the seeds of species other than the seed crop present in submitted sample and constituting the seed lot by number. Seed crop refers to all the botanical varieties and cultivars of the species considered for seed production.

2. APPARATUS

Physical purity work board, forceps, cups and specimen of objectionable weed seed (OWS).

Physical purity work board (Diaphanascope): The work board is made up of wooden flat smooth board of 55 cm length, 27 cm width and 10 cm height. Reflected light provided from the source at the base covered with opaque glass sheet. A magnifying glass (10x) is fixed on a stand helping the analyst to separate and observe the seeds from front side (Fig. 1).

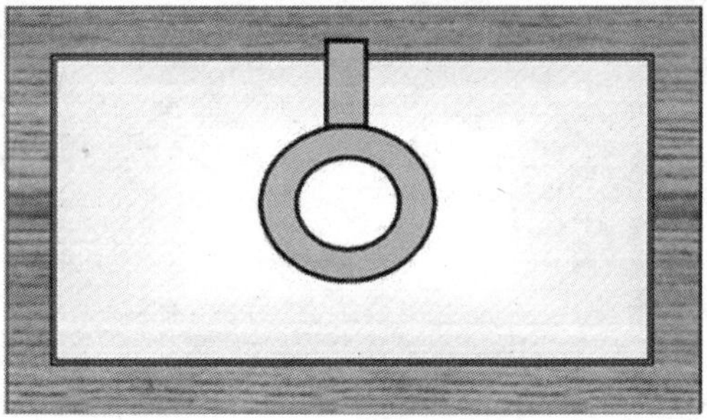

Fig. 1 Diaphanoscope

3. PROCEDURE

The sample is spread uniformly over the physical purity work board and the following categories of seeds are sorted out with the help of a pair of forceps by visual observation and separated in different cups.

3.1. Other Crop Seed (OCS)

Seed of any other crop plant excluding the crop under consideration (Table 1).

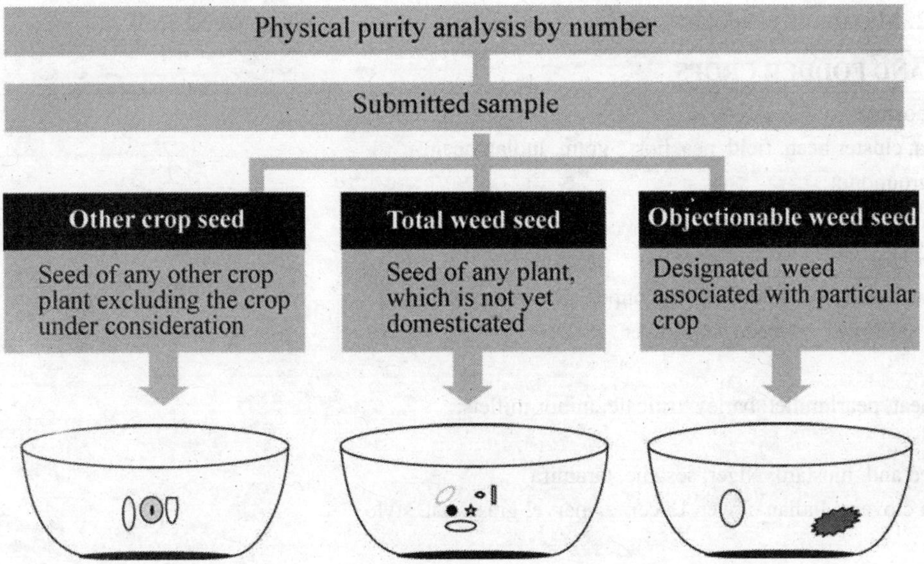

Fig. 2. Procedure of physical purity analysis by number

Table 1. Maximum seed certification standard for number for other crop seed

FIELD AND FODDER CROPS	**F**	**C**
Castor, groundnut, Indian bean, rajmash, safflower, sunflower	Nil	Nil
Chickpea, field pea , kidney bean, rice bean	-	5
Soybean, horse gram	-	10
Maize (composite, synthetic and open pollinated varieties), sorghum	5	10
Grass pea, lentil, moong, pigeonpea, urid	5	10
Cotton, teosinte	5	10
Barley, oat, pearl millet, rice, triticale, wheat, minor millets	10	20
Linseed, niger, rape seed and mustard, sesame, taramira	10	20
Egyptian clover, Indian clover, Lucerne, marvel grass, stylo	10	20
Jute cluster bean	10	20
Buffel, dharaf, Dinananth grass, guinea and setaria grass	20	40
VEGETABLE CROPS		
Brinjal, bitter gourd, bottle gourd, Indian summer and winter squash, lettuce, Pumpkin, ridge gourd, snake gourd, water melon	Nil	Nil
Okra	Nil	5
Amaranth, asparagus, cabbage, capsicum, carrot, celery, chilli, cauliflower Cucumber, little	5	10
gourd, long melon, musk melon, onion, parsley radish, spinach Sugar beet, tomato, turnip	5	10
Fenugreek	10	20

3.2 Total Weed Seed (TWS)

Seed of any plant which, is not yet domesticated (Table 2).

Table 2. Maximum seed certification standard for total number of weed seed

FIELD AND FODDER CROPS	F	C
Maize, teosinte	Nil	Nil
Chickpea, cluster bean, field pea, horse gram, Indian bean,	Nil	Nil
Castor, groundnut,	Nil	Nil
Cow pea, rajmash	Nil	10
Sorghum	5	10
Grass pea, kidney bean, moong, pigeonpea, rice bean, urid	5	10
Linseed, safflower, soybean, sunflower	5	10
Cotton	5	10
Rice, wheat, pearl millet, barley, triticale, minor millets,	10	20
Lentil	10	20
Rape seed and mustard, Niger, sesame, taramira	10	20
Egyptian clover , Indian clover, Lucerne, marvel grass, oat, stylo	10	20
Jute	10	20
Buffel, Dharaf, Dinanath, guinea and setaria grass	20	40
VEGETABLE CROPS		
Brinjal, Cucurbits, okra, tomato	Nil	Nil
Asparagus, Cabbage, capsicum, carrot, celery, Chinese cabbage, cauliflower, Chilli Lettuce, onion,	5	10
parsley, spinach, sugar beet, turnip	5	10
Amaranth, Fenugreek, radish	10	20

3.3. Objectionable Weed Seed (OWS)

These are the defined and designated weeds associated with particular crops (Table 3) as per Indian Seed Certification Standard (1966).

Table 3. Maximum seed certification standard for number of objectionable weed seed

Crop	Objectionable weed	F	C
Rice	Wild rice (*O. sativa* var. *fatua*)	2	5
Wheat	Wild morning glory (Hiran khuri; *Convolvulus arvensis)* Gulli danda (*Phalaris minor*)	2	5
Triticale	Wild morning glory(Hiran khuri; *Convolvulus arvensis)* *Phalaris minor*	2	5
Rape seed and mustard	Mexican prickly poppy(Satyanashi; *Argimone mexicana)*	5	10
Taramira	Mexican prickly poppy(Satyanashi; *Argimone mexicana)*	5	10
Safflower	Wild safflower (*Carthamus oxyacantha)*	Nil	Nil
Sunflower	Wild *Helianthus*	Nil	Nil
Egyptian clover	Chicory (Kasni; *Cichorium intybus)*	5	10
Napier Grass	Canada thistle(*Cirsium arvense)*	0	0
	Convolvulus arvensis	0	0

	Johnson grass (Saru; *Sorghum halepense*)	0	0
	Quack grass (Thitch / couch grass; *Agropyron repens*)	0	0
Oat	Wild oat (*Avena fatua*)	2	5
	VEGETABLE CROPS		
Bitter gourd	Balsam apple (mokha; *Momordica balsamina*)	Nil	Nil
	Bhat karela (*M. cochinchinensis*)	Nil	Nil
	Jangli karela (*M. diocia*)	Nil	Nil
Cucumber	*Cucumis hardwickii*	Nil	Nil
Long melon	*Cucumis prophetarum,*	Nil	Nil
	Weed melon (Takmek) (*Cucumis melo* var. *agrestis*)	Nil	Nil
	other non-desert forms of *C. melo*	Nil	Nil
Musk melon	*Cucumis prophetarum,*	Nil	Nil
	Weed melon (Takmek) (*Cucumis melo* var. *agrestis*)		
	other non-desert forms of *C. melo*		
Snap melon	*Cucumis prophetarum,*	Nil	Nil
	Weed melon (Takmek) (*Cucumis melo* var. *agrestis*)		
	other non-desert forms of *C. melo*		
Snake Gourd	*Trichosanthes cucumerina* (Jangli chichinda or Rambel)	Nil	Nil
	T. palmata	Nil	Nil
	T. lobata	Nil	Nil
Watermelon	Wild water melon (Indrayan; *Citrullus colocynthis*)	Nil	Nil
Okra	Wild okra	Nil	Nil
	A. fioulneus		
	A. manihot		
	A. moschatus		
Amaranthus	Wild amaranthus (Kantewali chaulai; *A. spinosus*)	10	20
Fenugreek	Senji (*Melilotus* spp.)	2	5
Lettuce	Wild lettuce (*Lactuca scariola*)	2	5

In the event (Fig. 2), the number of weed seed from any category exceeds than the maximum seed certification standard (Tables 1, 2, 3) the sample is liable to be rejected.

3.4. Objectionable parasites at seed level

These are the defined parasitic weeds associated with particular crop (Table 4).

Table 4. Maximum seed certification standard for number of parasitic weeds at seed level

Crop	**Objectionable parasitic weed**	**F**	**C**
Lucerne	Dodder (*Cuscuta* spp.)	5	10
Napier grass	Dodder (*Cuscuta* spp.)	0	0

4. RESULT

The result of the purity analysis is reported by number of seeds present in the prescribed quantity (g) of submitted sample

Number						Size of the submitted sample
OCS		WS		OWS		
Number	Name	Number	Name	Number	Name	
Total						

■■■■

Chapter 22

PHYSICAL PURITY ANALYSIS BY WEIGHT

1. The working sample is separated into pure seed, other seed and inert matter to determine the physical purity of a seed lot by weight in terms of presence of the other matter than the crop seed present in the working sample. The physical purity percentage is determined by weight.

- **(i) Pure seed:** It is the seed of an allotted crop produced by a grower, which includes all the botanical varieties and cultivars of that particular species. The seed of allotted crop is considered as pure even if it consists of immature, shrivelled, diseased, germinated, under sized or piece of seed unit larger than half of the original size provided identified authentically as of that crop.
- **(ii) Other seed:** Seed of any other crop or weed mixed in the sample.
- **(iii) Inert matter:** All the material present in the sample not defined and considered as seed e.g., plant parts other than the seed, soil particles, insect, nematode galls, seed of the allotted crop with less than half of its original size any appendages of seed not defined as part of the seed (eg., glume of wheat) are considered as inert matter (Fig. 1).

2. MATERIAL REQUIRED

Physical/precision balance, purity work board, spatula, forceps, cups, seed sample.

3. PROCEDURE

The working sample of desired weight is distributed evenly over the physical purity work board. Pure seed, other seed and inert matter are removed with the help of spatula and forceps and kept in different cups. No pressure or magnification is applied for separation except some grasses.

3.1. During physical purity analysis the seed of any crop from family Leguminoceae and Cruciferae without seed coat is considered as inert matter (Fig. 1).

3.2. Multiple seed unit (one fertile floret is attached with sterile or fertile floret) of grasses are weighed separately without breaking the attachment and reported in percentage.

3.3. Insects are considered as inert matter. Insect damaged seed is examined thoroughly and when the extent of damage is more than half of the seed then the seed and insect both are considered as inert matter.

3.4. Husk less seed from the sample of rice and sunflower are removed and counted to work out the percentage of husk less seed. The maximum permissible limit for husk less seed is 2%.

$$\text{Husk less seed (\%)} = \frac{\text{Number of husk less seeds}}{\text{Total number of seeds in sample}} \times 100$$

3.5. Use of seed blower : It is used to remove the light weight inert matter from the seeds of grasses or vice-versa. Working sample is kept at the lower portion of the tube and the required uniform upward flow of air is regulated upto prescribed period of time. Lighter matter is separated from the sample by air flow and settled down in the partition provided in the tube of the blower. The tube is removed and the inert matter is collected. All the fractions of seed samples are weighed at the end of the physical purity analysis.

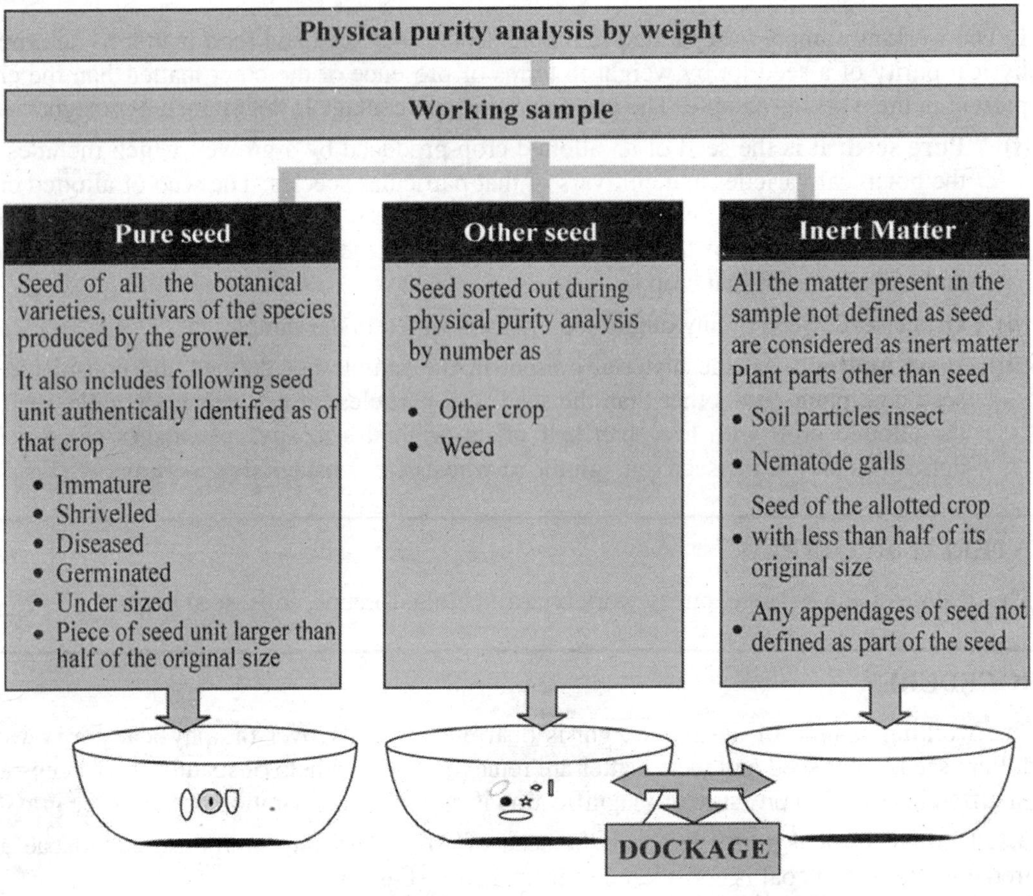

Fig. 1. Components of physical purity analysis by weight

4. CALCULATION

Physical purity analysis is based on the sum of the weight of different components and not on the original weight of working sample.

$$\text{Purity (\%)} = \frac{\text{Weight of pure seed (g)}}{\text{Pure seed (g)} + \text{Other seed (g)} + \text{Inert matter (g)}} \times 100$$

5. RESULT

The result of the purity analysis is presented using one decimal place. Component less than 0.05% is reported as trace.

Component	Weight (g)	Remark
Pure seed		
Other seed		
Inert matter		
Total		
Physical purity (%)		

The lot with physical purity percentage more than the minimum seed certification standard (Table 1) is liable to pass.

Table 1 Minimum seed certification standard for physical purity percentage by weight

Field and fodder crops	F and C
Buffel grass, Dharaf grass, Guinea grass	80
Marvel grass, stylo	90
Dinanath grass, setaria grass	95
Groundnut	96
Sesame, rape seed and mustard, minor-millet, taramira, jute	97
Barley, maize (inbred, single cross hybrids, other type of hybrids, composites, synthetics and open pollinated varieties) oat, pearl millet, rice, sorghum, triticale, wheat	98 98
Chickpea, cluster bean, cow pea, field pea, grass pea, horse gram, Indian bean, kidney bean, lentil, moong bean, pigeon pea, rajmash, rice bean, urid bean	98 98
Castor, linseed, niger, safflower, soybean, sunflower	98
Egyptian clover, Indian clover (*Melilotus* sp), Lucerne, teosinte	98
Cotton, hybrid cotton, daincha, Jatropha, mesta, sugarcane, sun hemp	98
Napier grass (root cutting by number)	99.5
Vegetable crops	
Amaranth, carrot	95
Asparagus, spinach, sugar beet	96
Celery, parsley	97
Brinjal, capsicum, chilli, tomato, vegetable pea	98
Fenugreek, lettuce, onion, radish, turnip	98
Broccoli, cabbage, cauliflower, all cucurbits(baring pointed and little gourd), Wahga (*Cucumis melo* spp. Melo)	98
Okra	99
Pure living planting stakes of little and pointed gourd	99.5©/ 98(F)

Seed propagated medicinal plants	
Lemon grass & Citrunella grass (*Cymbopogon* spp.) (seed)	90
Anise (*Pimpinella anisum*), Babchi (*Psoralea corylifolia*), Basil (*Ocimum* spp), Bell-dona (*Atropa* spp), Chamomile (*Matricaria chamomilla*), Datura, Henpane (*Hyoscyamus* spp.), ipecac (*Cephaelis ipecacuanha*), oxglove (*Tilpushpi Digitalis*), Palmorosa grass (*Cymbopogon* spp), Periwinkle (*Catharanthus roseus*), Sarpgandha (*Rauvolfia serpentina*), Senna (*Cassia angustifolia*)	95
Isabgol (*Psyllium* spp)	97
Ashvagandha (*Withaniq somnifera*), Bhang (*Cannabis sativa*), Jojoba (*Simmondsia chinansis*), Musk Mallow (*Abelmoschus moschatus*), Rat-tail Radish	98
Ornamental	
African marigold, Ammi majus, Annual Carnation, Marigold, Opium (*Papaver somniferum*)	97
Annual Chrysanthemum, Aster, Ornamental Sunflower, Petunia, Snapdragon	98
Spices	**F and C**
Coriander, Cumin, Dill (*Anethum graveolens*), Fennel	95
Tree fruits	
Arecanut (Husked nut), Coconut (Husked nut), Oil palm (Husked seed), Papaya	98
Clonally propagated (tubers/bulbs/bulblets/ cormels/rhizomes)	
Aloe (*Aloe* spp. (seed aloe/rhizome), Chinese potato (*Solenostemon rotundifolius* (Tubers), Daffodil (bulbs/bulbils), Elephant Foot Yam (*Amorphophallus companulatus*), (Corms/carmels), Garlic (Bulbs), Ginger (seed ginger, rhizome), Gladiolus (Corms/Cormels), Greater Yam (*Dioseorea alata*) (Tubers), Hyacinth (Bulbs), Iris (bulbs/bulbils), Lesser Yam (*Dioscoreq esculenta*) (Tubers), Multiplier onion (potato onion) (bulblets), Narcissus (bulbs/ bulbils), Safed musli (*Chlorophytum* spp) (Tubers), Saffron (*Corcus sativus*) (Bulbs), Seed potato (*Solanum tuberosum*) (Tubers), Steriodal Yam (*Dioscorea*) (Tubers), Sweet Potato (Tubers), Tannia (*Santhosoma sagittifolium*) (Corms/Cormlets), Taro (*Colocasia esculenta*) (Corms/Cormels), Tube rose (*Polinathus*) (Bulbs), Tuberous begonia (tubers), Tulip (bulbs), White Yam (*Dioscorea* Tubers)	95
Clonally propagated: planting stakes/root slips/self rooted plants/plantlets- produced through tissue culture)	
Acid lime (self-rooted plants), Almond (self-rooted plants), Amla (self-rooted plants), Anthurium (off'-shoots), Apple (self-rooted plants), Apricot (self-rooted plants), Areca nut (self-rooted plants), Bael (self-rooted plants), Banana (Sword sucker plantlets), Ber (self-rooted plants), Black pepper (planting stakes), Cardamom Small & large (self-rooted plants/plantlets), Clove (self-rooted plants), Cashew (self-rooted plants), Cherry (self-rooted plants), Chow chow (Planting stakes), Cirinamon (self rooted plants), Coconut (self-rooted plants), Custard apple (self-rooted plants), Date Palm (off shoots), Fig (planting stakes/self-rooted plants),Grape (planting stakes), Guava (self-rooted plants), Guggal (*Commjphora wightii*) (planting stakes), Hop (*Humulus lupulus*) (planting stakes), Jack fruit (self-rooted plants), Japanese mint (Mentha) (root slips), Jojoba (*Simmondsia chinensis*), Kewda (*Pandanus fascicularis* (Suckers), Kiwi fruits (planting stakes), Lemon (self-rooted plants), Lemon grass & Citrunella grass (*Cymbopogon*) (root slips), Liquorice (*Cylcyrrhiza glabra*), litchi (planting stakes(self-rooted plants), Little gourd (planting stakes), Long pepper (*Paper longum*) (Planting stake), Mandarin (self-rooted plants), Mango (self-rooted plants), Neem (*Azadirchta indica*) (self rooted plants), Nutmeg (self-rooted plants), Oilpalm (self-rooted plants), Olive (self-rooted plants), Passion fruit (planting stakes), Patchouli (*Progostemen patchouli*) (planting stakes), Peach (self-rooted	98

plants), Pear (self-rooted plants), Pecan nut (self-rooted plants), Perennial carnation (planting stakes), Pine apple (suckers/ rooted slips), Plum (self-rooted plants), Pointed gourd (planting stakes), Pomegranate (planting stakes/self-rooted plants), Sapota (self-rooted plants), Sweet Orange (self-rooted plants), Tamarind (self-rooted plants), Tapioca (planting stakes), Vetiver (*Vetivera zizanioides*) (root slips), Walnut (self-rooted plants)	
Ornamental	
Chrysanthemum (suckers/planting stakes), Damask Rose (*Rose damascena*), French Jasmine (*Jasminum*) (Root slips), Gerbera (sister clumps), Lavendur (*Lavenula stoechus*) (Planting stakes), Orchids (Off-shoots/pseudo bulbs), Rose (planting stakes), Rose-geranium (*Elargonium graveolens*) (Planting stakes), Rose-marium (*Rosemarium officinalis*) (self-rooted plants)	98

Chapter 23

GERMINATION

1. OBJECTIVE

To test the germinability of a seed lot.

Seed germination test is an index indicating its capacity of producing normal healthy seedling under optimum conditions. The emergence of essential structures of seedling from embryo i.e. root system, shoot axis, cotyledons, terminal bud and coleoptile (Gramineae) determine its potential to develop into a normal plant under favourable conditions.

2. APPARATUS

2.1. Substrata

The medium must be free from microorganisms, toxic substances, insects and foreign seeds. It should have good water holding capacity with pH 6.0-7.5. Substrata used for germination test are of the following types

2.1.1 Paper towel: Open and porous towels with sufficient strength are prepared from bleached wood.

2.1.2 Sand : It is a medium consisting of silica particles ranging from 0.05 to 0.08 mm.

2.1.3 Filter paper **:** It is a normal blotting paper with water holding capacity and sufficient strength.

2.2. Water

Distilled, deionized or sterilized water with 6 -7.2 pH.

2.3. Seed germinator

It is a chamber with facilities to manipulate temperature and photoperiod as per need with 100% humidity.

3. METHOD

Minimum 400 seeds from pure seed fraction of physical purity test are tested for germination in prescribed substratum (Table 1) as per following procedure

3.1. Paper towel method

The information pertaining to tested sample is written on the mid-portion of the paper towel prior to its proper moistening with distilled water. Moist towel is stretched on a clean Table and on the other side (without any information) of the paper towel 50 seeds are arranged on its half portion containing 5 rows each of 10 seeds. About 3 cm space is left on lower and right side of the paper towel. Seeds are covered with remaining half portion of the towel and the right and lower portion is folded upward to close both the ends. The paper towel is rolled from the right end, wrapped in a wax paper and the ends are tightened with rubber bands. This paper towel is placed vertically in a seed germinator with upward direction of open end (Fig 1). It is also known as Between Paper (BP) method.

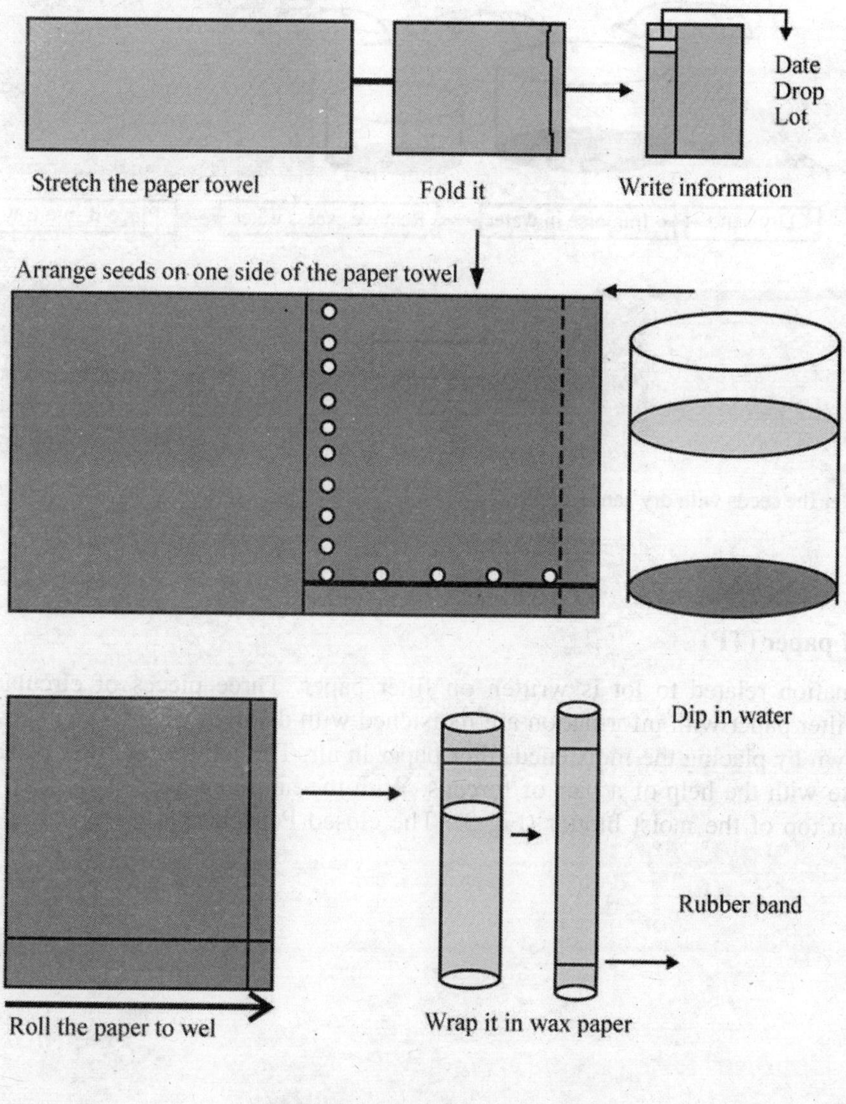

Fig. 1. Paper towel method

The seed can be arranged on entire towel in 10 rows each constituting 10 seeds keeping space of approximately 3 cm at lower and left and right margins. It is covered with other moistened paper towel and rolled prior to keeping in a seed germinator.

3.2. Sand (S)

Sterilized sand is immersed in distilled water. The excess water is removed by keeping the handful of this sand in air and placed in a tray of 6-9 cm depth. Counted seeds are planted on the levelled sand at proper distance and covered with a layer of dry sand (Fig. 2). The seeded tray is placed in a germinator.

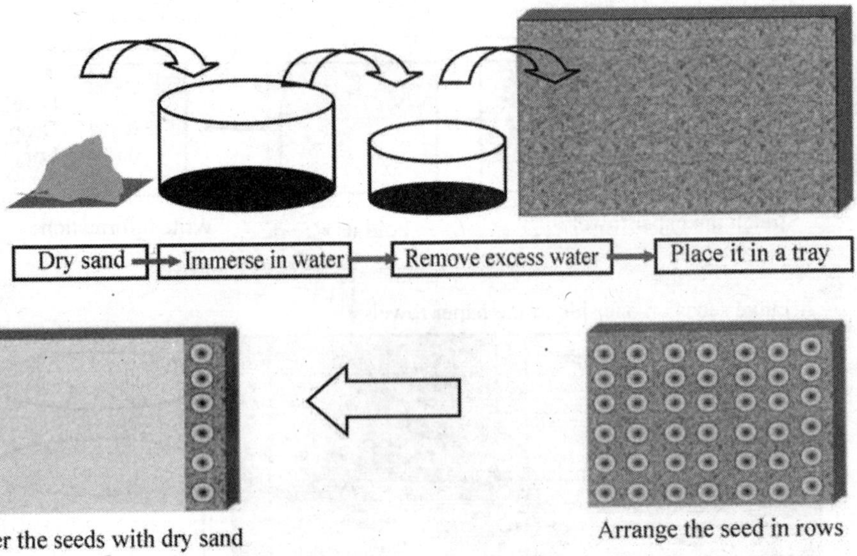

Fig. 2. Sand method

3.3. Top of paper (TP)

Information related to lot is written on filter paper. Three pieces of circular filter paper including filter paper with information are moistened with distilled water. The excess moisture is trickled down by placing the moistened filter paper in air. The moistened filter papers are placed in Petriplate with the help of a pair of forceps. With the help of a pair of forceps, 25 seeds are arranged on top of the moist blotter (Fig 3). The closed Petriplate is placed in an incubator or germinator.

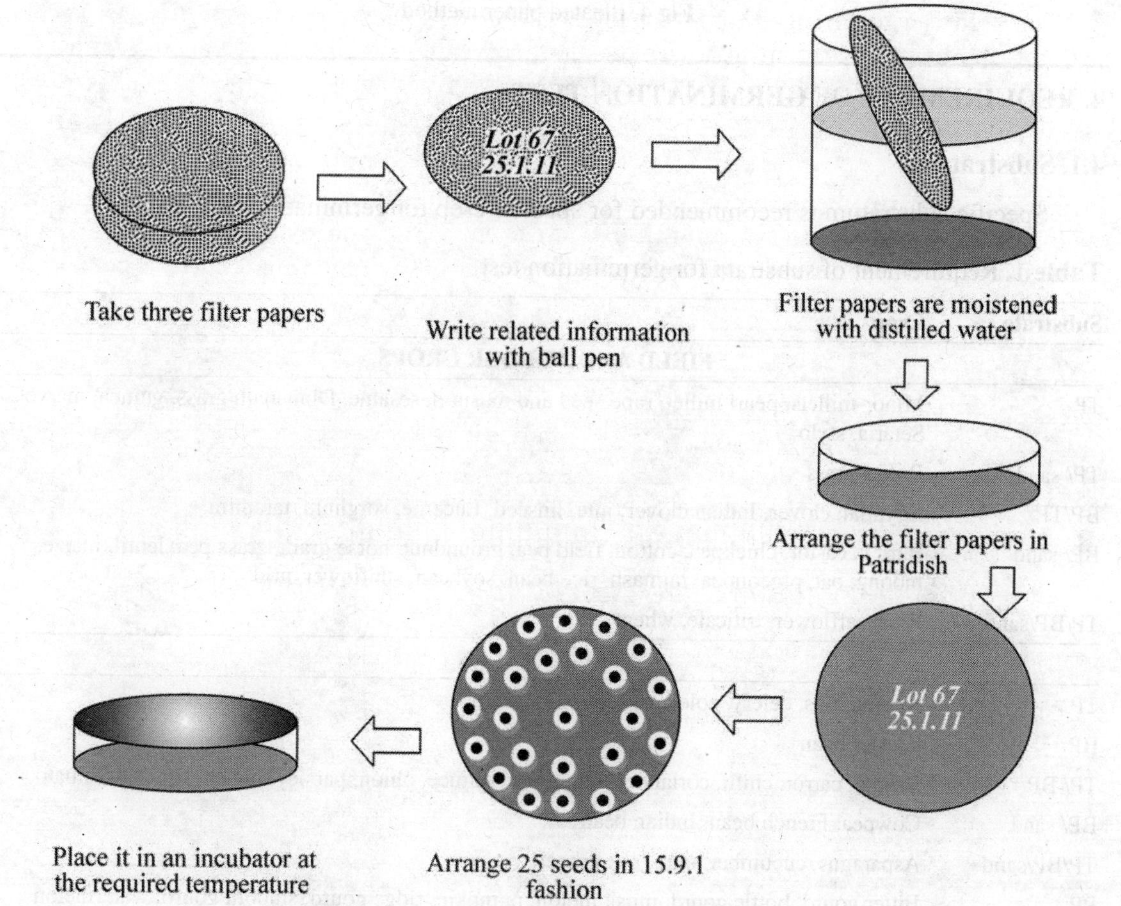

Fig. 3. Arrangement of seed in top of paper method

3.4. Pleated paper (PP)

Accordion like folded paper strip with 50 pleats is used as substratum. Generally 2 seeds are placed in one pleat (Fig. 4). This pleated paper is wrapped in wax paper and kept in germinator at required temperature for germination. This method is recommended mainly for the crops of the family Cucurbitaceae.

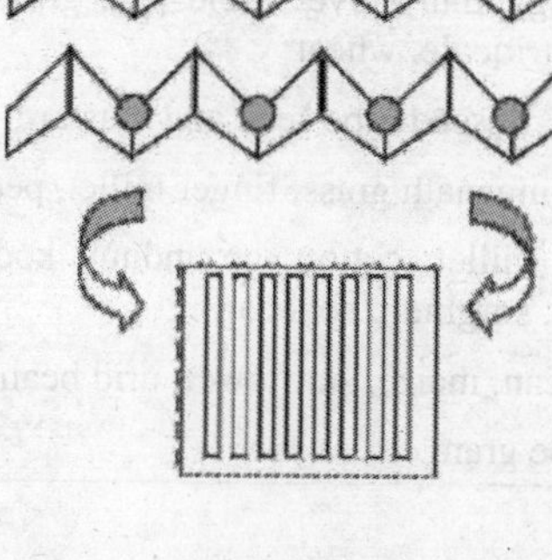

Fig 4. Pleated paper method

4. REQUIREMENT OF GERMINATION TEST

4.1. Substrata

Specific substratum is recommended for specific crop for germination test (Table 1).

Table 1. Requirement of substrata for germination test

Substrata	Crop
	FIELD AND FODDER CROPS
TP	Minor-millets, pearl millet, rape seed and mustard, sesame, Dinnanath grass, guinea, marvel, Setaria, stylo
TP/ sand	Buffel grass
BP/TP	Egyptian clover, Indian clover, jute, linseed, Lucerne, sorghum, taramira
BP/ sand	Barley, castor, chickpea, cotton, field pea, groundnut, horse gram, grass pea, lentil, maize, moong, oat, pigeonpea, rajmash, rice bean, soybean, sunflower, urid
TP/BP/sand	Rice, safflower, triticale, wheat
	VEGETABLES
TP	Amaranthus, celery, cole crops, cumin, turnip
BP	Cluster bean
TP/ BP	Brinjal, carrot, chilli, coriander, fenugreek, lettuce, onion, parsley, radish, spinach, tomato
BP/ sand	Cowpea, French bean, Indian bean
TP/BP/sand	Asparagus, cucumber, okra, sugar beet
PP	Bitter gourd, bottle gourd, musk melon, pumpkin, ridge gourd, smooth gourd, watermelon

4.2. Temperature

Normally seeds of *Kharif* crops are exposed to 25°C and *Rabi* crops to 20°C for germination (Table 2a and b).

Table 2a. Temperature (°C) requirement for germination test of field crops

Temperature		Crop
Range	**Constant**	**Field and Fodder Crops**
	20	Barley, Egyptian clover, field pea, grass pea, Indian clover, lentil, lucerne, oat, taramira, triticale, wheat
20-30	20	Chickpea, linseed rape seed and mustard
20-30	-	Castor, Dinnanath grass, finger millet, pearl millet, sesame
20-30	25	Common millet, cotton, groundnut, kodo millet, moong bean, pigeonpea, rice, safflower, sorghum, soybean
20-30	25, 20	French bean, maize, sunflower, urid bean
	30	Jute, horse gram

Table 2b. Temperature (°C) requirement for germination test of vegetable crops

Temperature		Crop
Range	Constant	Vegetable Crops
	15,10	Spinach
20-15		Onion
20		Lettuce
20-30		Asparagus, brinjal, celery, cluster bean, capsicum, okra, parsley, tomato
20-30	20	Amaranthus, cabbage, carrot, cauliflower, Chinese cabbage, coriander, fenugreek, knol khol, radish, sugar beet, turnip
20-30	25	Cowpea, cucumber, Indian bean, musk melon, pumpkin, summer squash, watermelon, winter squash
20-30	25,20	French bean
20-30	30	Bitter gourd, bottle gourd, sponge gourd
	30	Ridge gourd

4.3. Time

Time requirement for germination in different crops is as follows

4.3.1 First count

The first day on which germination is counted is known as first count (Table 3).

Table 3. Days to first count for germination test

Days	Crop
	FIELD AND FODDER CROPS
3	Common millet, Egyptian clover, Indian clover, jute, linseed, pearl millet, sesame
4	Barley, cotton, finger millet, horse gram, Italian millet, Lucerne, maize, pigeonpea, safflower, sorghum, stylo, sunflower, taramira, triticale, urid, wheat
5	Chickpea, field pea, grass pea, groundnut, lentil, moong, oat, rajmash, rape seed and mustard, rice bean, rice, soybean,
7	Buffel, castor, Dinnanath, kodo millet, marvel and Setaria grass
10	Guinea grass
	VEGETABLE CROPS
4	Amaranthus, bitter gourd, bottle gourd, cucumber, horse gram, Indian bean, lettuce, musk melon, okra, pumpkin, radish, ridge gourd, sponge gourd, sugar beet, summer squash, winter gourd
5	Chinese cabbage, cluster bean, cole crops, cowpea, fenugreek, French bean, tomato, turnip, water melon
6	Onion
7	Brinjal, capsicum, carrot, chilli, coriander, spinach

10	Asparagus, celery, parsley

4.3.2 Final count

The last day upto which the seed should germinate is considered as final count. Days for first and final count differ from crop to crop (Table 4).

Table 4. Days to final count for germination test

Days	Crop
	FIELD AND FODDER CROPS
5	Jute
6	Sesame
7	Barley, common millet, Egyptian clover, linseed, maize, pearl millet, rape seed and mustard, taramira, urid
8	Chickpea, field pea, finger millet, moong, rice bean, soybean, triticale, wheat
9	Rajmash
10	Groundnut, horse gram, Italian millet, lentil, lucerne, oat, pigeonpea, sorghum, stylo, sunflower
12	Cotton
14	Castor, grass pea, Indian clover, rice, safflower
20	Kodo millet
21	Marvel and setaria grass
28	Buffel, Dinnanath grass, guinea and grass
	VEGETABLE CROPS
7	Lettuce, turnip
8	Cowpea, cucumber, musk melon, pumpkin, summer squash, winter squash
9	French bean
10	Cole crop, Indian bean, opium, radish
12	Onion
14	Amaranthus, bottle gourd, brinjal, capsicum, carrot, chilli, cluster bean, fenugreek, ridge gourd, sugar beet, tomato, water melon
21	Celery, coriander, okra, spinach
28	Asparagus, parsley

5. TYPES OF GERMINATION

Germination is of following types

Hypogeal germination : It is the process of germination in which cotyledon(s) stay beneath the soil (Table 5 and Fig. 5). The epicotyl expands to raise the first true leaf out of the soil and the hypocotyl remains short and compact is known as hypogeal germination.

Epigeal germination : The process of germination in which cotyledon(s) get raised out of soil by extension of the hypocotyl and worked as leaf(ves) is known as epigeal germination (Table 5 and Fig. 5).

Table 5 Examples of hypogeal and epigeal germination

Cotyledons	Hypogeal	Epigeal
Dicot	Chickpea, fieldpea, pigeonpea	Oil seed crops, moong, urid
Monocot	Cereals	Onion

Vivipary : Germination occurs in fruits without completing the requirement of maturation (i.e., seed depends on plant for food and water) is known as vivipary e.g., mangroove. In this case, abscission layer is not formed between seed and plant.

Pre-harvest sprouting: Sprouting of mature seed (i.e., no relationship of seed with plant for nutrient or water) on the mature plant standing on the field due to high moisture is known as pre-harvest sprouting and it is different than vivipary. It occurs mainly due to no dormancy in the seed. In this case abscission layer is formed between seed and plant. Example sprouting in pods of mung bean, soybean, wheat etc.due to delayed harvesting coupled with high moisture content.

Hypo-epigeal germination : *Peperomia peruviana* a dicot species leaves one cotyledon beneath the soil as hypogeal germination while other one comes out of soil as epigeal germination (Fig 5).

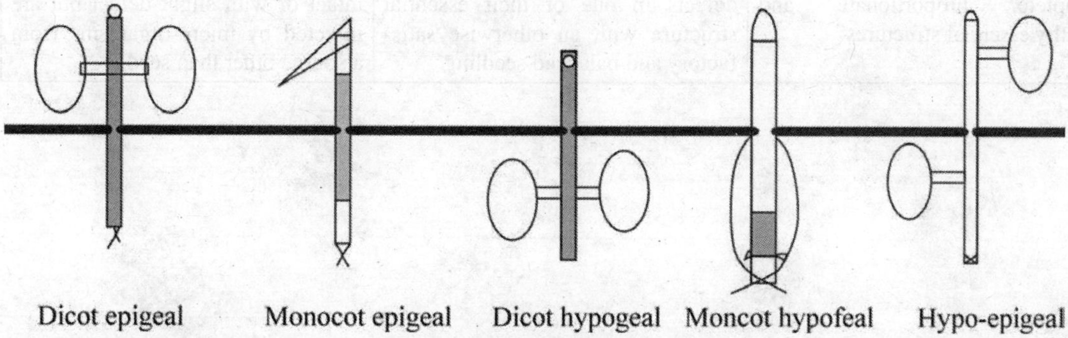

Fig. 5. Types of germination

***Precocious germination*:** Germination of seed inside the fruit before it begins to decay is known as precocious germination. Example green apple.

The first sign of germination is the increase in length and weight of the radicle. Whereas, in anaerobic condition (waterlogging) coleoptile growth of rice and barnyard precede radicle growth.

6. EVALUATION

Each seedling is critically evaluated on the day prescribed for first count. Seedlings with all their essential structures normal or badly decayed are counted and removed. Remaining seeds and seedlings are left as such and evaluated on the day of final count.

6.1 Normal seedling

Following seedlings are considered as normal.

6.1.1 *Intact seedling*

Seedlings with well developed, complete, proportionate and healthy essential structures are considered as normal seedlings (Fig. 6). Characters of normal seedlings are

i. *Well developed root system* : Long and slender primary root with numerous root hairs ending in a fine tip. In cereals several seminal roots (roots from embryonic tip instead of radicle) instead of primary root is also considered as normal root system.

ii. *Well developed shoot axis*: Straight, slender and elongated hypocotyl in seedlings exhibiting epigeal germination. Well developed epicotyl in seedling showing hypogeal germination and an elongated mesocotyl in seedlings of cereals.

iii. *Specific number of cotyledons* : One in monocot and two in dicot.

iv. *Green expanding primary leaves* : Single primary leaf in monocot and two in dicot.

Intact seedling	Seedlings with slight defects	Seedling with secondary infection
Seedlings with well developed, complete, proportionate and healthy essential structures	The seedlings exhibiting slight defects in one of their essential structure with an otherwise satis-factory and balanced seedling	Seedlings which are passed as intact or with slight defect but are infected by micro-organisms from a source other than seed.

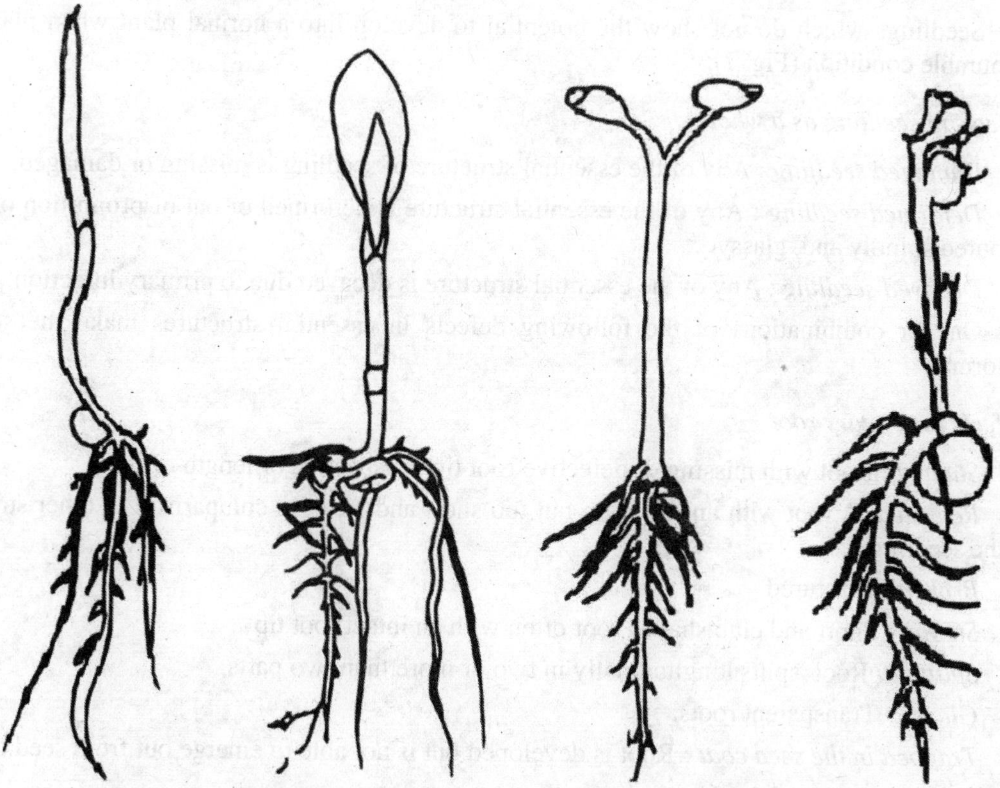

Monocot seedling **Dicot seedling**

Fig. 6. Intact seedling

6.1.2 Seedlings with slight defects

The seedlings exhibiting slight defects in one of their essential structures with an otherwise satisfactory and balanced seedling. Defects considered as slight are as follows :

(i) *Root* : Slight growth retardation in primary root, defective primary root with well developed secondary roots.

(ii) *Embryonic axis* : Hypocotyl, mesocotyl or epicotyl with limited damage. Only one cotyledon in dicot.

(iii) *Primary leaf* : Primary leaf with limited damage. In monocot small and loosely twisted coleoptile or coleoptile with half green leaf or splitting from the tip extending downward not more than one third of the length is considered as slight defect.

6.1.3 Seedling with secondary infection

Seedlings which are passed as intact or with slight defect but are infected by micro-organisms from a source other than seed.

6.2. Abnormal seeding

Seedlings which do not show the potential to develop into a normal plant when planted in favourable condition (Fig. 7).

Defects in seedling as a whole

Damaged seedling : Any of the essential structure of seedling is missing or damaged.

Deformed seedling : Any of the essential structure is deformed or out of proportion or white coloured spindly and, glassy.

Decayed seedling : Any of the essential structure is decayed due to primary infection.

One or combinations of the following defects in essential structures make the seedling abnormal.

Defects in primary root

Stunted : Root with missing or defective root tip irrespective of length of root

Retarded :A root with an intact tip but too short and weak in comparison to other structures of the seedling

Broken : Fractured

Stubby : Short and club shaped root often with an intact root tip

Splitted : Root splits longitudinally in two or more than two parts

Glassy : Transparent roots.

Trapped in the seed coat : Root is developed but is not able to emerge out from seed coat

Negative geotropism : Upward growth of root

Defects in shoot portion

Tightly twisted : Seedling twisted tightly around its main axis.

Looped structure : Seedling structure which forms a loop or circle instead of being more or less straight. Apart from these symptoms short and thick, broken, splitted, missing and constricted shoot axis are also considered as symptoms of abnormality.

Cotyledons : Swollen, deformed, broken, missing, discoloured, necrotic, glassy and decayed due to primary infection.

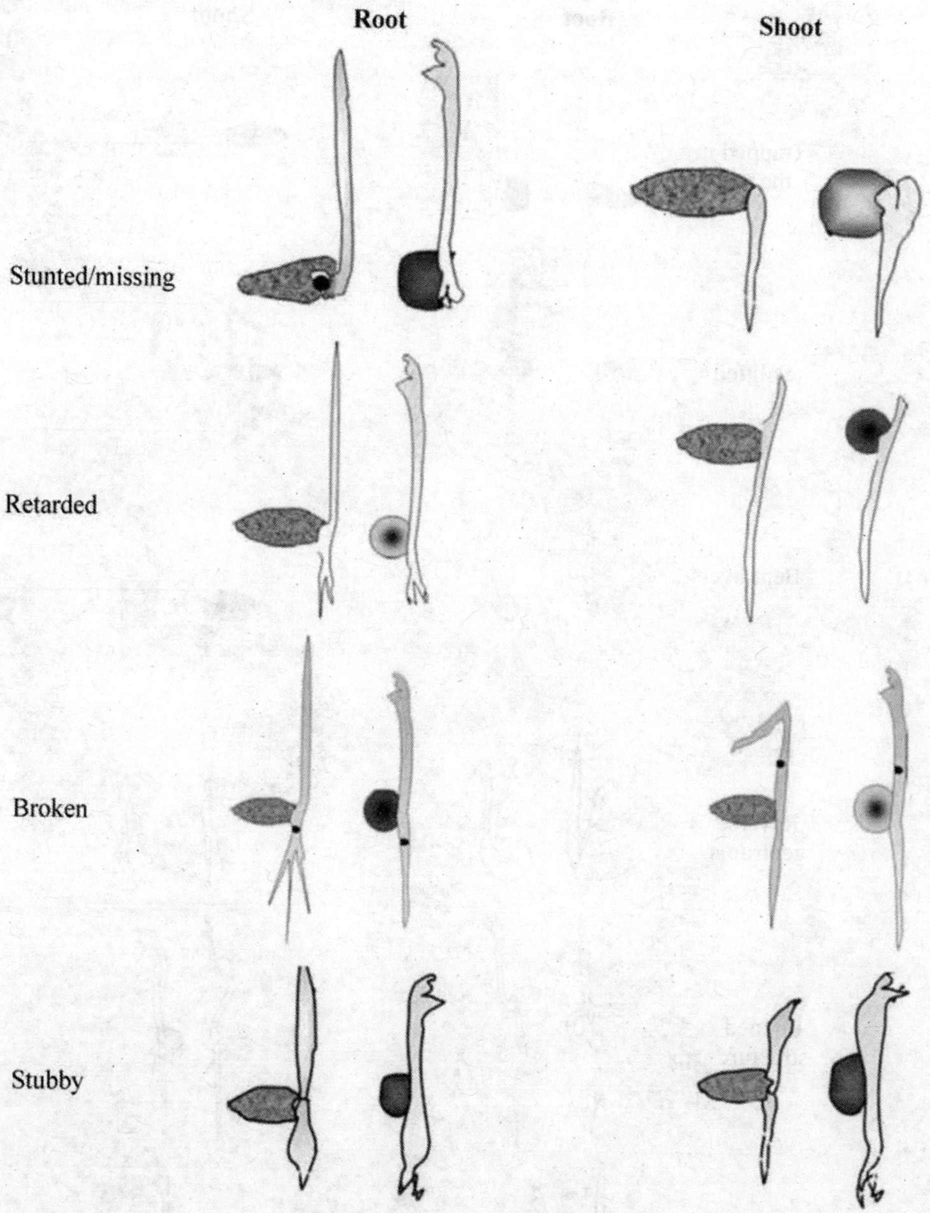

Fig. 7. Abnormal seedling

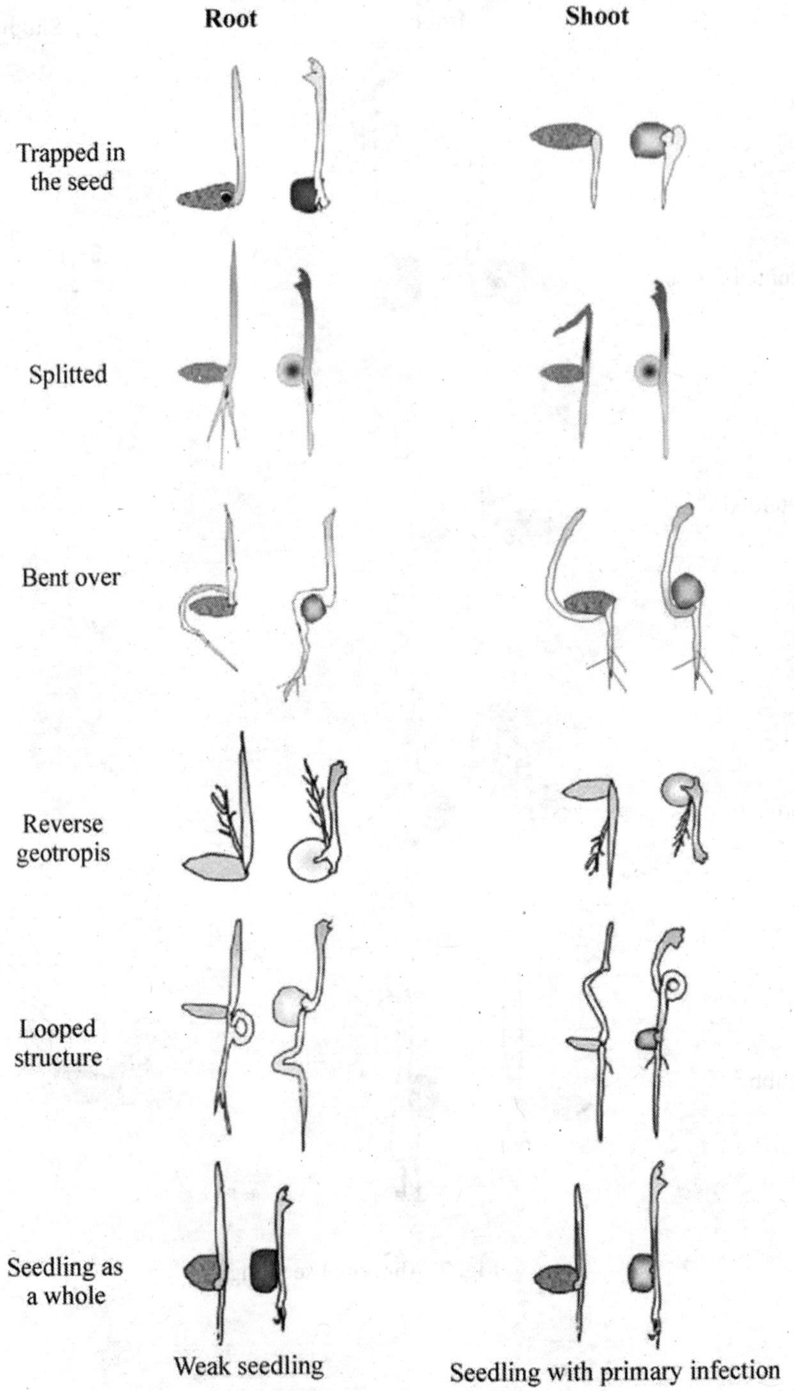

Fig. 7. Abnormal seedling

Primary leaves : Deformed, missing, damaged, necrotic due to primary infection and size less than 1/4 of the normal size.

Terminal bud : Deformed, damaged, missing, decayed due to primary infection and only 1-2 axillary buds with defective terminal bud.

6.3. Ungerminated seed

Seeds which do not germinate even up to the end of the test period (Fig. 8). These are of following types:

Hard seed : They remain hard at the end of the test period because of no water absorption. Hard seeds from family Leguminoceae and Malvaceae are considered as germinated, whereas, hard seeds of other families are considered as ungerminated abnormal seed.

Fresh ungerminated seed : Seeds which are neither hard nor germinated but remain clean, firm and apparently viable at the end of the test period. Pressure on seed with needle or dropping it from the height give the feeling of water absorption. This category is considered as ungerminated abnormal seed.

Dead seed : Seeds showing softness, oozing, discolouration, mouldy and decaying with no sign of seedling development. Pressure releases dirty water from the dead seed. This category of seed is considered as ungerminated abnormal seed.

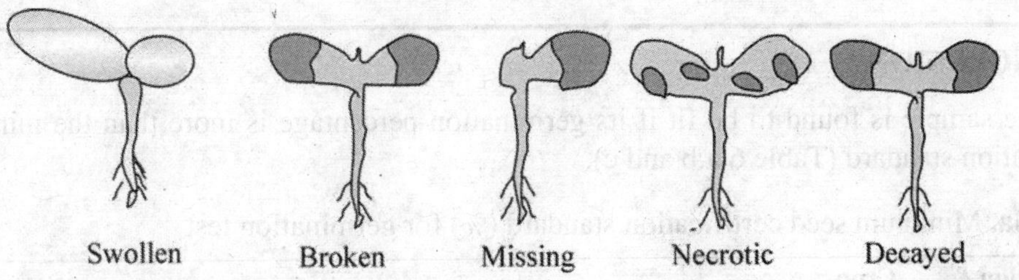

Types of abnormality in cotyledons

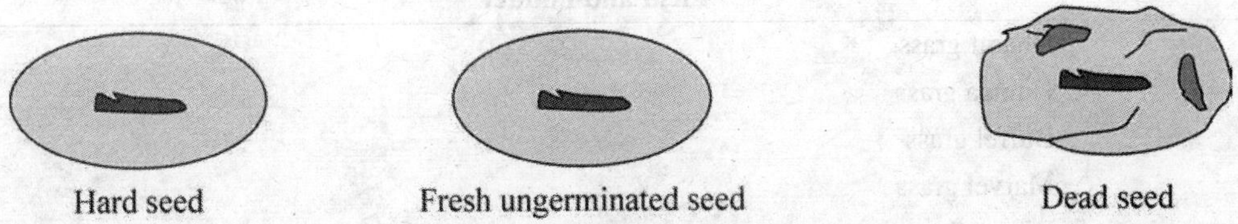

Types of ungerminated seeds

Fig. 8. Ungerminated seeds in test of germination

6.4. Multigerm seed unit

When more than one seedling is produced by a seed that seed is known as multigerm unit and only one seedling is counted for determination of germination percentage. This is the major problem of sugarbeet (*Beta vulgare*). Its seeds are prewashed for 2 hrs if genetically multigerm and 4 hrs if monogerm and then dried back at 25°C before germination test.

7. RESULT

Crop	Variety	Seed sample No.		
Date of placement	**Replications**	**Temp. °C**		
substrata		**Date of observation**		
Category	Replication	Total	Average	Remark
1. Normal seedling				
i) Intact				
ii) Slight defect				
2. Abnormal seedling				
3. Ungerminated seed				
i) Hard				
ii) Fresh				
iii) Dead				

8. CALCULATION

Germination is expressed as percentage by number of normal seedling and the sum of the normal, abnormal and ungerminated seeds must be hundred.

9. CONCLUSION

The sample is found to be fit if its germination percentage is more than the minimum seed certification standard (Table 6 a,b and c).

Table 6a. Minimum seed certification standard (%) for germination test

Foundation & Certified	Crop
	Field and Fodder
15	Dharaf grass
20	Guinea grass
30	Buffel grass
40	Marvel grass
50	Dinanath grass, setaria
65	Cotton, Indian clover
70	Castor, groundnut, Jatropha, rice bean, soybean, sunflower
75	Cotton (hybrid), daincha, grass pea, Indian bean, kidney bean, lentil, Mesta, minor millet, moong, field pea, pearl millet, pigeonpea, rajmash, sorghum, urid
80	Egyptian clover, horse gram, jute, linseed, Lucerne, maize (inbred) niger, rice, safflower, sesame, stylo, sunhemp, teosinte
85	Barley, chickpea, oat, rape seed and mustard, sugarcane, taramira, triticale, wheat
90	Maize (single cross hybrids, other type of hybrids, composites, synthetics and open pollinated varieties)

Table 6b. Minimum seed certification standard (%) for germination test

Foundation & Certified	Crop
Vegetables	
60	Capsicum, carrot, chilli, cucurbits, garden beet, spinach, sugar beet
65	Broccoli (*Brassica oleracea*. L), cauliflower, okra, parsley
70	Amaranth, asparagus, brinjal, cabbage, cauliflower, celery, cluster bean, fenugreek, knol-khol, lettuce, onion, radish, rat tail radish, tomato, turnip
75	Cowpea, French bean, Indian bean, vegetable pea
80	True potato seed
	Medicinal and, spices (Seed propagated)
25	Babchi (*Psoralea corylifolia*)
40	Ipecac (*Cephaelis ipecacuanha*)
50	Sarpgandha (*Rauvolfia serpentina*)
60	Basil (*Ocimum* spp), bell-dona (*Atropa* spp), chamomile (*Matricaria chamomilla*), datura, dill (*Anethum graveolens*), henpane (*Hyoscyamus* spp.), lemon grass & citrunella grass (*Cymbopogon* spp.), musk mallow (*Abelmoschus moschatus*), oxglove (*Tilpushpi digitalis*), palmorosa grass (*Cymbopogon* spp), periwinkle (*Catharanthus roseus*), wahga (*Cucumis melo* spp. melo)
65	Coriander, cumin, fennel
70	Ammi majus, ashvagandha (*Withania somnifera*), steroid sotanums (Keteri. *Solanum viarum* & other spp/vars.)
75	Jojoba (*Simmondsia chinansis*)
80	Senna (*Cassia angustifolia*)
85	Opium (*Papaver somniferum*)
90	Bhang (*Cannabis sativa*), isabgol (*Psyllium* spp)
	Ornamental
50	Annual chrysanthemum
60	Anise (*Pimpinella anisum*), aster
70	African marigold, marigold, ornamental sunflower , snapdragon
75	Annual carnation, petunia
	Trees
60	Papaya
90	Areca nut (Husked nut), coconut (Husked nut), oil palm (Husked seed)

Chapter 24

BIOCHEMICAL TEST FOR VIABILITY

It is a quick estimation of viability of a seed sample in general and of dormancy in particular.

Principle: A colourless solution of 2,3,5 triphenyl tetrazolium chloride (indicator) reacts with hydrogen released by the reduction process in living cell due to action of enzyme dehydrogenase, which catalyses the reactions in glycolysis and citric acid cycle. The malic acid dehydrogenase, carries out the reduction of the tetrazolium salt in living tissues. It produces a red, stable and non-diffusible substance triphenyl formazan to distinguish from the colourless dead ones. Red colour shows that the mitochondria of the cell are active.

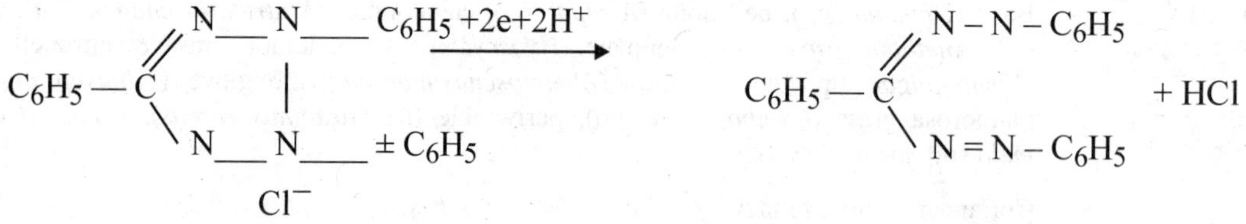

Tetrazolium *Formazan*

Reagent Tetrazolium salt (2,3,5 triphenyl tetrazolium chloride or bromide available in solid form) and Distilled water

Equipment: Conditioning media: blotter, paper towel or beaker, Dark amber coloured plastic/glass bottle; pH meter; Measuring cylinder; Electronic balance; Cutting and piercing device: razor blade, dissecting knives and needles; Staining dishes: watch glasses/ Petridishes; Medicine dropper: to remove tetrazolium solution after test period; Magnifying devices: hand lens (6X) and stereoscopic microscope.

Procedure

Preparation of solution One gram of tetrazolium salt is dissolved in 100ml of water to prepare one percent solution (w/v). The pH of the solution is adjusted at 6.5-7.5 and stored in a dark amber coloured bottle in dark and cool place to avoid reduction process (in light). For correct pH a buffered solution is prepared to dissolve tetrazolium by mixing two part of *solution-I* (9.078g KH_2PO_4 in 1000 ml water) with one part of *solution-II* (9.472g Na_2HPO_4 or 11.876g $Na_2HPO_4.2H_2O$ in 1000 ml of water). The seed bisected before staining requires dilute solution (0.25-0.5%) whereas seeds of pulses, oilseed and very small seeds (with lactophenol) require high concentration (1%).

Working sample: Four replications of 100 seeds (or 8 sub samples of 50 seeds each) each from the pure seed fraction of physical purity test or seeds observed as ungerminated at the end of the germination test are used for testing.

Preconditioning of seed: The seed is washed properly by clean tap water to remove the coat of any chemical present on the surface before conditioning. Conditioning is required to start activity of dehydrogenase enzyme, to facilitate penetration of tetrazolium solution and to soften the seed for sectioning and piercing. It may be achieved by following two procedures

Fast moistening: Seeds are fully immersed in distilled water for 12-15 hrs at 25ºC *e.g.*, cereals, castor, sunflower, safflower etc. Water should be changed in the event of more than 24 hrs soaking. Seed of some plant species (linseed) produce sticky mucilage that hampers testing of viability. These seeds are soaked in 1-2% solution of aluminium potassium sulphate ($AlK(SO_4)_2.12\ H_2O$) for five minutes.

Slow moistening: Seeds of dicot (e.g., leguminous crops) with thin and fragile testa are moistened at reduced rate of imbibition by wrapping in moist towel/ blotter paper and fracture in cotyledon to reduce mechanical injury. The seed wrapped in moist paper should be placed in germinator at 25°C to keep them saturated. The seed may be placed at high temperature (41°C) to reduce the time of preconditioning.

Hard seed: coat of hard seed should be scarified or incised at distal end of cotyledon to facilitate imbibition of water *e.g., Medicago, Trifolium*, etc.

Preparation of conditioned seed for staining: After preconditioning the coat of dicot seed is removed without any damage to cotyledon and embryo. Special precaution is advocated for groundnut and soybean. To facilitate entry of tetrazolium in the seed the seed coat and endosperm beneath fruit wall (sunflower, safflower etc) should be removed after soaking by sliding the seed between thumb and forefinger. Similar treatment is required for cotton and castor. The pericarp of cereals is not permeable to tetrazolium therefore the seed is bisected longitudinally or pierced with a needle at a non-essential part of the seed (Fig. 1). Bisecting the seed facilitates observation by exposing the embryo of monocot. Appendages present on seed as palea and lemma should be removed before staining. For critical observation to estimate germinability seeds are bisected by sharp razor through the seed coat and longitudinally through the midsection of embryonic axis.

Staining The conditioned seeds prepared for testing are completely immersed in 1% tetrazolium solution for 3 hr at 30-35°C. The rate of staining is increased with increase in temperature but in any condition it should not be more than 45°C. Light may affect the accuracy of the result therefore for staining it should be placed in dark.

The solution is decanted and excess reagent is removed with dropper. The seeds are then rinsed with water and evaluated.

To make embryo of small seeded cereals/ grasses visible without bisecting, seeds are arranged on a paper strip and the folded strip is immersed in the reagent.

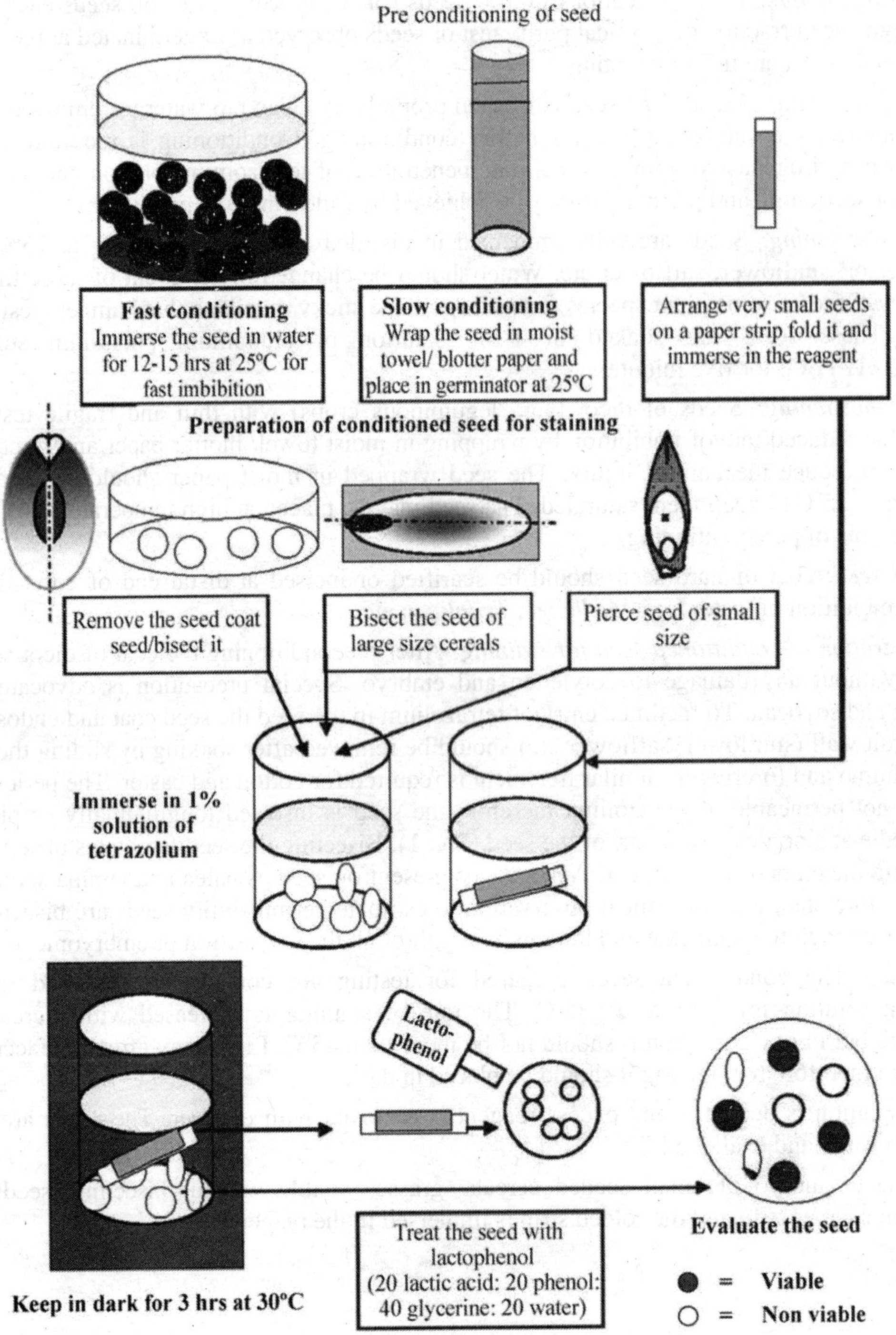

Fig. 1. Procedure for topographical viability test

Evaluation

After staining the tetrazolium is siphoned off by a medicinal dropper and rinsed with tap water. To avoid dehydration seeds are kept submerged in water. If seeds are not to be evaluated immediately then stained seeds can be stored in refrigerator upto 12 hrs before evaluation. The seeds are evaluated with the help of magnifying devices. Individual seed is evaluated as viable or dead on the basis of staining pattern in embryo followed by cotyledons (Fig. 2 and 3) with the help of magnifying device.

Tetrazolium in case of small seeds are siphoned off and the remaining tetrazolium is removed with the help of blotter after staining. Then the seeds are treated with **lactophenol** (20 lactic acid: 20 phenol: 40 glycerine: 20 water) so that the embryo is visible through lemma and palea for examination. The solution of lactophenol should be dispended with a medicine dropper as it is toxic both on contact and through inhalation. For 100 small seeds treatment with 2-3 drops of lactophenol for 10-30 minutes is sufficient to make hull transparent. The seeds are evaluated with the help of stereoscopic microscope.

Evaluation for viability and germinability requires a trained seed analyst with knowledge of seed and seedling structure. The seeds are evaluated to test viability of seed lot, estimation of germinability and seed vigour and to diagnose causes for non-viability by staining pattern of important part, intensity of colour, position of unstained tissues and extent of non viable tissues.

(A) Assessment on the basis of staining of embryo

Embryo completely stained	-	viable
Embryo unstained	-	non viable
Plumule or radicle unstained	-	non viable

(B) Assessment on the basis of staining of cotyledon

Complete staining	viable
No staining	non viable
Necrosis	evaluation on the basis of category "C"

(C) Assessment on the basis of staining in necrosis fashion

Unstained tissues at the attachment of the embryo	non viable
Unstained tissues are away and are not connected with embryo	viable

(D) Assessment on the basis of colour intensity

Faint red	vigorous seed
Dark red colour	weak seed (deteriorating)
Dark red fractured	non viable
White	non viable

The red colour is developed with more lustre and intensity on the outer surface in comparison to inner surface exposed by bisecting the seed because of lack of diffusion of tetrazolium solution. It should be considered during examination.

(E) Specific evaluation

Germinable seeds of cereals

In cereals only embryo, scutellum and aleurone layer show reaction with tetrazolium, other parts of endosperm remain completely unstained. The cells of endosperm are non-viable to store the food in complex form.

- Well-developed embryo with non-fractured normal cherry red stain. Scutellum with dark red dots against a lighter red background.
- Necrosis within the upper or lower ends of the scutellum.
- Radicle unstained but embryonic axis stained (exception-sorghum)

Non germinable seeds of cereals

- Whole embryo unstained
- Scutellum node unstained
- Major area of coleoptile unstained

Germinable seeds of pulses / oilseed crops

- Non fractured red coloured embryo and cotyledon
- Normal red coloured embryo with only one normal cotyledon
- Normal red coloured embryo with half or more than half of both the cotyledons attached to embryo are of red colour

Non germinable seeds of pulses/oilseed crops

- Embryo completely unstained
- Fracture at radicle or plumule with dark red line
- Plumule or radicle tip unstained
- More than 1/2 part of both the cotyledons attached to embryo are colourless
- Attachment of embryo to cotyledon is unstained

The embryonic axis of a dicot seed is made up of two types of tissues cortex and vascular cylinder (stele). Vascular cylinder connects radicle and plumule and supply food to both the tips from cotyledon. Its viability is essential for germination.

Advantages of the viability test

- No effect of environmental fluctuation that may affect germination test.
- Physical and physiological health of individual seed is determined.

- Estimation of viability and germinability is quick.
- Seed vigour can be estimated.
- The probable cause of deterioration may be determined.
- The test is simple that do not require specialized equipment.
- Rapid determination of viability helps to make decisions in a short notice.
- Accurate evaluation of seed viability even in dormant state.

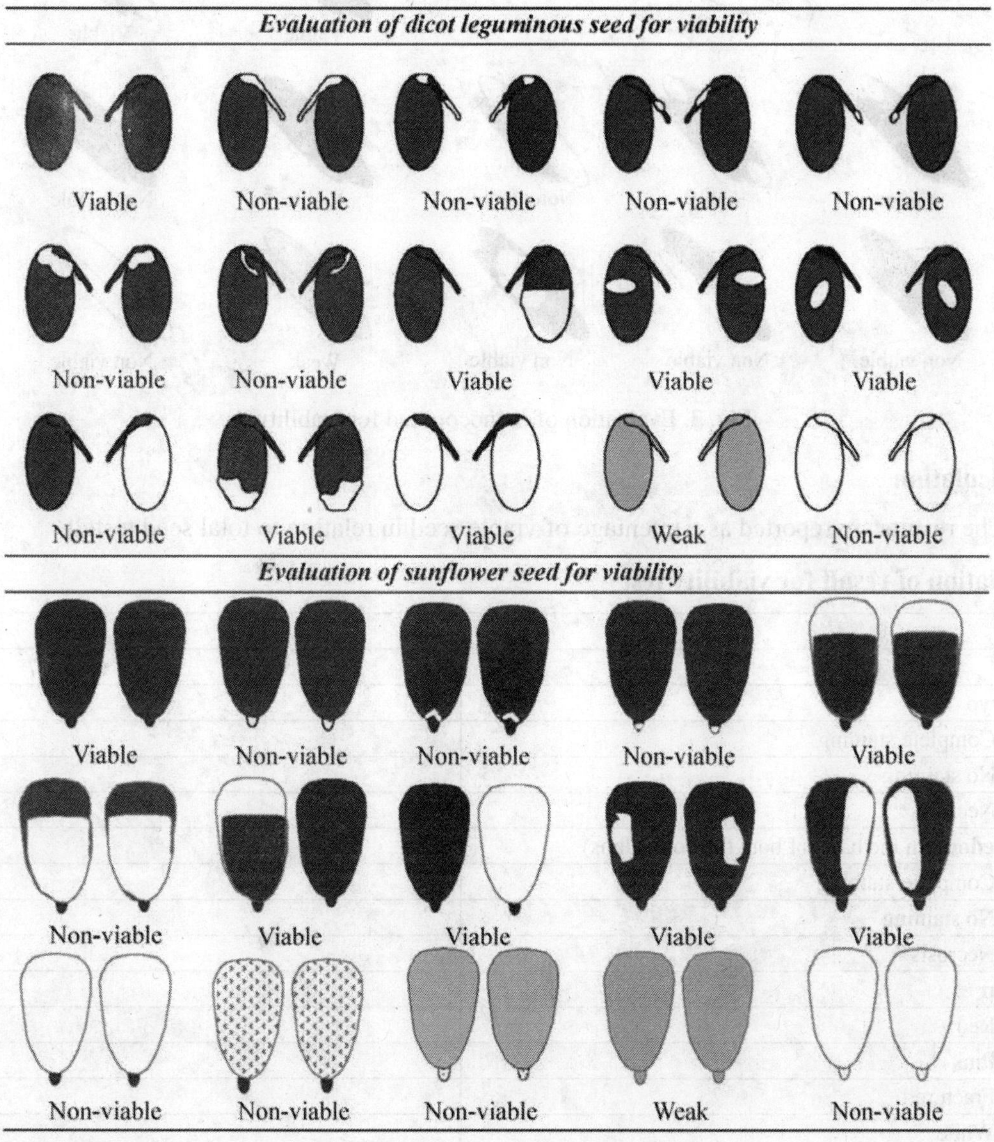

Fig. 2. Evaluation of monocot seed for viability

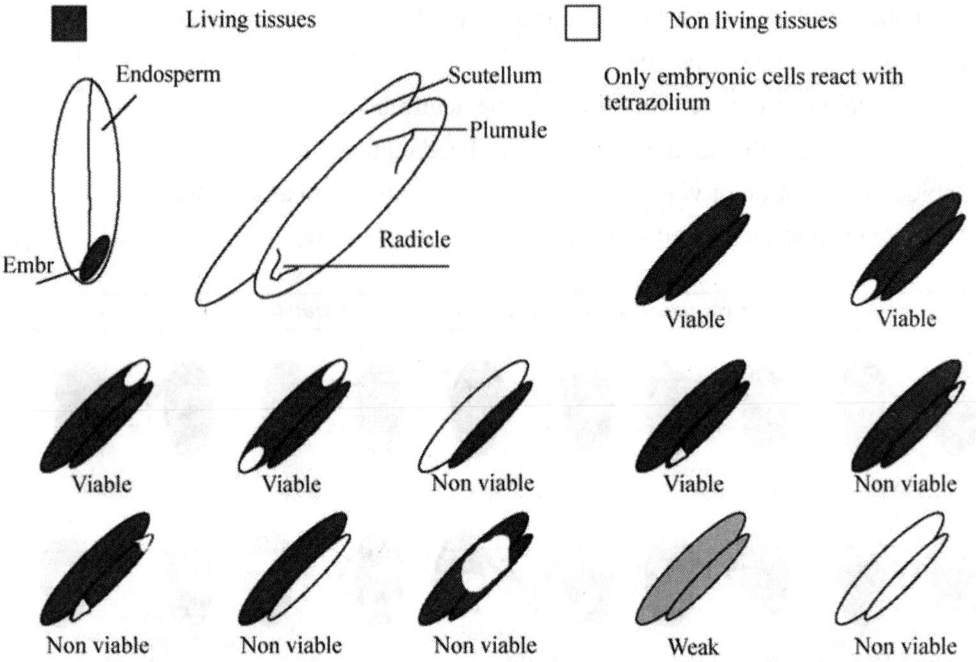

Fig. 3. Evaluation of monocot seed for viability

7. Calculation

The results are reported as percentage of viable seed in relation to total seed tested.

Tabulation of result for viability test

Seed No								
	1	**2**	**3**	**4**	**5**	**6**	**7**	**n**
Embryo								
Complete staining								
No staining								
Necrosis*								
Cotyledon (on the basis of both the cotyledons)								
Complete staining								
No staining								
Necrosis*								
Colour								
Red								
Pink								
Fractured								
White								

**Conclusion on the basis of expression in specific structure*

■■■■

Chapter 25

MOISTURE CONTENT

1. OBJECTIVE

To determine the moisture content of a seed sample.

2. OVEN DRY METHOD

The moisture content in the seed sample is the loss in weight when it is dried to reduce oxidation, decomposition and the loss of other volatile substances by removal of moisture. Seeds, which can withstand dehydration without damage, are known as **orthodox seed** e.g., seed of field crops. Where as seed, which can not withstand water losses and are stored at a relatively high moisture content in order to remain viable are known as **recalcitrant seeds** e.g., seeds of plantation crops like coffee, tea, cocoa, coconut etc.

2.1. Apparatus

Oven, desicator, analytical balance, grinding machine with uniform speed which does not cause heating to grinding material, sieve with mesh size of 0.5 mm, 1.0 mm and 4.00 mm and glass or metal containers of 8 or 10 cm diameter with a tightly fitted cover.

2.2. Procedure

Working sample: Moisture of the seed sample submitted in water-proof container is estimated by oven dry method.

Preparation of sample: Crops of larger seed size (Table 1) are ground before drying unless its high oil content makes it difficult. After grinding, the sample is passed through different sizes of sieves (Table 1). Pre drying before grinding is recommended for the sample having more moisture content than the minimum required percentage for moisture determination (Table 2). Pre dried material is exposed for 2 hrs before drying. For the container of 8cm diameter 4-5 g sample and for 10 cm diameter 10 g sample is weighed without exposing it for more than 30 second to the air.

Table 1: Requirement of grinding and other specification

Crop	Grinding	Mesh size	
		Pass through	Remain on
Rice, wheat, maize, sorghum, cotton	Fine	50% ground material is passed through 0.5 mm mesh	10% ground material remain on 1.00 mm mesh
Pea, chickpea soybean, grass pea	Coarse	50% ground material is passed through 4mm mesh	-

Table 2. Specification of pre-drying treatment

Crop	Pre drying is required if moisture is more than	Temperature required for pre drying (°C)	Duration
Maize	25%	70	2-5 hrs
Rice	13%	130	5-10 min.
Soybean	10%	130	5-10 min.

Moisture determination: Container and its cover are weighed (M_1). The sample is distributed evenly over the surface of the container and weighed with the cover of container (M_2). This container is placed in an oven at a temperature of 103 ± 2°C for 17 ± hrs or 130 ± 2°C for 2 ± 1 hr as per requirement of the crop (Table 3). The drying period starts when the temperature of the oven reaches at the required point. At the end of the prescribed period, container is placed for cooling in a desiccator for 30-45 minutes. After cooling, container is weighed with its cover and content (M_3). Determination of moisture content at low constant temperature (103 ± 2°C) is recommended at less than 70% relative humidity.

Table 3. Oven dry method recommended for different crops

Crop	Oven dry method
Chickpea, field pea, grass pea, maize, pearl millet, pigeonpea, rice, sorghum, wheat	High constant temperature (130 ± 2°C)
Castor, cotton, groundnut, linseed, rape seed and mustard, soybean, sesame	Low constant temperature (103 ± 2°C)

2.3. Calculation

The moisture content as percentage by weight is calculated to one decimal place by the following formula:

$$\text{Moisture (\%)} = \frac{(M_2 - M_3)}{(M_2 - M_1)} \times 100$$

Where

M_1 Weight of container and its cover

M_2 Weight of container, its cover and sample before drying

M_3 Weight of container, its cover and sample after drying and cooling

2.4. Conclusion

Seed lot with moisture content more than the minimum seed certification standard is recommended for more drying.

3. UNIVERSAL MOISTURE METER

Electrical conductivity of moist material is directly proportional to the amount of moisture content in it.

3.1 Equipment

Universal moisture meter: It is a machine with (i) Compression unit (ii) Moisture meter dial (iii) Thermometer (iv) Compression knob (v) Cups of different volumes.

3.2. Procedure

A representative sample of prescribed weight or volume (Table 4) is obtained and placed in the sample cup. It is fixed in the lower house of compression unit.

Meter is calibrated by pressing the button "cal" and "bell" with the help of calibration knob. Sample is compressed as per requirement with the help of compression knob and scale. At required compression the meter dial (M) is read by pressing the knob "Read" and bell.

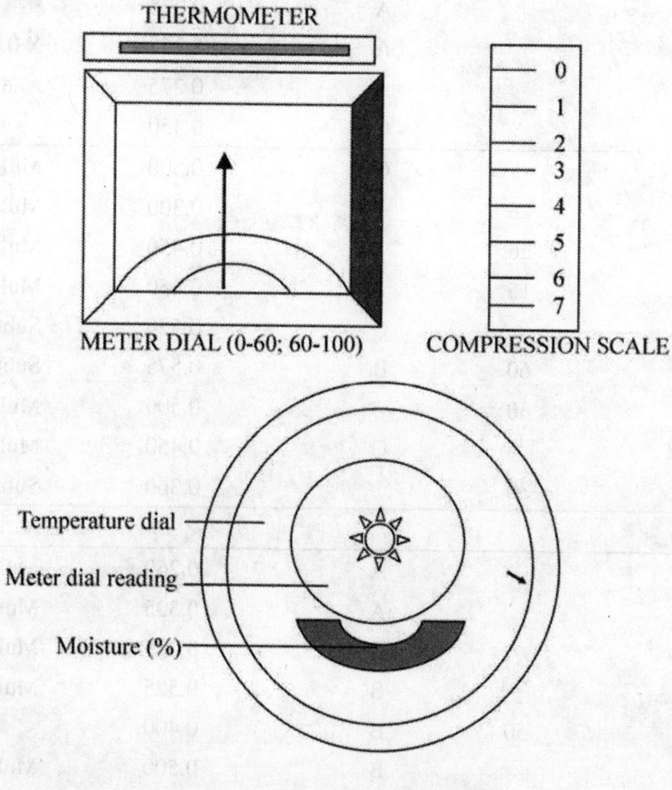

Fig. 1. Universal moisture meter.

Temperature (T) is observed by the thermometer fixed in between meter dial and compression chamber. The reading M and T are intercepted on the corelator dial (moisture meter dial) by turning the temperature dial (Fig. 1). On adjustment of both the reading mark of arrow on the outer reading of temperature dial indicates the moisture percentage. For some crops factor is also considered for estimation of moisture content (Table 4).

Table 4. Determination of moisture content by universal moisture meter

Crop	Sample size		Compression	Factor
	Weight (g)	Volume*		
Field and fodder crops				
Barley	50	B	0.600	
Maize	60	B	0.560	
Oat	30	B	0.400	
Pearl millet	60	B	0.500	
Rice	50	B	0.550	
Sorghum	50	B	0.675	
Wheat	30	A	0.275	Add 1%
Chickpea		C	0.500	Subtract 1%
Horse gram		A	0.275	
Lentil		A	0.250	X 0.7+3.5%
Moong & urid bean		A	0.275	Add 1.5%
Pigeonpea, field pea		C	0.450	
Castor		C	0.500	Multiplied by 0.5
Groundnut (pod)	25		0.300	Multiplied by 0.6
Groundnut (Kernel)	26		0.450	Multiplied by 0.56
Safflower	15		0.450	Multiply by 0.66
Sesame		C	0.550	Subtract 0.5%
Soybean	60	B	0.575	Subtract 2.5%
Sunflower	30		0.500	Multiplied by 0.6
Rape seed and mustard		C	0.450	Multiplied by 0.6
Cotton (linted)	30		0.360	Subtract 5%
Vegetable crops				
Cabbage		A	0.260	Multiplied by 0.6
Cowpea		A	0.325	Multiplied by 0.8
Coriander		C	0.325	Multiplied by 0.6
Cucumber		B	0.525	Multiplied by 0.8
Kidney bean	50	B	0.400	
Lettuce		B	0.500	Multiplied by 0.9
Okra		C	0.425	

Crop	Sample size		Compression	Factor
	Weight (g)	Volume*		
Onion		A	0.250	Subtract 2.5%
Tomato	25	B	0.250	Multiplied by 0.8
Turnip	25		0.200	Multiplied by 0.8
Watermelon		B	0.425	Subtract 3.5%

** A, B and C - Container size*

3.4. Conclusion

Seed lot with moisture content more than the minimum seed certification standard (Table 5) is recommended for drying.

Table 5. Minimum seed certification standard for moisture percentage

Crop	Sample in vapour proof container	Sample not in vapour proof bag
Field and fodder crop		
Castor, mustard, taramira	5	8
Groundnut, niger, sesame	5	9
Cotton	6	10
Rape seed	7	8
Jute, linseed, horse gram, rajmash, safflower, sunflower,	7	9
Egyptian clover, Indian clover, Lucerne	7	10
Soybean	7	12
chickpea, fieldpea, grass pea, kidney bean, lentil, moong bean, pigeonpea, rice bean, urid bean	8	9
Buffel, Dharaf, Dinanath, guinea, marvel, setaria and stylo grass	8	10
barley, forage sorghum, maize, minor millets, oat, pearl millet, sorghum, teosinte, triticale, wheat	8	12
Rice	8	13
Vegetable		
Radish, Rat tail radish, turnip	5	6
Cole crops	5	7
Cucurbits (all)	6	7
amaranth, asparagus, brinjal, capsicum, chilli, fenugreek, lettuce, onion, tomato, TPS	6	8
Carrot, celery, parsley	7	8
French bean	7	9
Cluster bean, cowpea, Indian bean, spinach, sugar beat	8	9
Okra	8	10

■■■■

Chapter 26

GENETIC PURITY

Objective

Determination of genetic purity of a seed lot.

Genetic purity of a seed lot is determined on the basis of distinct morphological characters of the variety expressed at seed, seedling and plant level by comparing its submitted sample with authentic sample under identical environmental situation.

Procedure

1. Determination of genetic purity on the basis of expression of distinguishing traits at seed level.
2. Grow-Out-Test for determination of genetic purity on the basis of plant morphological characters.

---

1. GENETIC PURITY TEST AT SEED LEVEL

1.1. Objective

To estimate the number of genetic impure seeds

The seeds of other varieties present in the submitted sample are visually observed on the basis of readily apparent differences in the stable and established seed characters of the variety under test. Expression of these characters can be influenced by environmental fluctuation, under such situation affected seeds are not classified as genetic impurity.

1.2. Apparatus and other requirements

(i) Physical purity work board, spatula, forceps, Petridishes and notebook. (ii) Authentic seed sample of the variety under test.

1.3. Procedure

The authentic seed sample is examined by visual observation for the expression of morphological traits at seed level. The submitted sample is uniformly spread over the surface of the physical purity work board and observations on the characters are made. Seeds exhibiting differences in distinguishing characters are sorted out and separated in Petridish (Fig 1). The seed sorted out as genetic impurity are compared with authentic seed for expression of morphological trait. The confirmed genetic impure seeds are counted and reported in number. Maximum permissible limit for genetic impure seeds in different crops is presented in Table 1.

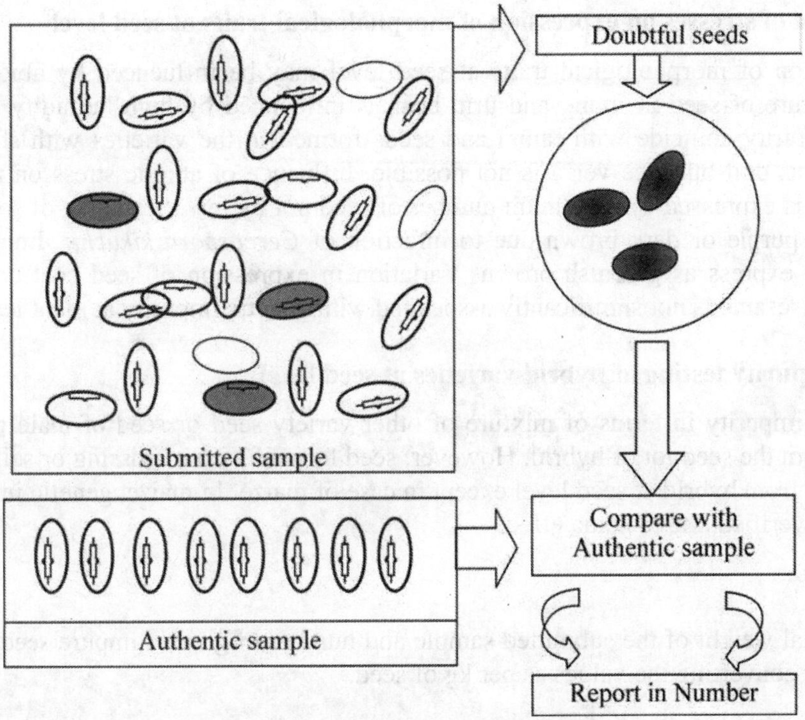

Fig. 1. Procedure of testing genetic purity at seed level

1.4. Basis for identification of genetic impure seed

Genetic impure seeds are mainly identified based on expression of morphological traits at seed level.

Crop	Characteristics
Linseed	Seed : colour
Rape seed and mustard	Seed : colour
Sesame	Seed : colour
Soybean	Seed : shape, ground color of testa, lusture, peroxidase activity in seed coat, hilum color and colour of hilum funicle
Maize	Endosperm colour
Rice	Grain length, Lemma and Palea colour and Tip of lemma
Sorghum	Shape of seed dorsal view and profile view, Size of mark of germ, Texture of endosperm, Colour of vitreous albumen, Grain lusture
Chickpea	Seed: colour, shape, testa texture, ribbing and puckering
Field pea	Shape, surface, cotyledon colour and testa mottling
Lentil	Seed : size (weight of 100 seeds) (g), coat colour, testa mottling and Cotyledon colour
Mung and Urid bean	Seed: colour, lusture and shape
Pigeonpea	Seed shape, colour, colour pattern

1.5. Influence of stresses on expression of morphological traits at seed level

Expression of morphological traits at seed level may be influenced by abitoic and biotic stresses. Lusture of seed in mung and urid bean is influenced by high humidity at the time of maturity (maturity coincide with rains) and seeds formed in the varieties with shiny seed coat may express as dull but vice versa is not possible. Influence of abiotic stress on morphological traits of seed is expressed by maximum number of seed not by few. Seed coat of soybean may be expressed as purple or dark brown due to infection of *Cercospora kikuchii*. Immature seed of soybean may express as greenish brown. Variation in expression of seed coat colour in lentil, chickpea and sesame is not significantly associated with genetic impurity at plant level.

1.6. Genetic purity testing of hybrid varieties at seed level

Genetic impurity in terms of mixture of other variety seed or seed of male parent may be sorted out from the seed lot of hybrid. However, seed formed by out crossing or selfing cannot be distinguished from hybrid at seed level except in case of maize. In maize, genetic impurity at seed level may be verified due to xenia effect.

1.7. Result

The actual weight of the submitted sample and number of genetic impure seeds sorted out is reported after converting the values as per kg of seed.

1.8. ODV Test

Testing of genetic purity at seed level by sorting out genetic impure seed with mentioning the name of the variety to which the seed sorted out as genetic impure represents. Therefore, it is known as Other Distinguishing Variety (ODV) Test. In this the name of the variety to which genetic impurity represents has to be mentioned. Whereas, genetic purity test at seed level performed in India, the number of impure seed is reported without mentioning the name of the variety to which genetic impurity represents.

Table 1. Permissible limits of genetic impurity at seed level

Crop	Maximum permissible number	
	Foundation	Certified
Barley	10	20
Castor (variety and hybrid)	5	10
Chickpea	5	10
Clover	10	20
Cluster -bean	10	20
Cowpea	5	10
Field pea	5	10
Grass pea	10	20
Horse gram	5	10
Indian	10	20

Crop	Maximum permissible number	
	Foundation	Certified
Indian bean	5	10
Jute	10	20
Kidney bean	5	10
Lentil	10	20
Linseed	10	20
Maize (inbred and single cross)	5	20
Maize(Composite synthetic and hybrid)	10	20
Marvel grass	10	20
Moong	10	20
Niger	10	20
Oat	10	20
Pigeonpea	10	20
Rajmash	5	10
Rape seed and mustard	10	20
Rice (variety and hybrid)	10	20
Rice-bean	10	20
Sesame	10	20
Setaria grass	10	20
Sorghum (varieties and hybrids)	10	20
Soybean	10	40
Stylo	10	20
Taramira	10	20
Urid	10	20
Vegetables		
Amaranth	10	20
Bitter gourd	5	10
Carrot	10	10
Fenugreek	10	20
Lettuce	10	20
Okra	10	20
Ridge gourd	5	10
Sponge gourd	5	10
Watermelon	5	10

2. GROW OUT TEST

2.1. Objective

To estimate the genetic purity of a seed lot on the basis of expression of morphological characteristics of the plants produced by the seeds under test.

The seed sample of a seed lot is sown in controlled conditions with authentic sample. Genetic purity is determined on the basis of variation in the expression of distinguishing, stable and uniform morphological characteristics at plant level with reference to authentic sample.

2.2. Sampling

Submitted sample: The submitted sample for grow out test is drawn simultaneously with the submitted sample for other tests. The size of submitted sample varies according to the crop species (Table 2).

Table 2. Recommended size of sample submitted for grow out test*

Crop	Size of sample (g)
Genera with seed size similar to Pearl millet	100
Genera with seed size similar to *Beta vulgaris*	250
Sorghum, rice, wheat and other genera of similar seed size	500
Maize, cotton, groundnut, soybean and other genera of similar seed size	1000
Seed potato, sweet potato and other vegetatively propagated crops	250 tubers/planting stakes/roots/ corms

* If genetic purity has to be worked out both at seed and plant level then the size of submitted sample is doubled.

Working sample: The size of working sample mainly depends on the test weight and germination percentage of the crop to observe the permissible offtype plants prescribed as minimum seed certification standards (Table 3) in the optimum population i.e., minimum 400 plants.

Table 3. Number of plants required for grow out test

Maximum permissible off type (%)	Number of plants required
0.10	4000
0.20	2000
0.30	1350
0.50	800
1.00 and above	400

4. LOCATION OF THE GROW OUT TEST

The grow out test is conducted in the area where crop can express its marker characters without any variation due to the influence of environment even in off-season or under controlled conditions in glass house.

Authentic sample of the variety under test: It is the official sample of the variety obtained from the originating institute and grown in identical situation with sample under test for comparison.

5. LAND REQUIREMENT

In field, test is conducted on land which is free from other crops, weed and volunteer plant seeds with adequate fertility and irrigation facility.

Raising of the crop: Standard and recommended cultural practices are adopted for raising of the crop in two replications but the row length, plant to plant, row to row and plot to plot distances have to be altered as per recommendations (Table 4). Proper care should be taken to avoid admixture and to maintain optimum plant population. Subsequent thinning and transplanting is not recommended. Authentic sample is grown with same cultural practices at suitable interval.

Table 4. Recommended specifications to raise the crop for grow out test

Crop	Space/distance (cm)			Length (m)
	Plant to plant	Row to row	Plot to plot	Row
Hybrid cotton	10	45	45	5
Wheat, barley, oat	02	25	50	6
Chickpea, moong, urid	10	30	60	6
Sorghum	10	45	60	6
Pea, cowpea, soybean	10	45	90	6
Pearl millet	10	60	90	6
Rice				
Early to medium	15	20	45	6
Late	25	30	60	6
Maize	25	60	90	10

6. OBSERVATIONS

Each and every plant is examined throughout the growing season with emphasis on the expression of stable, uniform and distinguishing marker characteristics and time of their expression. The morphological characteristics of different crops with stable and uniform expression are discussed in the chapter entitled NDUS test. The plants showing deviation in expression of the characters against the control are tagged and examined thoroughly to confirm their genetic purity. The number of offtype plants and total population is counted and recorded.

7. CALCULATION

Percentage of genetic purity is worked on the basis of number of offtype and total plant population upto first decimal place. Result is interpreted by using the reject number for prescribed standards with reference to sample size (Table 5).

Table 5. Reject number for prescribed standards and sample size

Genetic Purity (%)	Reject number for sample size (number of plants)			
	100	**400**	**800**	**2000**
99.5	00.5	02	008	010
99.0	01.0	08	016	020
95.0	05.0	24	048	100
90.0	10.0	44	088	200
85.0	15.0	64	128	300

8. REPORTING RESULT

Result is reported as percentage of genetic purity. The impurity is not reported by name.

9. CONCLUSION

The sample with genetic purity less than the MSCS is not accepted for sowing purpose.

10. FIELD OF APPLICATION

At present this test is recommended only for verification of cotton hybrid with minimum 90% genetic purity. Out of 10% offtypes maximum 1.50% plants of other varieties and segregants are permitted and remaining offtypes are ear marked for selfed plants. The total offtype should not exceed 10% i.e. 10 in 100 with 44 as reject number out of minimum 400 plants under GOT.

Pre and Post Control Test: It can be performed as Pre Control Test as well as Post Control Test. For testing of genetic purity of Foundation and Certified category of all the varieties and hybrid seed the test is performed as Pre Control Plot. Whereas, **Post Control Test** is adopted for testing of Breeder seed. In the event of genetic impurity, the producer used the particular seed for raising Foundation category is informed to rectify the particular problem by rouging based on the informed expression of distinguishing traits.

Genetic purity of Genetically Modified (GM) Crops: As per notification number 1567 (E) dated 5.1.2005, the genetic purity in terms of quantum of gene express of *Bascillus thuringiensis* (Bt.) Protein (Toxin) as 90 per cent in *Bascillus thuringiensis* cotton seed lot for labelling of *Bascillus thuringiensis* Cotton Seed. Laboratory of Central Institute of Cotton Research (CICR), Indian Council of Agricultural Research (ICAR), Nagpur is notified as the Central Seed Laboratory to carry out the functions of ascertaining the presence or absence of Cry 1AC gene *(Bacillus thuringiensis as per SO 1107 E 8.5.2008)* in cotton seeds and acts as a referral laboratory for *Bacillus thuringiensis* in cotton as per notification number 1300 (E) dated 12.11.2003. Procedure of testing is discussed in separate chapter.

■■■■

Chapter 27

BIOCHEMICAL TESTS FOR VERIFICATION OF GENETIC PURITY

1. OBJECTIVE

Verification of genetic purity of cultivars based on variation in expression in response to biochemical tests at seed level.

2. FLUORESCENT TEST

- Seeds of oat are exposed to UV light at 360- 380 nm wave length on a dark background.
- Varieties are classified in two groups on the basis of fluorescent effect i.e., fluorescent or non- fluorescent (Anon, 1985).

3. PHENOL COLOUR TEST

Variety of cereals particularly wheat, rice and sorghum are verified through involvement of the enzyme tyrosinase using phenol as a substance (Walls, 1965).

Reagent and equipment: 1% freshly prepared solution of carbolic acid (V/V in water), Petriplate, beaker, pipette, incubator, filter paper, forceps and distilled water.

Procedure

- Seed treated with fungicide are rinsed with methanol prior to soaking.
- Seeds are placed in a beaker and immersed in distilled water.
- Two sheets of filter paper are arranged in a Petriplate.
- Seeds are arranged on filter paper with the help of a pair of forceps keeping the hump portion of the seed upward.
- 1% solution of carbolic acid is applied in each Petriplate with the help of pipette till 3 to 4 part of the seed is covered.
- After application of carbolic acid the Petriplate is covered with lid, immediately to avoid evaporation of carbolic acid.
- The Petriplate is placed in an incubator for required period at prescribed temperature (Table 1).

Table 1. Details for phenol colour test

Crop	Soaking period (hrs)	Concentration of phenol (%)	Duration of test (hrs)	Temperature (°C)
Wheat	16	1	4	30
Rice	24	1	24	40

- Seeds are observed after prescribed period for development of colour and its intensity to distinguish the genetic impure seed.
- Intensity of the colour developed on the seed is observed as no reaction (-) deep olive (+), light brown (++), brown (+++) and black (++++) (Fig 1).

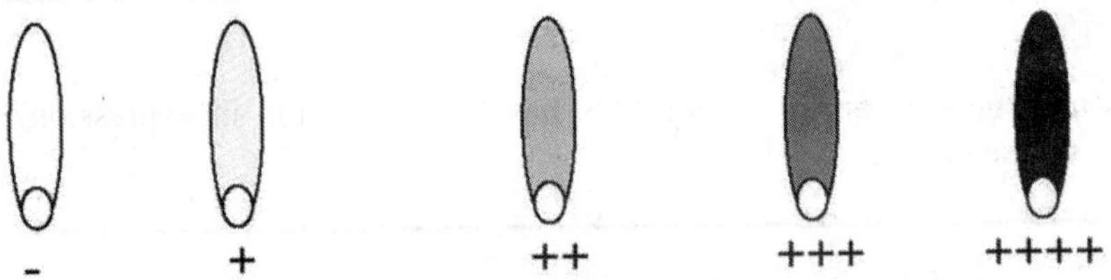

Fig. 1. Classification based on the intensity of the colour developed by phenol test

- Seed, pure at genetic level expresses similar intensity of colour in response to the test. Similarly, any seed showing variation in expression indicates genetic impurity.
- Systemic fungicides like Vitavax are not removed by methanol that results in uneven staining. Therefore, a final inspection should be made after 24 hrs of staining to confirm the result.
- **The test is approved for DUS testing of wheat varieties.** 20 grains are tested for distinctness whereas, 100 grains for homogeneity with at least two example varieties as control.

4. SODIUM HYDROXIDE TEST

Verification of red wheat seed in adversely damaged white wheat due to unfavourable environmental conditions is difficult by visual observation. These seeds can be differentiated by sodium hydroxide test (Payne, 1988).

Reagent and equipment: Methyl alcohol, sodium hydroxide, distilled water, beaker, Petriplate and pH meter.

Procedure

- Methyl alcohol solution of 95% concentration is prepared in water.
- Sodium hydroxide solution of 5 N is prepared with the help of pH meter.
- Seeds are soaked in methyl alcohol solution for 15 minutes.
- Seeds are air dried for 30 minutes, after soaking.
- These dried seeds are then soaked in sodium hydroxide solution for 2 minutes.

- The seeds of white wheat exhibits bright yellow colour, while red wheat expresses dark brown colour.

5. KOH BLEACH TEST

Genetic purity of sorghum varieties can be verified on the basis of dark pigmentation present on the tegmen of the seed coat of some varieties due to tannic acid. This character can be confirmed by KOH bleach test (Payne and Koszykowski, 1980).

Reagent and equipment: Potassium hydroxide, bleach (NaOCl) and beaker.

Procedure:

- 5.25% solution of bleach is prepared in water.
- 1: 5 (weight / volume) solution of KOH is prepared with 5.25% bleach solution.
- Seeds are soaked in KOH - bleach solution for 10 minutes.
- Seeds are gently swirled in the solution.
- Seeds are placed on filter paper for air drying.
- Varieties are verified based on dark or light colour developed on the seed.

The test is approved for DUS testing. For observation of the content of tannin, the EEC norm is applied. Annex III to Commission Regulation (EEC) No 1908/84 of 4 July 1984 (OJ No. L 178, 5. 7. 1984, p. 22) inserted by Commission Regulation (EEC) No. 2159/84 of 26 July 1984 (OJ No. L 197, 27. 7. 1984, p. 18), and amended by Commission Regulation (EEC) No. 2281/86 of 22 July 1986 (OJ No. L 200, 7. 1987, p. 7).

6. KOH TEST:

Seed of wild rice (*O. sativa* var. *fatua)* an objectionable weed is of red colour but some of its seeds become nearly colourless. Identification of colourless seed of red rice in normal rice seed lot is difficult. KOH test is recommended to verify presence of objectionable weed seed in the seed lot of rice (Rosta, 1975).

Reagent and equipment: Potassium hydroxide, distilled water, test tube and pipette.

Procedure

- Seed suspected as wild rice are sorted out from the sample.
- KOH solution of 2% is prepared in water.
- Seed sorted out as of objectionable weed are placed in a test tube.
- Two drops of 2% KOH solution is added in each test tube.
- Colour of the KOH solution is observed after 10 minutes.
- Development of red color in the solution confirms presence of wild rice seed. Whereas, no change in the color of KOH solution confirms that the seed is of normal rice.

7. PEROXIDASE ACTIVITY TEST

Presence or absence of peroxidase enzyme in seeds of crops from family Leguminoceae is under genetic control therefore, variety may be verified based on variation in the expression with the reaction of hydrogen peroxidase (Buttery and Buzzell, 1968). **The test is approved for DUS testing of soybean.**

Reagent and equipment: 0.5% (V/V) guaicol, 0.1% (V/V) hydrogen peroxide, razor, test tube and dropper.

Procedure

- Preparation of seed: Coat of 20 seed sorted as genetic impure is removed with the help of a sharp razor. No part of embryo or cotyledon should remain attached with the separated seed coat.
- Seed coat is placed in test tube in such a way that one test tube contains coat of only one seed.
- Guaicol solution of 0.5% concentration is prepared in water.
- In each test tube containing seed coat, 10 drops of 0.5% guaicol solution is added.
- After 10 minutes, one drop of 0.1% hydrogen peroxide (H_2O_2) is added in each test tube.
- The change in the colour of the solution is observed immediately after addition of hydrogen peroxide.
- Development of red color in the solution shows presence, while no change shows absence of the activity of peroxidase enzyme.
- Seed present with no change, in a variety with presence of peroxidase activity or vice versa confirms genetic impurity.

8. COPPER SULPHATE AMMONIA TEST

Seed of *Melilotus alba* is of white colour whereas, *M. officinalis* is of yellow colour, but the differentiation in a seed lot on the basis of seed coat color and expression of other morphological traits of the seed is difficult. Seed of both the species have different expression for copper sulphate ammonia test (Maxon and Hurst, 1983).

Reagent and equipment: Copper sulphate, ammonia, Stopper bottle and Petriplate.

Procedure

- Preparation of solution: A solution of tetra amine copper sulphate is prepared by dissolving 3 g $CuSO_4$ in 30 ml NH_4OH_2 (4.8%) in a Stopper bottle. In the event of precipitation in the solution more quantity of $CuSO_4$ is added.
- Seeds are soaked in the solution for 20 minutes.
- Both the species have different expression, seeds of *M. alba* exhibits olive or yellow green, while *M. officinalis* shows dark brown to black colour.

9. COLEOPTILE : ANTHOCYANIN COLOURATION

- 20 seeds of wheat are placed on a moistened filter paper in Petriplate for germination.
- The germinated seed with 1cm coleoptile is shifted to artificial light of 15,000 lux at 15^0 to 20ºC.
- After one week when leaf is within coleoptile (stage 9-10) the colour of coleoptile is observed for presence or absence of anthocynin pigmentation by naked eye.
- The test is approved for DUS testing of wheat.

10. SEEDLING COLOUR

Development of anthocyanin pigment on the stem of the cereals, soybean and pigeonpea is under the control of genes and a variety can be verified based on pigmentation.

Reagent and equipment: 1% solution of HCl or NaCl, filter paper, pipette, Petriplate, germinator with fluorescent light and UV lamp.

Procedure:

- Three layers of filter paper are placed in a Petriplate.
- It is moistened with 1% solution of NaCl or HCl to intensify the colour in a Petriplate.
- Seeds of cereals are arranged with the help of a pair of forceps on the moistened filter paper.
- Petriplate is placed in a germinator for germination in presence of light.
- After 4-7 days seedlings are illuminated for 1-2 hrs in UV light prior to observation.
- Genetic impurity is verified based on the variation in expression of pigmentation on coleoptile as purple or green.
- Seeds of crops from family Leguminoceae are planted in a sand medium (1/3 sand, 1/3 soil and 1/3 vermiculture) for germination.
- The tray is kept under fluorescent light (450 μ Em^{-2} S^{-2}) for germination of seed.
- Hypocotyl colour of each seedling is evaluated 7-10 days after germination to verify the cultivar.

11. REACTION OF SEEDLING ROOT TO UV LIGHT

- Verification of rye grass seed as of annual or perennial can be done by fluorescent test of the seedling roots (Genter,1929).
- Varieties of annual rye grass (*Lolium multiforum*) exhibit a positive response due to presence of alkaloid annuloline, while varieties of perennial rye (*L. perenne*) are non fluorescent (Dales, 1953).

Reagent and equipment: UV lamp, paper towel, germinator and distilled water.

Procedure

- Seeds of rye grass are arranged on a moist filter paper eliminating the chances of root entanglement with each other.
- This is kept for germination in dark.
- 5-10 days old seedlings are examined under UV light at 300-400 nm wavelength for fluorescent effect on root.
- Varieties are verified as of positive response (annual rye grass) or negative response (perennial rye grass).
- Similar test has been recommended for verification of soybean varieties (Grabe, 1957).

12. AMMONIA FLUORESCENT TEST

Differentiation between *Festuca rubra* and *F. ovine* at seed level on the basis of morphological characters in a seed lot is difficult. Their seedlings can be differentiated on the basis of Ammonia fluorescent test (Linder, 1958).

Reagents and equipment: Ammonium hydroxide, distilled water, Petriplate, filter paper, pipette, germinator and UV view finder.

Procedure

- Three pieces of filter paper are moistened and placed in a Petriplate.
- Seeds are arranged on the Petriplate in such a way that the roots are not entangled with each other during germination.
- The Petriplate is placed in a germinator at alternate temperature (10-30^{0}C or 15-25^{0}C) in weak light (1000 lux) for 14 days.
- On 14th day 0.5% solution of ammonium hydroxide is sprayed on the seedlings.
- After spray seedlings are observed under UV light.
- Root of *F. rubra* shows fluorescent effect of yellow green, while *F. ovina* exhibits blue green colour.

13. AMYLOSE CONTENT IN ENDOSPERM

Verification of varieties based on amylose content in the endosperm. **The test is approved for DUS testing of rice varieties.**

Reagents and equipment: 95% ethanol, 1N Sodium hydroxide, 1N acetic acid, 2 ml iodine solution (0.2 g iodine and 2.0 g potassium iodide in 100 ml of aqueous solution), UDY cyclone mill, volumetric flask, waterbath, spectrophotometer.

Procedure:

- 25 grains of milled rice are grounded in a UDY cyclone mill of mesh size 60.
- 100 mg powder is placed in a volumetric flask (100 ml).

- 1 ml 95% ethanol and 9 ml 1N sodium hydroxide are added to it.
- It is heated on a boiling water bath to gelatinize the starch.
- After gelatinization it is cooled at room temperature.
- After cooling, distilled water is added.
- To determine the amylase content, 1 ml of 1N acetic acid and 2 ml of iodine solution are added in the volumetric flask containing 5 ml starch solution, to make up the volume by distilled water.
- Contents are kept stand still for 20 minutes after shaking.
- Absorbance of the solution is measured with a spectrophotometer at 620 nm.
- Amylose content is determined with the help of conversion factor.
- Based on amylase content the varieties are classified as

14. EXPRESSION OF WHITE CORE IN THE ENDOSPERM OF POLISHED GRAINS

Chalky white spots appear on the grain with starchy endosperm. Extension of chalky white region to the edge of the ventral side and towards the centre of the endosperm is known as *white core*. Soft textured white spot present on the middle part of the embryonic (ventral) side is termed as *white belly* whereas, a white spot on the dorsal side as long streak is called *white back*. Degree of chalkiness confirms the quality of milled rice based on the presence of white core, white belly and white back. **The test is approved for DUS testing of rice varieties.**

■■■■

Chapter 28

DETECTION OF GENETICALLY MODIFIED CROPS

Objective

Detection of genetically modified organism through **Protein based methods/ Immunoassays**

Immunoassay is analytical measurement system that uses antibodies as test reagents. Antibodies are gamma globulin proteins present in blood. Its function is to identify and neutralize foreign objects. When all the antibodies are same because they are formed by identical immune cell then it is known as monoclonal antibodies and when formed by different immune cell then as polyclonal antibodies. The formation of antibody in the body is stimulated by introduction of some substances known as antigen. A laboratory technique that makes use of the binding between an antigen and its homologous antibody in order to identify and quantify the specific antigen or antibody in a sample is known as immunoassay. The sample that has to be analysed is termed as analyte. In immunoassays of GM the antigen is considered as analyte. Prerequisite of immunological detection methods is the availability of highly specific antibodies directed against the protein to be detected.

Principle

Binding of antigen (Ag) with antibody (Ab) is detected with the help of enzyme that converts a colorless substrate (chromogen) to a colored product. Change of colour indicates the presence of Ag:Ab binding.

High specificity of the immunological reaction allows accurate recognition of an antigenic substance even in the presence of interfering compounds. It is of two types

- Enzyme-Linked Immunosorbent Assay (ELISA)
- Lateral flow strip method

Enzyme-Linked Immunosorbent Assay (ELISA) Method

It is a protein-based test method that uses antibodies specific to the protein of interest. ELISA detects and quantifies the amount of specific protein in a sample in the presence of other proteins. In ELISA test one antibody binds with the specific protein whereas the second antibody amplifies. This antibody binds to an enzyme and forms a colored product.

Materials: Fresh young leaf tissue

Equipment: Pipette and microliter plate

Microliter plate is a flat plate with 6, 24, 96 or 384 multiple circular wells of 10 nanolitres to several milliliters capacity arranged in a 2: 3 rectangular matrix.

Chemicals: Antibodies to be tested, enzyme labeled specific antibody, antigen and10 ml Tween 20® detergent

Procedure: Fixed volume of homogenate test sample prepared from fresh young leaf tissue is loaded in the well of ELISA plate with the help of micro pipette. In each row of micropipette a blank is accommodated. It is kept for half an hour to fix the antigens of the sample to the wall of the well. After specified time the well is rinsed with buffer to remove unbound part of the sample from the wells so that on the wall of the well only the required antigens remain fixed.

In each rinsed, well fixed volume of enzyme linked anti-body is poured for firm binding between enzyme linked anti-bodies and antigens fixed on the wall of the well. After half an hour the wells are again rinsed with washing buffer to remove unbound enzyme linked anti-bodies.

These washed wells are loaded with a specified amount of ELISA substrate. It is incubated for specified time to start the reaction and generate color.

Result: The well with colour confirms presence of transgenic.

Lateral Flow Strip Method

Principle: Antibodies specific for the expressed protein are coupled to a color reagent, and incorporated into the nitrocellulose strip. The strip is then placed in an eppendorf tube that contains a plant tissue with the transgenic protein. A colored antibody sandwich is formed. This colored sandwich flows to the other end of the strip through a porous membrane that contains two captured zones, one specific for the transgenic protein sandwich and another specific for untreated antibodies coupled to the color reagent. A positive result contains two lines, while a negative result just has one.

Materials required: Leaf sample

Equipment: Lateral flow strip, eppendorf tubes, mortal pastel

Chemicals: Antibody reagent and colour reagent,10 ml Tween 20®

Lateral flow strip: Lateral flow strip works on capillary action and uses gold-labeled antibodies for detection of GMOs. Antibodies are 'labeled' with detectable substances viz., gold particles or enzymes which can catalyze reactions in the solution to give the color for the ELISA and to provide a means for visualizing and quantifying the presence of the target protein.

The antibody binds to Bt protein in solution with other proteins. These antibodies are purified and attached to a 'solid phase' which filters on the test strip. The 'capture antibody' would be the one which is attached to the filter on the test, the solid phase. The 'detector antibody' binds to a different site on the Bt protein than the capture antibody, and its role is to provide visualization which is the final step and is obtained by the presence of pink/ red bands on the lateral flow strip due to the concentration of the gold linked to the first antibody.

A typical test strip consists of the following components:

Sample pad : It is an adsorbent pad on which the test sample is placed.

Conjugate or reagent pad: This pad contains antibodies specific to the target analyte labeled with nanometer sized colloid gold formed coloured particles when joint to GM protein. The antibody-protein complex moves up the strip upto the second zone of antibodies.

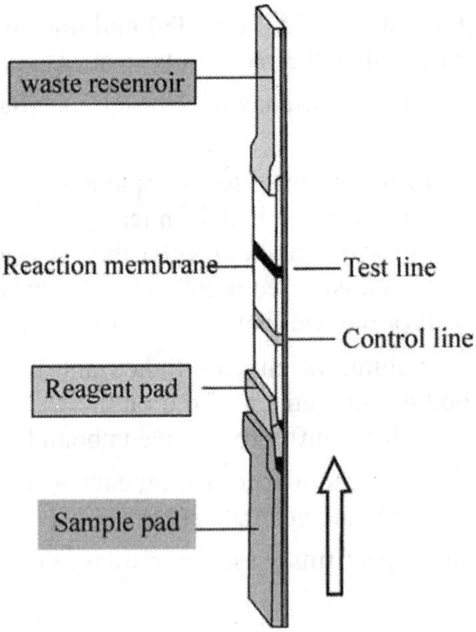

Reaction membrane (Immobilized control zone): It is a hydrophobic membrane made up of nitrocellulose or cellulose acetate. On this membrane anti-target analyte antibodies are immobilized.

The complex concentrates into this immobilized antibody zone where the gold becomes visible as a red band.

It also binds a control complex that is present in the extraction solution and also produces a visible line. It contains antibodies specific for the conjugate antibodies.

Wick or waste reservoir: an absorbent pad to draw the sample across the reaction membrane by capillary action and collect, it is provided as waste reservoir.

The components of the strip are usually fixed to an inert backing material and may be presented in a simple dipstick format or within a plastic casing with a sample port and reaction window showing the capture and control zones.

Procedure

Small piece of young leaf tissue is taken from plant and chopped in water and transferred to eppendorf tube. It is mixed with the protein extraction solution provided with the kit. Protein from the sample comes out of the cell in dissolved or suspended form in the solution.

The base i.e., sample pad of the lateral flow test strip is dipped into the eppendorf tube containing prepared sample in extraction solution. The sample migrates upward in the strip by capillary action.

With the upward movement, the sample reaches in reaction membrane after passing reagent pad that contains first antibodies, labeled with colloidal gold. This labeled antibody binds to the GM protein, if present in the sample and forms antibody – protein complex.

Interpretation

The final reaction includes a colour change in the vial, where the strip/paddle is placed. The gold linked to the first antibody (which is still stuck to the Bt protein) will become concentrated and a pink color appears on the filter stick. Gold is a pinkish color in its natural state. If no Bt protein is present, no color appears. A positive control filter is also used to make sure everything has been added and the color reagents are working. This filter gives a dark color in all tests done correctly. If this filter stays white, something is wrong with the procedure or the test strip. The presence of a single line (control line) on the membrane indicates a negative sample and the presence of two lines indicates a positive sample using antibodies and color reagents incorporated into a lateral flow strip. However, if an excess of unlabeled target analyte is not present, a weak line may be produced in the capture zone, indicating an inconclusive result.

Result

Visibility of both control and test line indicates presence of target GM protein.

Objective

Detection of genetically modified organism through **DNA based method**

It is based on detection of the specific sequence of transgenic DNA genetically engineered into the crop. Presence of transgene is tested by Polymerase Chain Reaction (PCR) technology based on million or billion fold amplification of specific target DNA of transgene with synthetic oligo-nucleotide primers. An event-specific presence of a DNA sequence is unique for detection of GM. Generally it is preset in between transgene and the organism's original DNA.

Procedure

The polymerase chain reaction (PCR) is a technique to isolate and amplify a fragment of DNA, via enzymatic replication. It makes millions of copies of the target genetic sequence (DNA). The targeted genetic sequence pairs with primers i.e., custom designed complementary part of DNA and trigger a chain reaction. DNA replication enzymes use the primers as starting points for doubling the target sequences. The process is repeated over and over again by sequential heating and cooling until doubling and redoubling to multiply the target sequence several million-fold. The copies of identical fragments are purified in a slab gel, and observed under UV light in the presence of dye.

■■■■

Chapter - 29

SEED HEALTH

Objective

Determination of health status of a seed sample.

1. OUT BREAKS OF DISEASES OF SEED BORNE NATURE

Indian agriculture has witnessed following outbreaks of plant diseases of seedborne nature during past decades in different ago-ecological zones of the country under varied farming systems.

- Substantial losses were experienced in pearl millet hybrids were experienced in southern part of country due to downy mildew (*Sclerospora graminicola)* pathogen during 1970-76, 1983, 1984 and 1987-88
- During 1956, ergot disease of pearl millet (*Claviceps fusiformis*) appeared in epidemic form in south western part. There are several known endemic region where plant pathogens of seed borne nature have been well established and under favorable conditions pose a threat to successful seed production.
- Marathawada region of Maharashtra is identified endemic region for sunflower downy mildew disease (*Plasmopara halstedii);* Nillgiri Hill region is known for potato golden nematode (*Heterodera rostochiensis*) and Darjeeling region for wart of potato (*Synchytrium endobioticum*).
- Out breaks of late blight (*Phytophthora infestans*)of potato has always been experienced in hilly tracts of the country.
- Wheat bunt (Karnal bunt) (*Neovossia indica*) and rice bunt (*Neovossia horrida*) are of the major concern in the state of Punjab, Haryana and Uttar Pradesh.
- Ear cockle (*Anguina tritici*) is endemic in north-western part (near Satna & Chhattapur districts) of Madhya Pradesh.
- Ripe rot of chili (*Colletotrichum capsici*) is a wide spread problem in central part of Madhya Pradesh.

Role of seed borne pathogens

Seed is the vehicle of disease causing agents viz., fungi, bacteria, virus, and nematodes. Pathogen may be externally seedborne, internally seedborne or present as admixture or concomitant contamination. Infection of seed borne pathogens results in seed rot, seedling decay and seedling deformities, pre- and post –emergence mortality and abnormalities, distortions, discolorations, reduction in seed size and ultimate alteration in nutritive composition.

Effect of seed borne pathogens

- Loss in germination and vigour
- Development of plant diseases
- Discoloration and shriveling
- Biochemical changes in seeds – protein/oil/carbohydrates
- Alteration in physical properties of seeds
- Poor plant population resulting reduced crop yield

Seed Certification: It is a legally approved system to maintain quality of seeds during seed production, post harvest operations and distribution. It includes field inspection, seed quality tests and pre- & post –control plots.

Threshold value: It is defined as the level of infection on or in the seed that will significantly affect disease development and results in economic losses.

Tolerance value: The term is applied to the permissible level of seed contamination or infection, beyond which the economic loss may occur.

Objective of Seed Certification: To ensure the supply of quality seed to the farmers and to protect the interests of seed traders.

Legislation

- The Indian parliament passed the Seed Act on December 29, 1966.
- Seed Rules under the Act were notified in September, 1968.
- The Act was implemented with effect from 18 October 1969.
- In 1983 the Ministry of Agriculture, Government of India, passed the Seed (Control) Order, accordingly Seeds have been declared as an Essential commodity, under the essential Commodity Act, 1955.
- Accordingly, labeling of seed is compulsory for the purpose of selling in the market, whereas, seed certification is voluntary.
- Certification began, as a result of Seed Act, 1966 that empowered the State and Central Government authorities to stop distribution of uncertified seeds to growers.
- The Central Seed Certification Board was setup in 1972.
- On the recommendations of Seed Review Team, the State Governments were advised to set up the Certification Agencies.
- Minimum Seed Certificate Standards were formulated and a manual known as Indian Minimum Seed Certification Standards was published in 1981.
- The manual of 1971 was revised in 1988 by Tunwar and Singh.
- Book on *Seedborne Diseases Objectionable in Seed Production Programme and their Management* was brought out by Bhale MS & Khare, D in 2001.

- A supplement list was brought out in 2002 Khetrapal and later in 2006 by Kumar and Vari.
- Book on *Seedborne Diseases of Field Crops* by Bhale & Gaur (2012) provides detailed status of seedborne pathogens.

Historical Importance: In different crop plants regular selection for maintenance of seed quality along with consideration of seed health has been a major criterion throughout the world.

1990s Potato seed certification was practiced for checking the tuber borne and other viruses in Germany

1913 Potato certification was introduced in 28 states of US and in Provinces of Canada

1920 In US and Canada, certification of cereals was combined with seed treatment for the effective control of plant diseases that results in reduction of incidence of wheat bunt (*Tilletia leavis*) ,*Urocystis occulata* in rye, and loose smut of oat (*Ustilago avenae*).

1930 Certification of *Phaseolus vulgaris* based on strict possible field inspections resulted in reduction of bacterial blight and anthracnose in New South Wales and Victoria Province of Australia.

In France, upon embankments of seed companies for seed health testing of bean for halo blight pathogen, substantial reduction in losses could be achieved.

Field Standard: A seed production plot is inspected by visual observation for the percent disease incidence. The seed production plot with the disease incidence within the prescribed limit is acceptable for seed production. It the disease incidence is higher than the prescribed limits then the plot is not accepted (passed). The results are expressed as *"% plant infected"*.

Seed Standard: A number of diseases have been identified for which seed standards have been worked out based on per cent seed infection. The result is expressed as *"% seed infection by number"*. The analysis of seed standards is done by a qualified seed analyst in lab.

As per Indian Seed Act some seedborne diseases are designated as objectionable for specific crops (Table 1) and seed of those crops are certified after testing for seed certification standard (Table 1) on the basis of number of infected seed prescribed in number of observed seed. Infected seeds are identified on the basis of symptoms and reported in percentage.

Table 1. Designated objectionable diseases in seed production programme as per Indian Minimum Seed Certification Standards (Tunwar and Singh, 1988) at seed stage and their certification standard

Crop	Disease	Maximum infected seed by number (%)	
		Foundation	Certified
FIELD AND FODDER CROPS			
Wheat	Karnal Bunt	0.050	0.25
	Seed Gall, Tundu	0.00	0.00
Rice variety and hybrid	Kernel Smut (Bunt)	0.10	0.50
Sorghum	Ergot	0.02	0.04
Pearl millet	Ergot	0.02	0.04

Crop	Disease	Maximum infected seed by number (%)		
		Foundation		Certified
Triticale	Karnal Bunt	0.05		0.25
	Ergot	0.02		0.04
Forage sorghum	Ergot	0.02		0.04
Sudan grass	Ergot	0.02		0.04
VEGETABLES				
Sweet Potato	Black Rot	0.0		0.0
	Scurf	0.0		0.0
	Fusarial Wilt	0.0		0.0
	Internal Cork	5.0		5.0
	Storage Rot	0.0		0.0
Multiplier onion	Bacterial Brown Rot	0.0		0.0
	Bacterial Soft Rot	0.0		0.0
	Fusarial Basal Rot	0.0		0.0
Potato		F1	F2	
	Phytophthora Late Blight, Fusarial Dry Rot, Macrophonmina Charcoal Rot	1.0	1.0	1.0
	Sclerotial Wet Rot	0.0	0.0	0.0
	Common Scab	3.0	3.0	5.0
	Black Scurf	5.0	5.0	5.0
	Total	5.0	5.0	5.0

Seed of pearl millet from fields having ergot infection even within the prescribed limits should be subjected to floatation treatment with brine solution to become eligible for certification.

Per cent seed infection of ergot in sorghum is reported by number, having ergot, sclerotia, seed entirely or partially modified as sclerotia or broken sclerotia.

In the presence of even single potato seed tuber infected by common scab in Foundation I, II and Certified seed lot, the entire seed lot is treated with approved fungicide before declaration of lot fit for certification. Seed lot having infected tubers more than the prescribed limits are certified even after dressing with appropriate pesticide.

1.1 Methods employed for the detection of fungal pathogen associated with seed

i. Inspection of dry seed with unaided eye, hand lens or under low power of stereoscopic microscope.

ii. Inspection of seed after soaking in water or NaOH solution.

NaOH seed soaking method: This method is used for detection of karnal bunt of wheat and kernel smut of rice pathogen. Seeds are soaked in 0.2% sodium hydrochloride solution (2g NaOH/1000ml water) for 24 hours at 25ºC. Seeds are examined visually on white paper with aided light. Bunted seeds become jet black and can be separated. Diseased seeds can be confirmed for the presence of teliospore.

iii. Microscopic examination of suspension obtained by seed wash.

iv. Incubation test
   - Standard blotter method
   - Standard agar method
   - Peptone Penta Chloro Nitro Benzene (PCNB) method: The basic culture media, peptone is amended with terramycin (50ppm) and 1000 ppm of penta chloro nitro benzene (PCNB) and 2% agar agar. It is highly selective medium for *Fusarium* species.

v. Examination of the symptoms developed on seedling grown in soil or sand

vi. Growing on test carried out in green house, environmental controlled chamber or in the field.

vii. Test based on the biochemical reaction (serology test)

1.2. Methods employed for the detection of bacterial pathogen associated with seed

i. Inspection of dry seed with unaided eye, hand lens or under low power of stereoscopic microscope.

ii. Microscopic examination of suspension obtained by seed washing by isolation technique on bacteriological cultural media.

iii. Tests based on biochemical and physiological reaction on the isolated bacteria.

iv. Test based on Enzyme Linked Immuno Sorbant Assay (ELISA).

ELISA Test: It is an immuno assay system involving polyclonal and monoclonal antibodies, making use of the inherent mechanism in living system to produce specific antibodies that interact with foreign protein substances. The entry of an antigen into a living system triggers the production of antibodies by lymphoide tissues. Antibodies have high affinity for very specific chemical structure, therefore they bind only to predetermined structure of an antigen.

Direct ELISA (DAS- ELISA) and indirect ELISA protocols are used for clinical diagnosis of seed borne pathogens, identification and quantification of food constituents, microbes and toxic metabolite produced by them. The specificity of an antigen- antibody interaction is extremely sensitive. In ELISA the reaction between antibodies and antigen is amplified by an enzyme conjugated to the primary antiserum as in double antibody Sandwich method or to secondary antibodies as an indirect ELISA. The reaction is quantified by absorbence value at a particular wavelength read in ELISA reader.

1.3. Methods employed for detection of virus pathogen associated with seed

i. Inspection of dry seed with unaided eye, hand lens or under low power of stereoscopic microscope.

ii. Examination of symptoms developed on seedling grown in soil/sand under insect proof chamber.

iii. Examination of symptoms on indicator plants by infectivity test. Inoculation of plant sap and close observation on the development of symptoms on a particular indicator plant under insect proof and environmentally controlled chamber.

iv. Electron microscopy and serological tests including advanced techniques eg., ELISA, Polymrease Chain Reaction (PCR), Dot Immuno Binding Assay (DIBA) etc.

Table 2. Seed health methods employed for detection of associated fungus

Crop	Disease	Method
Paddy	Kernel smut	Visual inspection of planting material, examination of suspension obtained by seed washing, NaOH soak
Wheat	Ear cockle	Visual inspection of planting material, examination of planting material by dissection and staining
	Tundu	Visual inspection of planting material, test based on biochemical and physiological reaction
	Karnal bunt	Visual inspection of planting material, examination of suspension obtained by seed washing, NaOH seed soak
Triticale	Karnal bunt	Visual inspection of planting material, examination of suspension obtained by seed washing, NaOH seed soak
	Ergot	Visual inspection of planting material
Sorghum	Ergot	Visual inspection of planting material
Pearl millet	Ergot	Visual inspection of planting material
Sweet potato	Storage rot	Visual inspection of planting material, standard agar method
	Scurf	Visual inspection of planting material, standard agar method
	Wilt	Visual inspection of planting material, standard agar method, examination of symptoms developed on seedlings in soil or sand
	Black rot	Visual inspection of planting material, standard agar method
	Internal cork	Visual inspection of planting material, examinat88ion of seedlings on the basis of these symptoms developed in insect proof chamber, electron microscopy and serological test
	Nematode	Visual inspection of planting material, examination of planting material by dissection and staining
Potato	Late blight	Visual inspection of planting material, standard agar method
	Dry rot	Visual inspection of planting material, standard agar method, peptone-PCNB method
	Wet rot	Visual inspection of planting material, standard agar method
	Common scab	Visual inspection of planting material
	Black scurf	Visual inspection of planting material, standard agar method
Multiplier onion	Bacterial soft rot	Visual inspection of planting material, examination of suspension obtained by seed washing-standard agar method, test based on biochemical and physiological reaction, ELISA
	Bacterial brown rot	Visual inspection of planting material, examination of suspension obtained by seed washing-standard agar method, test based on biochemical and physiological reaction, ELISA
	Fusarium basal rot	Visual inspection of planting material, standard agar method- peptone-PCNB method

2. INFESTATION OF OBJECTIONABLE PARASITIC WEED PLANT SEED

Some crops have designated objectionable parasitic weeds at seed level, in the event of more number than the prescribed standard the sample is not accepted (Table 3).

Table 3. Maximum seed certification standard for infestation of seed with objectionable parasitic weeds

Crop	Parasitic weed plant seeds	Standard (%)	
		Foundation	Certified
Sunflower	*Orobanche cumna*	0	0
Lucerne	Dodder (*Cuscutta* spp.)	5	10

3. INFESTATION OF NEMATODES

Some crops have designated objectionable nematodes at seed level, in the event of more number than the prescribed standard the sample is not accepted (Table 4).

Table 4. Maximum seed certification standard for infestation of seed with objectionable nematodes

Crop	Parasitic nematodes	Standard (%)	
		Foundation	Certified
Sweet potato	Root knot	0	0
Wheat	Ear cockle	5	10

Detection of nematode pathogen associated with seed

- Inspection of dry seed with unaided eye, hand lens or under low power of stereoscopic microscope.
- Examination of seed and / or plant material by dissection technique under microscope.
- Examination of seed and / or plant material by dissection and staining method and observation under microscope.
- Examination of soil washing and extraction of nematode from soil and plant material.

4. INFESTATION OF OBJECTIONABLE INSECT PESTS

Some vegetable crops multiplied by its vegetative parts have designated objectionable insect pest at seed level and in the event of more number of insects than the prescribed standard the sample is not accepted (Table 5).

Table 5. Maximum seed certification standard for infestation of seed with objectionable insects

Crop	Insect	Standard(%)	
		Foundation	Certified
Colocasia	Scale insect	0.0	0.0
	Mealy bug	0.0	0.0
Lesser yam	Scale	0.0	0.0
Sweet potato	Wireworm	1.0	1.0
	Weevil	0.0	0.0
Tapioca	Scale insect	0.0	0.0

5. INSECT INFESTATION

Objective

Determination of insect infestation in the seed sample.

Submitted sample of Foundation and Certified categories are observed visually for insect infestation. Seed sample with insect infestation more than the seed certification standards (Table 6) is not accepted during certification.

Table 6. Maximum seed certification standard for insect infestation

Crop	Insect infestation (%)	
	Foundation	Certified
Maize and legumes	1.0	1.0
Other than maize and legume	0.5	0.5

■■■■

Chapter 30

IDENTIFICATION OF OBJECTIONABLE SEED BORNE DISEASES

30.1 OBJECTIVE

Verification of diseases at field level

LOOSE SMUT OF WHEAT

Causal pathogen: *Ustilago segatum* var. *tritici* (= *U. tritici*)

Significance: In India, it is prevalent in areas having cool humid and semi-humid climate during flowering period. In India average losses ranges from 3 to 4 per cent. The disease is exclusively internally seed-borne (embryo borne) in nature

Symptoms: Few earheads are converted into dark olive brown black powdery mass. The dark spore mass can be seen through the wall of the boot several days before heading. The black powdery mass (smut spores) is initially covered with a delicate membrane which soon ruptures and releases the spores. After breaking up the membrane, spores are dispersed by wind leaving behind the naked rachis. The loose smut affected seedlings are often weaker. Relative humidity (90%) and temperature (20-25°C) are required for teliospore germination.

DOWNY MILDEW OF SUNFLOWER

Causal pathogen: *Plasmopara halstedii.*

Significance: In India, first observed in Marathawada region (Lattur and Beed) during 1984, the disease is now recorded in Karnataka, Madhya Pradesh and Andhra Pradesh with increasing concern. Incidence of the disease from 5 to 60% in substantially large area is reported from

Maharashtra while in Bangalore up to 38% was recorded. Because of systemic infection, practically no seed formation takes place in the undersized flower heads.

Symptoms: Cultivars, moisture, temperature, age of the host and level of inoculum are the factors which influence infection and severity of symptom expression. Under cool and moist condition, the infected seed produces infected seedling, which often killed by the whitish downy growth of the fungus on the cotyledons. In systemic infection, the characteristic extreme stunting. Venial chlorosis, downy growth, thickening of foliage, erect head and root growth reduction and root gall formation may be observed.

DOWNY MILDEW (GREEN EAR) OF PEARL MILLET

Causal pathogen: *Sclerospora graminicola*

Significance: Several epidemics have been observed which resulted in substantial grain yield losses (upto 60 per cent) in F_1 hybrids during 1970, 1976, 1983, 1984 and 1988 under south Indian conditions. Usually the losses in grain yield are directly proportional to disease incidence in the field. The disease is soil and seed-borne in the form of mycelium in embryo and seed carried oospores. After transmission of the pathogen through infected seeds and infected leaf the disease establishes in a region and becomes soil-borne.

Symptoms: Systemic infection results in variation in symptoms. The most conspicuous symptoms are the transformation of floral parts into green leafy structure. Green ears are the most commonly and easily identified symptoms. Systemic symptoms generally appear on the second leaf and later on all the subsequent leaves and panicle. Under humid and moderate temperature conditions, the infected chlorotic leaf areas support a massive amount of sporulation as whitish downy growth of the fungus.

ERGOT OF PEARL MILLET

Causal pathogen: *Claviceps fusiformis* (= *C. microcephala*)

Significance: The disease is reported from all over the country. Quality of grains is also adversely affected. The ergot (sclerotia) contains *ergotin* alkaloid and assimilation of this poison in the blood system causes convulsions, nervous breakdown, shrinkage of cells and many more symptoms in cattle and human beings. Hence, the disease has an additional importance. The damage caused by the disease is weather dependent. The loss in grain yield is directly proportional to per cent infection, as infected ovary is completely transformed into a sclerotium. The pathogen is observed as concomitant contamination.

Symptoms: The disease is visible at flowering stage. The diseased spikelets (florets) exudates the droplets of pink or honey coloured sticky liquid which turns darker and cover large area of the inflorescence. The honey- dew contains spore mass. Under advanced conditions infected kernels are transformed into straight or curved (1-2 cm), cremate light brown horny structure with sclerotia projecting between the glumes as ergot. The floral ovary is transformed into sclerotial body, which attains a large size. It is visible due to size, colour and shape.

GRAIN SMUT OF PEARL MILLET

Causal pathogen: *Tolyposporium penicillariae*

Significance: The disease is seed and soil-borne. The disease is observed on earheads. The infection may involve scattering light-coloured sori among grains up to complete coverage of earheads. More often, the infection is noticed on one side and at the base of the ear head. The floret is transformed into a smutted grain. Hence, the loss is proportional to the floret infection. The losses up to 30 per cent have been reported.

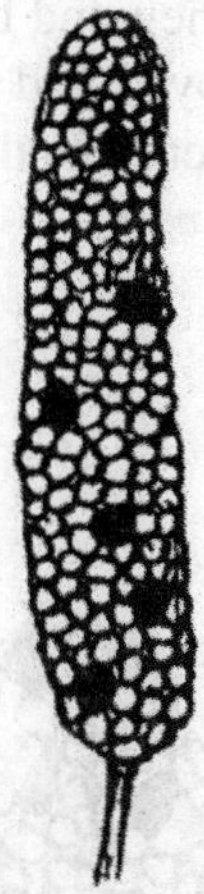

Symptoms: In an earhead the florets infection transform developing seed into sori (spore sac) containing black powder of smutted spores. These sacs are larger and greener as compared to no infected seeds in early stages. The infected seeds remain green whereas the normal seeds become creamy or grey coloured. As the sori mature they become dark brown and are easily broken, spilling out thousands of black spore balls of smut. Smut infection can also occur in inflorescence already infected by downy mildew, ergot or both. The smut sori are formed in place of grains. Sori are oval, bigger than seeds and projects beyond the glumes.

HEAD SMUT OF SORGHUM

Causal pathogen: *Sphacelotheca reiliana* (=*Sporisorium reilianum*)

Significance: With the advent of improved varieties and hybrids the comparative incidence of the disease is decreasing, however, it has been reported from Madhya Pradesh, Andhra Pradesh, Maharashtra and Uttar Pradesh. The disease is observed at flowering stage. The young plants are more susceptible.

Symptoms: The diseased earheads can be identified easily. Each infected head is modified into a large whitish gall. The galls are at first covered by a whitish grey membrane of fungal tissue, which breaks up often before the head emerges from the boot. Black brown spores of smut are exposed, embedded in long thin dark brown coloured filaments. The filaments are the vascular bundles of the diseased earheads. The wind blows the smut spores leaving the filaments. Since the entire earhead is completely destroyed and converted into smut spores, therefore it can easily be differentiated with other smuts where usually the individual florets express the disease symptoms.

KERNEL SMUT OF SORGHUM

Causal pathogen: *Sphacelotheca sorghi* (= *Sporisorium sorghi*)

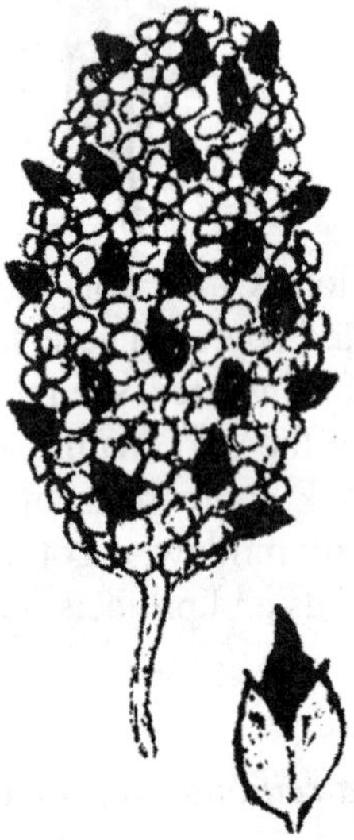

Significance: The kernel smut is the most destructive disease causing losses upto 25-30 per cent in affected regions. Since the individual floret is attacked, which is converted into a smut sorus (smut sac) the yield losses is practically less.

Symptoms: In a flower head, fungus attacks individual floret resulting in the transformation of each grain into a sorus. The size and shape of the diseased florets (grains) changes in spore-sac, according to the degree of susceptibility of a variety. However, usually the infection results in the development of oval dirty grey coloured grain with conical tip end. The sac is surrounded by unaltered glumes at the base. In infected florets the stigma is often not transformed while the stamens develop normally or sometimes absent or involved in the sorus. The stamens are represented by three conical protuberances from the side of the sorus. Varietal response to the fungal infection is very prominent. In some cases the grain turn slightly red and look like normal ones, however brown kernels are full of smutted spores (black powder). Therefore, the diseased seeds can be escaped from the eyes of seed analysts, hence critical observations are needed.

LOOSE SMUT OF OAT

Causal pathogen: *Ustilago segatum* var. *avanae*

Significance: Loose smut and covered smut are the two important smuts of oats. The situation becomes complicated when the interspecific hybridization takes place and several intermediate types are developed. Therefore, one hybrid may produce loose smut on one oat variety and covered smut on another. While another combination may produce the same type of smut on both varieties. The disease is exclusively internally seedborne. It has a potential to cause heavy losses.

Symptoms: In a panicle, the grains are replaced by olive brown to brownish black powdery mass of smut spores. In severe case the glumes and awns are also converted in black powder (Fig. 45). The spore mass is covered with delicate whitish grey membrane. In a plant all the panicles are smutted, however, occasionally one panicle may escape. Smutted plants are comparatively shorter. The thin whitish grey membrane breaks and disintegrates soon after the oat panicle emerges in loose smut disease while it does not break and persists till harvesting in covered smut. Breakage of membrane is the main distinguishing character. The spore mass is liberated by wind.

LOOSE SMUT OF BARLEY

Causal pathogen: *Ustilago segatum* var. *nuda* (=*Ustilago nuda*)

Significance: It is exclusively a seed-borne disease and in most of the features quite similar to loose smut of wheat.

Symptoms: The smutted head usually appears few days earlier than heads of healthy plants. The dark brown powdery spore masses (smut spores) are enclosed within a fragile, silvery membrane that soon ruptures releasing the spores. At the time of harvest, an erect naked spike of a smutted head is visible. The diseased seeds cannot be differentiated from normal ones.

CERCOSPORA LEAF SPOT OF SESAME

Causal pathogen: *Cercospora sesami*

Significance: The disease is a widespread problem and observed wherever the crop is grown extensively. In India, it has been reported from all the major sesame growing tracts. Under high humidity (more than 60 per cent) complete defoliation takes place which ultimately results in reduced yield and seed quality.

Symptoms: Initially the disease appears as small, minute, circular leaf spots and later increase in size up to 10 mm. Under favourable conditions of high humidity (more than 60 percent) and moderate temperature (about 25°C) the disease spreads to petioles, stem and pods producing linear dark coloured deep-seated lesions. The spots may coalesce into irregular blotches. Capsule shows more or less circular, brown to black lesions ranging up to 7 mm with deformed and shrivelled seeds.

HALO BLIGHT OF GREEN GRAM

Causal pathogen: *Pseudomonas savastanci* pv. *phaseolicola*

Significance: One infected seed per 5 kg healthy is capable of giving rise the epidemics under favourable conditions. The pathogen also affects French bean, lima bean, scarlet runner, soybean and kudzu vine. Heavily affected seed do not germinate. The disease is seedborne in nature, which survives in plant residues.

Symptoms: Symptoms appear as translucent spots (lesions), water soaked on the under surface of the leaves. These spots/lesions enlarge and run together. As they develop, the central portion becomes dry. Affected leaves turn yellow and yellowish green "halo" surrounds the small, dead leaf lesions. The halo effect tends to disappear when the air temperature rises. The spots become angular, reddish brown to brown and dry. Infected leaves or even entire plant may wilt and die quickly with the upper leaves crinkled and mottled. Whitish light green exudates may also be observed on the lesions developed on pods and stem, under severe conditions, indicating the infection of Pseudomonas while if the ooze colour is yellow, it may be due to Xanthomonas sp.

JUTE CHLOROSIS

Causal pathogen: The etiology of the causal agent (a virus pathogen) is unknown. Many workers have referred Chlorosis as morphological imperfection.

Symptoms: The jute plants may be chlorotic from the very early stages or may become chlorotic later at any time during the vegetative period. The major symptoms include yellow and light yellow patches on the leaf lamina (surface) leading to variegated appearance hence the term Chlorosis is applied. The affected plants exhibit stunting (reduced growth) compare to normal healthy ones. Chlorotic plants have lesser surface roots. In mild form of infection, no harmful effects have been noticed; however in severe cases the leaves become crinkled and brittle. The plant remains stunted.

ANTHRACNOSE OF CHILLI ES

Causal pathogen: *Colletotrichum capsici* (= *C. dematium*)

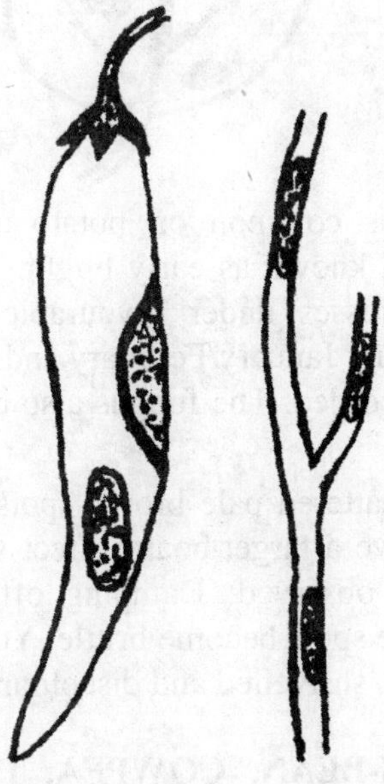

Significance: It is a wide spread problem limiting the profitable cultivation and seed production throughout the major chilli growing regions of India. Fruit rot upto 32 per cent and die back upto 29 per cent have been noticed. Seedling decay and seed rot up to 21 per cent ware recorded under Central Indian conditions. The market value and nutritive value is degraded in infected fruits resulting in the poor quality of seed.

Symptoms: Seed rot, seedling decay, leaf spot stem blackening and fruit rot symptoms are incited by the fungus. The fruit rot is conspicuous at the time of fruit maturity, which initially appears as small black slightly sunken circular spots on the skin, which later spread in the direction of long axis. The spot is lighter brown in colour with a thick and sharp black outline. The discoloured, straw coloured skin of chilli fruit contains numerous pinkish black fungal bodies known as acervuli. In die back phase, tender tip's necrosis progress downward results in withering of entire branch or tip. Under humid conditions the irregular spots on the stem or fruit develop rapidly and form irregular necrotic black lesions. The dead twig turn into grey white or straw colour with numerous black specks and acervuli scattered over. Infected seed become rusty and shrivelled.

LEAF BLIGHT OF CHILLIES

Causal pathogen: *Alternaria solani*

Significance: The pathogen is common on potato and tomato. The disease is widely distributed all over the country and known as early blight. The pathogen can attack the crop at any growth stage causing huge losses under favourable conditions. Under Central Indian conditions the disease is severe during January/February and the August/September months. Fruit Loss up to 24 per cent has been recorded. The fungus also causes post harvest fruit decay. Fruit loss up to 15 per cent is observed.

Symptoms: Small isolated scattered pale brown spots on leaves appear with deep green-bluish fungal growth. The spots give a target-board effect with concentric ridges. Seed rot and seedling blight at nursery bed is observed. Damping off seedlings take place under moist conditions. Under dry conditions the spots become brittle. Young tender tip drying and fruit rot is observed. Infected seeds are smaller, shrivelled and discoloured and have low germinability.

ANTHRACNOSE OF FRENCH-BEAN, COWPEA, INDIAN BEAN AND CLUSTER BEAN

Causal pathogen: *Colletotrichum lindemuthianum*

Significance: The disease is temperature and wet weather dependent. The affected seeds and pods have lower marketability. Usually the disease increases with the age of the crop. The infected seeds have low germination or produce weak plants.

Symptoms: The pathogen is able to attack on all above ground plant parts. The disease is more severe in cool and wet weather. Black sunken cankerous spots/lesions with light grey central portion appear on developing French bean and cowpea pods. Infected seeds have lesions (spots) of different shades of brown colour. Similar spots appear on cotyledons with sunken deep brown to black with ash coloured centre having pink mass of conidia. Black and necrotic spots on and along veins on the under side surface of leaves may be recorded. Blackening on stem and petiole can be seen.

ASHY STEM BLIGHT OF COWPEA

Causal pathogen: *Macrophomina phaseolina*

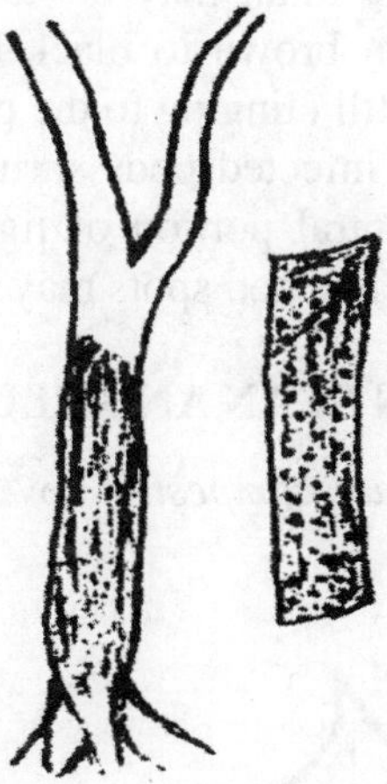

Significance: Pathogen is responsible for ashy stem blight of cowpea. The disease is also known as charcoal rot or ashy grey stem. Losses up to 40 per cent have been observed under Central Indian conditions.

Symptoms: The mature plants show yellowing of leaves and within 2-6 days, entire plant dries up. The disease appears as seed rot, seedling decay and root rot. Seedling death is noticed due to development of black sunken cankerous region near cotyledons. Darkening of stem at soil base may be visible. On infected root, symptom appears as if the charcoal powder is sprinkled. The charcoal powder consists of fruiting bodies (pycnidia) or sclerotia embed in the host tissues, which can be seen by pealing-off the epidermis. The infected plants are easily uprooted.

ASCOCHYTA BLIGHT OF COW-PEA AND FRENCHBEAN

Causal pathogen: *Ascochyta phasiolorum*

Significance: Infection may occur at any stage and at any plant part of the crop. Relatively of lesser importance the disease may be a serious problem causing leaf blight, and pod spot under favourable weather conditions of cool climate and ample moisture on leaves. The pod infection reduces the marketability and seed quality largely. Severely infected seeds do not germinate.

Symptoms: Major symptoms include seed rot, seedling mortality, stem-infection, pod-infection (spots) and seed deformities. Plant may be attacked at any stage of growth. Initially small purple spot appear later on turn brown to black become zonate with a definite margin. Infected leaves soon dry, hang down still clinging to the plant. Browning or blackening of stem is common with elongated lesions. The infected pods shrink. The seeds are deformed, shrivelled, small and sometimes stained. The central portion of light coloured spots may contain fruiting bodies, embedded in the plant tissues. The pod spots may extend to the underlying seeds.

BACTERIAL BLIGHT OF INDIAN BEAN AND CLUSTER BEAN

Causal pathogen: *Xanthomonas campestris* pv. *cyamopsidis* (= *X. axonopodis* pv. *cyamopsidis*).

Significance: The disease is more prevalent in cool and wet weather conditions. High temperature inhibits the disease expression and development Blight affected beans have low marketability due to spots on pod surface. Under severe conditions of infection seeds are not developed. The severely infected seed do not germinate. Under favourable conditions of ample moisture and low temperature, losses up to 40 per cent have been observed in our conditions. Infected seeds produce weak plants.

Symptoms: Bacterial pathogen induces the lesions on leaves, pods, stem and seed. The lesion develops as minute, small water soaked with light green area on leaves. The brown lesions enlarge with dead (necrotic) central area. Yellow halo around coalescing of lesions, result in larger dead areas as blight. On pods similar water soaked lesions are formed which soon enlarge into broad regular blotches. In humid weather, a yellow encrustation of *Xanthomonas* covers the surface of lesions. The margin of the spot or occasionally the entire spot has some shade of red, which gradually changes to brown. Under severe conditions of infection pod shrinks, seed remain shrivelled. Under favourable conditions of moisture, stem blackening and girdling may occur, which results in stem breaking during rain or strong wind.

ASCOCHYTA BLIGHT OF CLUSTER BEAN (GUAR)

Causal pathogen: *Ascochyta rabiei*

Significance: It is the most destructive disease problem of cluster bean (*Cyamopsis tetragonolobus*) for hill regions. The disease is widespread and favoured by low temperature and high humidity. Plant may be attacked at any stage of crop growth. Pod spots and seed deformities result in poor germination and marketability. Infected seed either do not germinate or produce weak plants resulting in substantial losses.

Symptoms: The symptoms induced by the pathogen include seed rot, seedling mortality, pod and stem infection and seed deformities. The initial symptoms appear near the tip of young shoots and later the disease spread rapidly on other aerial parts and produces brown spots. Fungal fruiting bodies (pycnidia) are produced in the form of concentric rings on leaves, stem, pods and seed.

BEAN COMMON MOSAIC OF FRENCHBEAN

Causal pathogen: Bean common mosaic virus (BCMV)

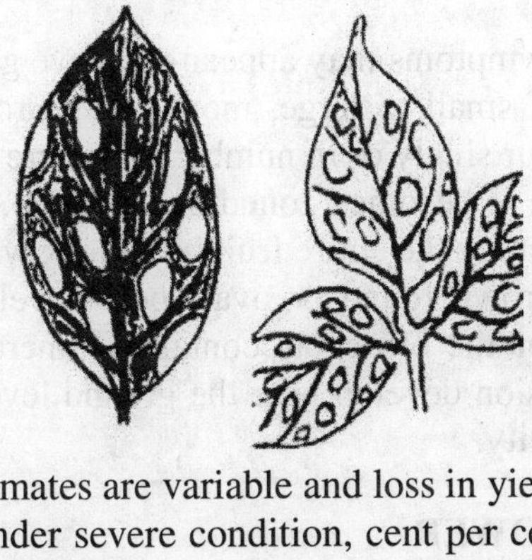

Significance: Crop loss estimates are variable and loss in yield is due to reduction in number of pod along with seed yield. Under severe condition, cent per cent losses have been reported in Latin America. The level of transmission ranges from 3 to 95 per cent. Several strains of the virus exist. The pathogen is seedborne. The extent of damage depends upon climatic condition host susceptibility, infective viral strain and the time of infection to the crop.

Symptoms: The virus is seed transmitted and infected plants exhibit mottling, leaf curling, reduction in height (stunting) and deformities (malformation) of primary leaves. Dark-green and light-green patches result in severe mosaic symptoms on trifoliate leaves is a common symptom.

The infected pods remain underdeveloped. It also delays the seed maturity. In general, three types of symptoms i.e., mosaic, systemic infection (necrosis) and hypersensitive reaction are noticed. In hypersensitive reaction, the infection by virus is followed by death of the surrounding tissues and restriction of viral movement by plant itself. Systemic necrosis or black rot syndrome is characterised by a vascular necrosis appearing first as an apical die back followed by plant death.

PHOMOPSIS BLIGHT AND FRUIT ROT OF EGGPLANT

Causal pathogen: *Phomopsis vexans* (= *Diaporthe vexans*).

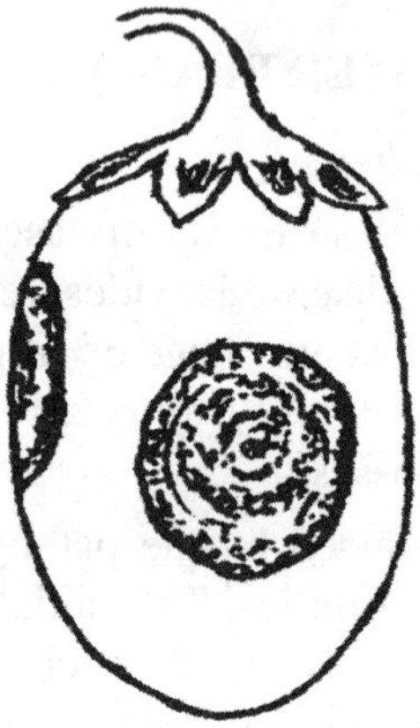

Significance: The fungus may attack any above ground plant part, from seedlings to fruits. It is a common and destructive disease observed in transit to market resulting in decay of fruits. The infection on leaf, stem, seedling and fruit often reduces the production, marketability and seed quality. Under Central Indian conditions the disease incidence upto 30 per cent is recorded. It is calculated that with the increase of one per cent disease incidence there is a reduction upto 2 kg yield of eggplant fruit.

Symptoms: The disease symptoms may appear on above ground parts from seedling to fruit. Symptoms on fruits consist of small to large, more or less round, tan to light brown, slightly sunken lesions (spots) may occur singly or in number anywhere on the surface. The decaying spot turn chocolate brown with age it becomes zoned and enlarge to a diameter of 2-3 inches. The decay is soft and spongy; ultimately the entire fruit turn black with rings on affected fruit area and shrivelled. On leaves, small brown round or oval spots develop usually on lower leaves with definite margins. The centre of the old spots contains numerous black pycnidia. On the stem sunken, girdling canker like lesion develop near the ground level. The fungus is also responsible for seed rot and seedling mortality.

BLACK ROT OF CAULIFLOWER

Causal pathogen: *Xanthomonas campestris* pv. *campestris*

Significance: The disease is also known as bacterial blight and stump rot. The disease is severe in regions having cooler temperature (22° to 25°C) and plentiful rains or dew. The plant may be infected at any stage of the growth. After establishment in the soil the disease become extremely serious. It is experienced that large percentage of crop plants becomes incapable of producing marketable heads of cabbage and cauliflower heads. The storability of infected heads is greatly reduced.

Symptoms: Disease can easily be identified in the field, as a V-shaped infected area develops on leaf lamina that extends inwards from the edge. Diseased areas turn yellow to brown and dry out. Leaf veins become dark brown or black. Black rot may appear in the form of numerous minute brown specks. The infected lower leaves wilt, dry and droop prematurely. The bacterium may move through veins and cause water soaking inside the head. Because of seedborne infection seedling results in mortality and dropping-off the cotyledons.

BLACK LEG OF CABBAGE

Causal pathogen: *Phoma lingam*

Significance: The disease may appear in severe form in the regions with wet seasons and continuous cropping. The disease is also a problem in cauliflower and other crucifers. Severely infected seeds are shrivelled, which do not germinate.

Symptoms: The symptoms appear as oval to linear sunken light brown coloured cankers near the soil line and with the advent of time it girdles the complete stem. The infection results in black brown cankers with purple margin and affected plants soon wilt. Numerous fungal bodies can be observed on decayed (necrotic) area. The infected seeds give rise to the diseased seedlings. Tremendous number of spores is produced on the cotyledons of prematurely killed seedlings.

Spots with ashy grey centres develop on leaves containing scattered fruiting bodies. Severely infected seeds are shrivelled, which do not germinate.

BACTERIAL SOFT ROT OF CABBAGE, CAULIFLOWER, BROCCOLI, KNOLKHOL AND MULTIPLIER ONION

Causal pathogen: *Erwinia caratovora* subsp. *caratovora*

Significance: Bacterial soft rot is the most widespread and destructive disease of vegetables. Carrot, onion, potato, parsnip, turnip, celery, lettuce, spinach etc. and large number of other plants are attacked by the bacterium. The bacterium can attack the crop at any stage in the field, garden, transit and storage in market. The disease is more prevalent under storage conditions. Substantial loses (more than 50 per cent) may occur if the storage conditions are congenial for the disease development.

Symptoms: The disease starts on leaves, stem and underground parts as small, water-soaked, translucent spots, which rapidly enlarge. Because of infection the host tissue softens and become stinky, mushy or watery. The infected portion emits strong repulsive smell due to entry of other secondary bacterial pathogen. Soft rot is slimy or watery because the causal bacteria produce enzymes which dissolve the middle lamella. Due to dissolution of middle lamella the tissues become loose and change their form and structure.

BASAL ROT OF MULTIPLIER ONION

Causal pathogen: *Fusarium oxysporum* f.sp. *cepae*

Significance: Disease is devastating under favourable conditions. However, no precise estimates are available under our conditions. The disease is seed and soilborne in nature.

Symptoms: Xylem tissues are attacked by the fungal infection resulting in wilt symptoms. The fungus invades the bulbs and induces rotting. At seedling stage, post emergence mortality is very severe. The affected tissues become brown due to the toxic metabolites produced by the fungus, known as fusaric acid. Affected plants can easily be identified due to paleness and yellowing. The rotten bulbs become watery and white mycelial growth can be observed on affected parts. Seeds turn brown, shrivelled and deformed due to the infection.

EARLY BLIGHT OF TOMATO

Causal pathogen: *Alternaria solani*

Significance: The disease is not dependent on weather. It is observed both in cool and warm areas with equal intensity. The fungus is also responsible for early blight in potato, chilli and eggplant. The disease may appear at any stage of crop growth. Under severe outbreaks losses upto 70 per cent are common. The early blight causes the leaves to drop-off, expose the fruit to sunscald, and contribute to the development of anthracnose. The disease also causes the fruit drop and rot. Infected seed produce weaker and diseased plants.

Symptoms: The most common symptoms are seed rot, leaf spot and fruit rots. The disease appears as small brown spots on the lower and older leaves as isolated and scattered spots. The spots often covered by grey-blue fungal growth. The spot enlarge until they are one fourth or half an inch in diameter and show concentric rings giving a target board effect. This results in defoliation. Early blight may cause a dark, sunken, leathery lesion around the attachment of the fruit with stem. The fungus often covers the fruit spot as mouldy growth. Flower drop and infection is also common and results in young fruit drop.

LEAF SPOT OF TOMATO

Causal pathogen: *Stemphylium solani*

Significance: It is not a common widespread problem as compared to *Alternaria* leaf spot/ blight. The disease is favoured by hot wet weather. No precise scientific estimates are available for the losses caused by the disease. Plants may be attacked at any stage from seedlings to maturity. Severe infection results in leaf drop, exposing the fruits to sun scold and other fungal diseases.

Symptoms: The disease appears as scattered minute circular black spots on leaves. The central necrotic black portion of the spots turns into greyish brown. Central portion may be cracked, ragged or fall out. Fruiting of the pathogen occur on both the surfaces. Some of the spots may be surrounded by yellow halo. Larger spots may collapse. The spots caused by *Stemphylium* are limited to leaf lamina only, rarely present on petiole, never on stem and fruit.

MOSAIC OF TOMATO (TMV)

Causal pathogen: Tobacco Mosaic Virus (TMV)

Significance: TMV has the widest host range and very common destructive problem. Practically every plant has infection of one or other virus in the field and often it is overlooked. The viral infected plants yield lesser and less number of fruits. Sometimes the fruits are deformed. Under severe conditions losses upto 50 per cent have been observed. The disease is both internally and externally seed-borne in nature and readily spread by contact. The symptoms are influenced by temperature, day length, and age of plant, susceptibility of cultivar and strain of virus

Symptoms: Leaves show dark green mosaic or mottle. The green parts continue to grow while lighter green area has slower growth rate. Therefore, leaf gives a rough appearance. The most obvious symptoms are a mottle on the leaves, which may vary from a slight patchy chlorosis to bright yellow and green. There may be some wilting in the early stage with recovery as temperature rise. Some strain may exhibit stripping from pale green to black coloured.

LEAF SPOT OF PARSLEY AND CELERIAC (TURNIP ROOTED CELERY)

Causal pathogen: *Septoria petroselini*

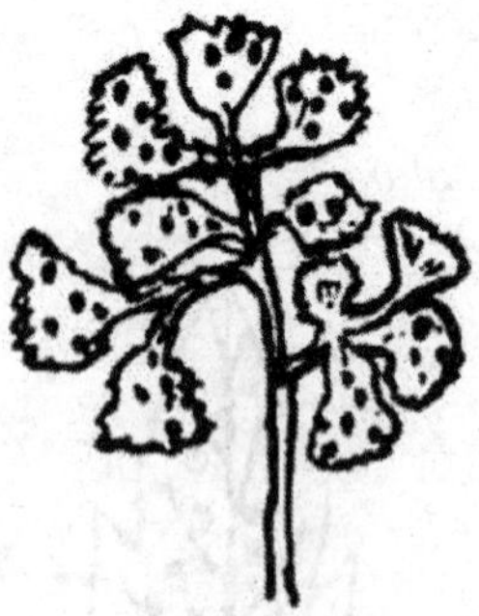

Significance: The fungal disease is the most destructive problem. The fungal infection reduces the marketability. The infected leaves are spoiled. Early infection destroys the entire crop long before harvest time. The disease is seed-borne and pathogen can remain viable in plant refuse in the soil.

Symptoms: The fungus causes sub-circular brown spots with grey centres. The spots begin with small specks with slight darker margin. The central portion of the spot consists of numerous black fungal fruiting bodies. It can easily be seen by unaided eye.

LEAF BLIGHT OF CELERY

Causal pathogen: Septoria apicola

Significance: The blight and leaf spot phase greatly reduces the quality of celery crop by the rusty brown lesions. The greatest damage results in loss of foliage. Spotted and blighted foliage have unacceptable consumers' choice and reduced marketability.

Symptoms: The symptoms are best seen on the older leaves where groups of fungal fruiting bodies (pycnidia) can be seen on water soaked spots which ultimately turn brown. These are surrounded by a distinct chlorotic halo. The spots are usually very numerous and eventually coalesce thus, the leaf shrivels and collapses.

ROOT ROT OF CELERY

Causal pathogen: *Phoma apicola*

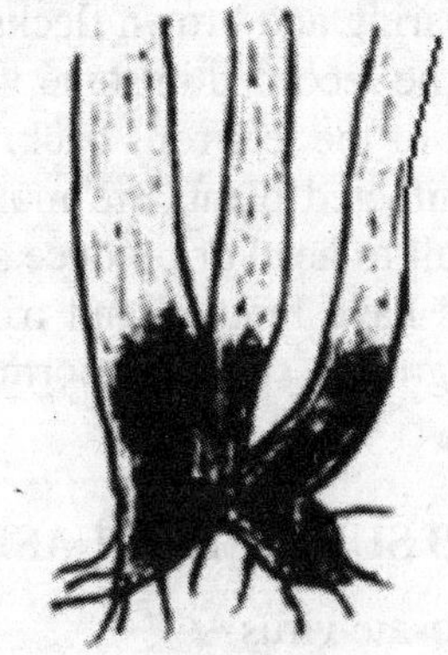

Significance: The root rot is a serious problem of the temperate- zone. Entire plant dies due to the infection under cool weather. The disease is seed and soilborne. Severely infected seeds do not germinate or produce weak seedlings.

Symptoms: The infected pale looking plants exhibit wilting and drooping of leaves in the field. The infected roots turn brown black and large spots (lesions) develop. Under advance situations the epiblema of root tissue cracks, exposing the discoloured below tissues. Secondary invasion of micro-organisms result in complete rotting of the basal portion. The major root infection is usually recorded just below the soil. A black zone may be observed on the roots.

MOSAIC OF LETTUCE

Causal pathogen: Lettuce virus 1(*Marmor lactucae* Holmes)

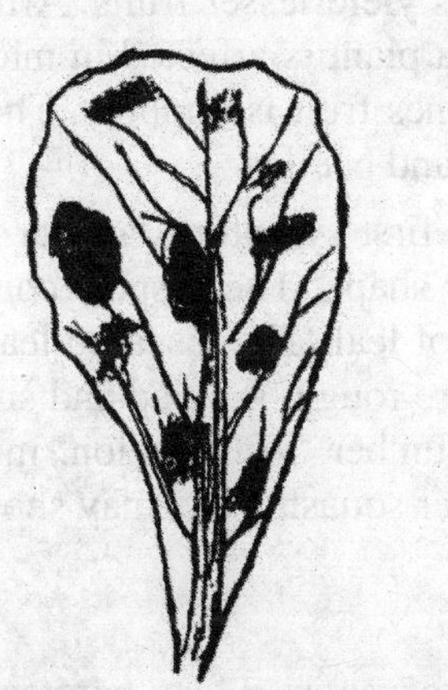

Significance: Under cloudy weather conditions the diseases is more destructive. The disease reduces the marketability of leaves. The extent of loses may vary depending upon the aphid vector population. Diseased plants yield less leaves and sterile heads. The disease is seedborne and transmitted through species of Aphids, which include *Myzus persicae, Macrosiphum euphorbiae* and *Aphis gossypii*.

Symptoms: Severe stunting of seedlings and young plants is due to seedborne infection. The first true leaves are often irregularly shaped and slightly bilobed followed by pale green and mottling. Mottling is most noticeable in cool cloudy weather, especially when lettuce leaves are viewed by transmitted light. Vein clearing and brown flecks along the smaller vein-lets are clear. Leaf blade tissue death/necrosis may be recorded in some varieties. The leaf margin may become excessively ruffled, distorted. The tip of the leaf rolls back. In old plant symptom, the light green to yellow mottling may disappear. Infected plants are more prostrate than healthy ones. Under sever conditions the infected plants fail to head or produce small loose heads of poor quality. The seed stalks of mosaic infected plants have mottled and frilled leaves. The seed stalk is usually stunted. The plants infected at early growth stage are normally less productive, yellow in colour and produce sterile heads.

MOSAIC OF MUSK MELON AND SUMMER SQUASH

Causal pathogen: Cucumber mosaic virus

Significance: Several strains of mosaic infect the crop. Early plant infection leads to heavy losses. Systemically affected vines yield lesser fruits. Affected fruits have low marketability due to size, rough skin and colour. If a plant is affected in mid season the growth remain normal and produces healthy fruit but sometimes fruit is knobbly. The virus is seedborne and transmitted by several vectors, chiefly by aphids and beetles.

Symptoms: On older plants, first symptoms appear on younger leaves with alternate green and yellow patches with irregular shape. These spots contain viral particles and enlarge rapidly ultimately covering entire lamina of leaf. The diseased leaves are mottled, deformed curled down wards and smaller. Young fruits are rough, mottled and small in size. The virus is responsible for warts on pale green fruits of cucumber. Watermelon, muskmelon and winter squash fruits are often stunted mottled while summer squash fruits may show warts and rings.

WILT OF SWEET POTATO

Causal pathogen: *Fusarium oxysporum* f.sp. *batatas*

Significance: The disease is also known as vascular wilt, surface rot and stem rot. Losses due to the disease usually range from 10-15 per cent. In storage the fungus causes the surface rot and provides the entry to another organisms resulting in substantial losses. The disease is vine, tuber and soil-borne in nature

Symptoms: The young leaves of plant, first become dull green, turn yellow between veins, wilt and remain attach to the plant. Older leaves drop off. The vascular bundles of infected stem become dark brown to black. Stem discolouration may extend upto several centimetres from the roots and easily seen when the outer layer of the stem is peeled away. During dry weather infected stem may swell and split near the ground line. Invasion to the tubers (potatoes) may take place from infected vascular bundles of infected vines.

BLACK ROT OF SWEET POTATO

Causal pathogen: *Ceratocystis fimbriata*

Significance: The disease is also known as black shank, black shoot and black root. The disease may appear in plant, seed- bed. The disease frequently develops on potatoes, which have been washed just before transportation to market or store. The infected/diseased potatoes have bitterness in taste. Single infected potato is capable of rotting the lot in storage under favourable conditions.

Symptoms: Disease appears as small, almost circular, slightly sunken, firm, dark brown spots or lesions. These lesions become greenish-black to black as they enlarge turning grey black with age. The flesh beneath the skin is black and under favourable conditions, it extends deeper in tubers. Small black spots (lesions) appear on the stem portion near the point of attachment.

30.2 OBJECTIVE: VERIFICATION OF DISEASES AT SEED LEVEL

KARNAL BUNT OF WHEAT AND TRITICALE

Causal pathogen: *Neovossia indica* (= *Tilletia indica*)

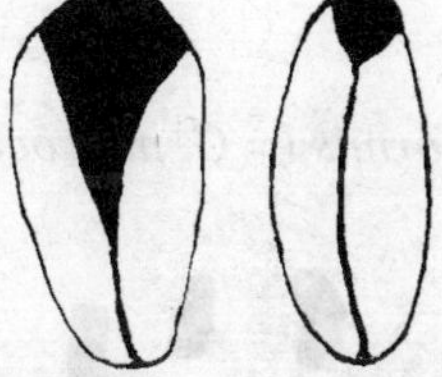

Significance: The disease normally appears sporadically causing losses from 0.3 to 0.5 per cent, however, maximum losses up to 20% have been recorded under epidemic conditions. The disease also adversely affects the seed vigour, seed germination, seed weight and quality of wheat flour. Karnal bunt infected wheat grain lots are not accepted by many countries, which is a cause of concern in wheat export.

Symptoms: The diagnostic symptom of the disease is blackening of the seeds at germinal end and becomes slightly swollen which, gives a silvery and shiny appearance. The grains are partially black and contain powdery mass of teliospores of the fungus. In heavily infected seed, whole seed is converted into black powdery mass, except its pericarp and alueron layer (seed wall). The dorsal side of the seed remains unaffected while blackening spreads along the groove. Diseased seeds are lighter in weight, discoloured and deshaped. Infected seeds omit a bad odour similar to rotten fish due to the presence of a chemical-trimethylamine.

BUNT OF RICE

Causal pathogen: *Tilletia barclayana* (=*Neovossia barclayana* = *N. horrida* = *Tilletia horrida*)

Significance: Once considered of minor importance is now a potential threat for healthy rice seed production especially in north Indian conditions. Primarily the disease is seed-borne and carried by seed, thereafter it establishes in the soil.

Symptoms: In a panicle, few grains are converted into black powdery mass of bunt spores. The infected seeds often show greenish black discolouration, which can be detected during visual observations of the seed lot. Disease appears at the time of crop maturity. In severe infections rupturing glumes show beak like fungal out growth. With the bursting of glumes the black powder of teliospores scatter on to the other seeds and leaves. It is visible as black powder. The spore mass is produced in the ovary.

ERGOT OF PEARL MILLET

Causal pathogen: *Claviceps fusiformis* (= *C. microcephala*)

Significance: The disease is reported from all over the country. Quality of grains is also adversely affected. The ergot (sclerotia) contains *ergotin* alkaloid and assimilation of this poison in the blood system causes convulsions, nervous breakdown, shrinkage of cells and many more symptoms in cattle and human beings. Hence, the disease has an additional importance. The damage caused by the disease is weather dependent. The loss in grain yield is directly proportional to per cent infection, as infected ovary is completely transformed into a sclerotium. The pathogen is observed as concomitant contamination.

Symptoms: The disease is visible at flowering stage. The diseased spikelet (florets) exudates the droplets of pink or honey coloured sticky liquid which turns darker and cover large area of the inflorescence. The honey- dew contains spore mass. Under advanced conditions infected kernels are transformed into straight or curved (1-2 cm), cremate light brown horny structure with sclerotia projecting between the glumes as ergot. The floral ovary is transformed into sclerotial body, which attains a large size. It is visible due to size, colour and shape.

ERGOT OF SORGHUM

Causal pathogen: *Claviceps sorghi*

Significance: Recent studies indicate that ergot is a serious limiting factor in the production programme of hybrid seed. Particularly if seed set in male sterile line is delayed due to lack of viable pollen caused by non-synchronous flowering of male sterile and restorer lines. In areas, where hybrid seed are in great demand but in short supply the damage is two folds, directly through loss in seed yield and indirectly through rejection of ergot sclerotia contaminated seeds. The pathogen causes losses to male fertile cultivars. The adverse effect of ergotin alkaloid in the sclerotial bodies mixed with the seeds has very well documented. The disease is seed-borne and recorded as concomitant contamination.

Symptoms: The first field symptoms of the disease are commonly conspicuous from blossoming stage of the crop. The infection is marked by the exudation of pink drops from infected ovary and known as sugary disease. The sticky pink droplets contain several fungal conidia. Under humid conditions the pink sugary mass exudates from infected ovaries is attacked and colonized by several other saprophytic fungi and convert the conidial mass into black mass. As the infection and colonization of the ovaries progresses, the fungal stroma under warm conditions is transformed into a hard over seasoning structure called sclerotia.

WILT OF SWEET POTATO

Causal pathogen: *Fusarium oxysporum* f.sp. *batatas*

Significance: The disease is also known as vascular wilt, surface rot and stem rot. Losses due to the disease usually range from 10-15 per cent. In storage the fungus causes the surface rot and provides the entry to another organisms resulting in substantial losses. The disease is vine, tuber and soil-borne in nature

Symptoms: The young leaves of plant, first become dull green, turn yellow between veins, wilt and remain attach to the plant. Older leaves drop off. The vascular bundles of infected stem become dark brown to black. Stem discolouration may extend up to several centimetres from the roots and easily seen when the outer layer of the stem is peeled away. During dry weather infected stem may swell and split near the ground line. Invasion to the tubers (potatoes) may take place from infected vascular bundles of infected vines.

SCURF OF SWEET POTATO

Causal pathogen: *Monilochaetes infuscans*

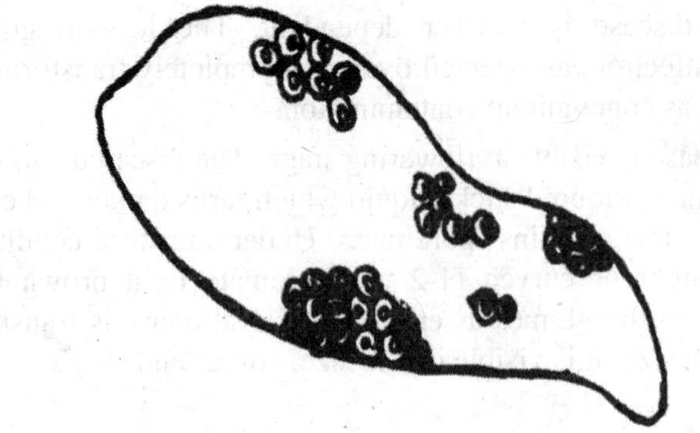

Significance: The disease is also known as rust, black skin or soil stain or Jersey mark. The disease is more prevalent in heavy and alkaline soils Practically the disease doesn't cause any loss but cause a rusty brown to black stain on the skin that adversely affect the marketability. Badly infected tubers may become shrivelled, spongy.

Symptoms: Brown small speck on the root appears as first symptom. Scurf is easily identified by a rusty brown to an almost black discolouration of the potato skin. The lesions may be a few to indefinite in number. The infection is superficial. In severe conditions tubers become corky and leathery. Due to moisture loss from skin, cracks may develop and secondary organism gets the chances to enter.

STORAGE ROTS OF SWEET POTATO

Causal pathogen: *Ceratocystis fimbriata* (black rot pathogen), *Rhizopus stolonifer* (soft rot), *Fusarium oxysporum* (surface rot), *Diaporthe phaseolorum* (dry or end rot), *Diplodia theobromae* (Java black rot).

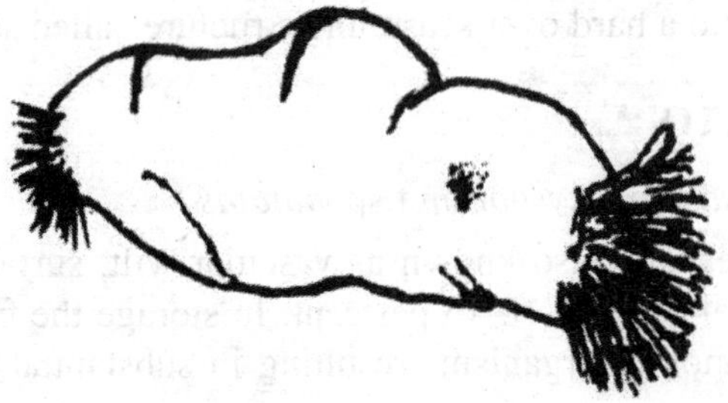

Significance: Huge losses due to storage rot fungi have been observed throughout the world. The rots not only cause the decay of the tubers but also adversely affect the marketability and taste.

Symptoms: The black rot causing fungus (*C. fimbriata*) attack both healthy and injured flashy roots. Symptoms include small circular, depressed, dark brown spots. Minute, black, flask shaped, bristly fruiting bodies (perithecia) are formed abundantly on the infected host part. The flesh below the skin or infected parts is blackened. It extends further deep.

The fungus (*R. stolonifer*) induces soft rot in wounded potatoes. Two types of symptoms i.e., soft rot and ring rot have been reported. Soft rot usually develops on any end of the potato resulting in spongy soft watery rot. Infected tubers become dark brown with mild odour. Straw coloured liquid ooze out from the cracked skin. The tubers become shrunken and brittle. In ring rot, the infection starts at any point in between the end-points and forms rings on the surface.

The surface rot causing fungus (*Fusarium oxysporum*) is severe in potatoes, which are improperly sorted, cured and stored. The fungus causes surface rot, which appear as small, circular, shallow, grey brown lesions. The area immediately around the lesion shrinks. The entire root shrivels, dries and mummifies. Moisture favours the disease development.

The dry or end rot causing fungus (*Diaporthe phaseolorum*) attacks the flashy roots in the field. The storage rot symptoms include brown rot which is coal black with small pimple like mounds. The tuber slowly dries up.

The Java black rot (*Diplodia theobromae*) symptoms include brownish infected tissues, later turning coal black, dry, hard, brittle and difficult to break. Infected parts contain large number of black fruiting bodies.

PHYTOPHTHORA BLIGHT OF TARO

Causal pathogen: *Phytophthora colocasiae*

Significance: Disease is favoured by high moisture, cloudy conditions and more prevalent in poorly drained soils.

Symptoms: Infection results in the development of small, dark, water soaked, round spots on the leaves of colocassia. Under high humidity (80-90 per cent) the spot rapidly increases and cover large area of the leaf. Under comparative dry conditions the leaf spot, blight become brittle and dry, and finally drop-off exhibiting shot hole symptoms .The periphery of the spot is zoned in different shades of brown, green and yellow. This is the main diagnostic character for identification of the disease in field and has quite resemblance with late blight symptoms in potato.

LATE BLIGHT OF POTATO

Causal pathogen: *Phytophthora infestans*

Significance: The development and spread of the disease is extremely fast and unless the prophylactic measures are adopted, it cannot be stopped. The problem is more acute in areas with cooler and moist humid conditions resulting in dry or wet rot expression of the disease. Under favourable conditions of temperature and humidity, the disease is capable of causing cent per cent losses. The disease is tuber and soilborne.

Symptoms: Potato tubers get infection under field condition. The skin of tuber becomes discoloured It turns initially brown then purple. The internal matter rots and turns black. Complex infection due to bacterial and fungal infection results into pulpyness of tuber. The tuber is rotted due to downward systemic infection. The soft slimy decay is generally the results of the bacterial infection, which often follow development of late blight pathogen. In moist atmosphere, white tufts of mycelium and sporangiophores of the fungus appear on the surface of tubers.

COMMON SCAB OF POTATO

Causal pathogen: *Streptomyces scabies*

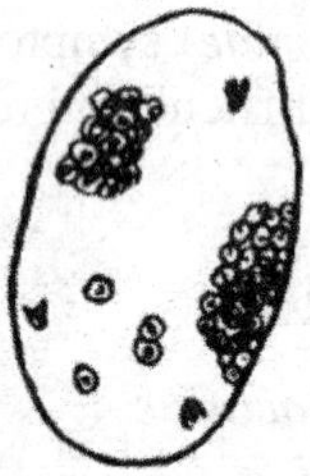

Significance: The shallow and deep scabbing on potato skin results in lower market quality. The infection leads to disfigure potato ultimately poor customer appeal. Losses up to 16 per cent have been experienced in severely affected soils especially in Uttar Pradesh and Punjab.

Symptoms: Scab appears as small, brownish specks or lesions on the skin of potato tuber. Later these specks become corky rough, may become slightly raised. Some lesions may be sunken and quite deep-pitted symptoms are observed

BLACK SCURF OF POTATO

Causal pathogen: *Rhizoctonia solani*

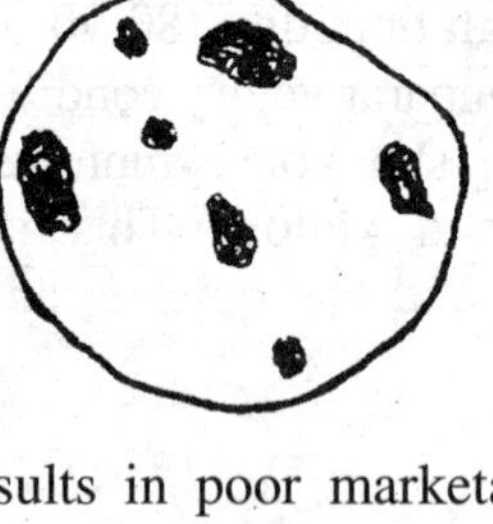

Significance: Black scurfing results in poor marketability. Fungal infection reduces the carbohydrate quality of tubers.

Symptoms: In stem canker, when the seed tuber sprouts, the growing tip is attacked by the fungus, which results in pre-emergence mortality. A dormant bud near the base may spout but it may also be killed or may give rise to stunted and yellowish seedling. The delayed emergence and reduced vigour of the seedling causes reduction in yield. The aerial green tubers are formed on the axis of branches and petioles. Under cool and wet conditions, development of grey fungal growth is observed on the base of stem. Normally the tuber infection is skin deep and looks as the crust of soils adhering to the tubers.

RED ROT OF SUGARCANE

Causal pathogen: *Colletotrichum falcatum*

Significance: It is the most destructive disease in the tropics and subtropics. Severe epidemics have occurred in northern region of India leading to substantial losses in yields and quality of the canes.

Symptoms: Earliest symptoms are loss of colour and drooping of third or fourth leaf that become shrivelled and the rind are lighter in weight. It can be broken easily. Upon longitudinal splitting of the diseased canes, the pith looks red coloured. In advanced cases red colour is replaced by dirty brown with white bands. Fungal fruiting bodies are formed near the nodes, on red patches with ash coloured centre which are observed at maturity of the crop.

SMUT OF SUGARCANE

Causal pathogen: *Ustilago scitaminea*

Significance: The smut is a primarily problem of wild canes and of those varieties which nearly approach the wild as compared to improved and thick varieties. Negative correlation exists between percent smut and yield. The disease appeared in early stage leads to the total loss as compared mid and late season of crop growth.

Symptoms: Due to systemic infection affected plants produce whip like black long and curved shoot. On the whip, the black powdery mass (smut spores) is initially covered with thin silvery membrane that releases spores upon rupturing. Affected plants have slender and thin canes. Primary infection is through seed sets and secondary through dissemination of fungal spores during crop growth season.

WILT OF SUGARCANE

Causal pathogen: *Cephalosporium sacchari*

Significance: The disease has been spread due to unrestricted movement of diseased seed canes and improper certification measures. it is a major problem in northern India, however, epidemics have been noticed in Tamil Nadu too. The disease adversely affects the germinability, yield and juice quality. Losses up to 56% have been reported.

Symptoms: The earliest symptom is stunting and unthrifty appearance. Root system does not reveal the damage or discoloration which can be attributed to the disease. The most striking symptoms is yellowing, drying and withering of top portion of the plant. The cane become light and hallow. When cane is split in early stages the lowest internodes have brick red colour There are no white transverse bands are seen as in the case of red rot disease. Gradually the pith becomes hollow and there is usually a copies greyish white, fluffy growth of hyphae is noticed.

GRASSY SHOOT OF SUGARCANE

Causal pathogen: Mycoplasma like Organism

Significance: First reported from Maharashtra in 1919, the problem is wide spread in major sugarcane growing areas of the country. Loss in juice quality, and yield reduction is common.

Symptoms: Profuse tillering and grassy appearance of stools is the chief symptom. The infected leaves become narrow and small, like grass leaves. The canes in infected plants are thin with short internodes giving bunchy appearance to clump. The leaves exhibit straight, long white or light green or yellow streaks. The lower nodes produce large number of grassy shoots.

RHIZOME ROT OF GINGER

Causal pathogen: *Pythium aphanidermatum*

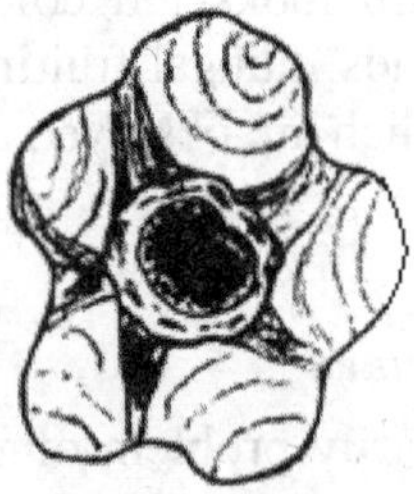

Significance: It is a wide spread problem all over the ginger growing areas. Under severe state of conditions, the pathogen is cable of causing losses up to 80% under fields where as about 90% rotting of rhizomes has been reported in storage conditions. In ill-drained fields cent per cent losses are not uncommon.

Symptoms: The young shoots of infected rhizomes show the damping-off symptoms. In early stage leaves turn yellow and cholorosis proceed downwards. This results withering and death of leaf. The central portion of leaf remains green under normal conditions. The base portion of the pesudostem exhibits translucent brown discolorations. The portion just above the ground level become watery and soft. The collar region becomes mass of decaying tissues enclosed by the comparatively tough rind.

PHYLLOSTICTA LEAF SPOT OF GINGER

Causal pathogen: *Phyllosticta zingiberi*

Significance: The disease is prevalent in central Indian condition. Due to infection the photosynthetic area is decreased that results in poor yield with low quality of marketable produce.

Symptoms: The disease can be identified on the basis of formation of minute colourless specks on the lamina. The specks later on turn in to oval to elongated irregular conspicuous spots with white centre and brown margins having extended yellow hallo. Numerous fungal fruiting bodies are observed on the on the lamina. In sever infection blighted appearance is noticed from a distance.

BACTERIAL WILT OF GINGER AND TURMERIC

Causal pathogen: *Ralstonia solanacearum* (*Pseudomonas solanacearum*)

Significance: The pathogen is seed and soilborne in nature. The crop is not eligible for certification if grown in land infested with the bacterium. Inspections are made to verify the presence of disease. The bacterium has the wide host range and able to infect several vegetables including potato, tomato, capsicum and egg plant.

Symptoms: The aerial and underground plant parts are affected. Sudden wilting and death of plants within 24-48 hours is the chief symptom. Brown vascular discoloration appears in the water conducting tissues. Due the production of enzymes by the bacterial pathogen, softening of rhizome is observed. The rhizomes become pulpy and watery under severe infection.

30.3 VERIFICATION OF DISEASES WITH NO SPECIFIED STANDARDS AT FIELD LEVEL

BUNCHY TOP OF BANANA

Causal pathogen: Banana Bunchy Top Virus

Significance: It is major wide spread disease problem in banana plantation in Kerala state.

Symptoms: The disease can be observed at any stage of growth. The leaves become typically a bunched and at the apex form a dense rosette. The young infected plants are usually stunted and show more erect leaves than normal. In a primary type of infection growth of the plant remain slow, plant is unable to attain the height of 2 feet and unable to produce fruits. In secondary infection plant exhibit the similar symptoms with lesser intensity. Older leaves show dwarfing, marginal necrosis and curling is common. Green, pale white streaks are observed on the leaves

PANAMA DISEASE OF BANANA

Causal pathogen: *Fusarium oxysporum* f.sp. *cubense*

Significance: Also known as Fusarium wilt is major wide spread disease problem in banana plantation world wide

Symptoms: The disease can be verified by the presence of faint yellow streaks in the petioles of the oldest lower leaves. The progressive yellowing kills the petiole. In non-yellowing type, no cholorosis is observed. Discoloured vascular strands with light yellow to dark brown are the distinguishing internal symptoms. Rootlets from rhizome are blackened and get decayed due to fungal invasion.

ANTHRACNOSE OF GRAPE

Causal pathogen: *Elsinoe ampelina*

Significance: It is major wide spread disease problem in grapevine yards throughout the world. The infected barriers are mummified and losses to the barriers are enormous and the development of the disease is weather dependent.

Symptoms: All the aerial parts are affected at any stage of crop growth. Typical symptoms include grey black spots with red brown margins which develop rapidly on growing leaves. Affected leaves become tattered and shot hole symptoms are developed. Brown to black spot are formed on stem. The sunken cankerous spot coalesce and stem are girdled and killed. The fleshes of infected berries remain firm. Infected berries are mummified in clusters.

■■■■

Chapter 31

IDENTIFICATION OF OBJECTIONABLE INSECT PEST

OBJECTIVE

Objectionable insect pest in seed certification and their identification

Some crops with vegetative reproduction have designated objectionable insect pest at seed level some at field level and in the event of more number of insects than the prescribed standard the sample is rejected (Table 31.1 and 2).

Table 31.1 Maximum seed certification standard for infestation of seed with objectionable insects

Crop	Insect	Scientific name
Colocasia	Scale insect	*Aspidiella hartii, Aspiditous destructor*
	Mealy bug	*Geococcus coffea, Planococcus dioscoreae*
Lesser yam	Scale	*Aspidiella hartii, Aspiditous destructor*
Sweet potato	Wireworm	*Cylas formicarius*
	Weevil	*Cylas formicaris*
Tapioca	Scale insect	*Aspidiella hartii, Aspiditous destructor*

Objectionable insect pest of crops at field level

Crop	Insect	Scientific name
Colocasia	Scale insect	*Aspidiella hartii, Aspiditous destructor*
	Mealy bug	*Geococcus coffea, Planococcus dioscoreae*
Lesser yam	Scale	*Aspidiella hartii, Aspiditous destructor*
Sweet potato	Wireworm	*Cylas formicarius*
	Weevil	*Cylas formicaris*
Tapioca	Scale insect	*Aspidiella hartii, Aspiditous destructor*
Sugarcane		
Gurdaspur borer		*Acigona steniellus*
Plassey borer		*Chilo tumidicostalis*
Mealy bug		*Sacchariphagus sacchari*
Scale insect		*Melanaspis glomerata*
Internode borer		*Chilo sacchariphagus indicus*
Stalk borer		*Chilo auricilius*
Top borer		*Scirpophaga excerptalis*

Scale insect of colocasia and lesser yam : Two types of scale insects viz., *Aspidiella hartii* cockerell and *Aspiditous destructor* Signoret (Hemiptera : Didspidae) affect the lesser yam, coolocasia and tapioca. Whereas *Melanaspis glomerata* is associated with sugarcane. Nymphs and adult are whitish yellow in colour, present usually in a group, on the tuber or on the aerial portion of the plant. Female insect remains attached to the host plant even with tuber under protection of whitish yellow scales. The problem is perpetuated by the use of infected tubers as planting material. To remove the insect from tuber, individual tuber should be scraped and should be treated with rogor 0.03% solution just before planting.

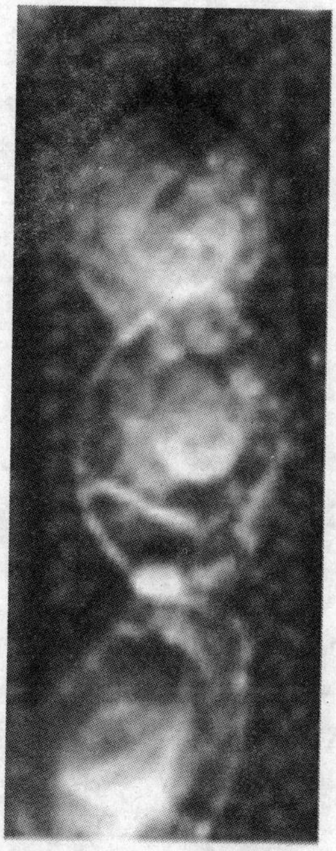

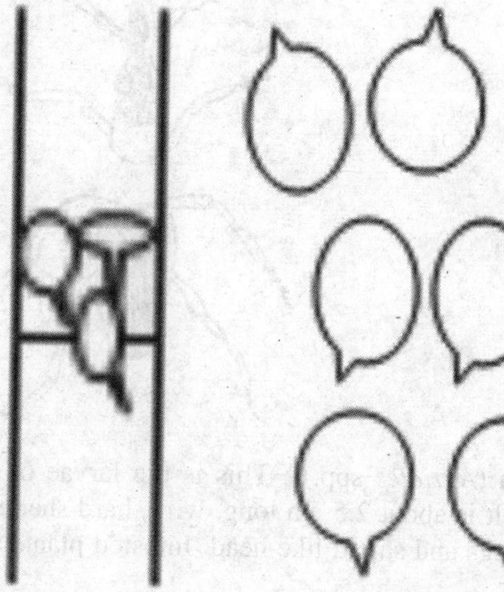

Mealy bug of lesser yam (*Geococcus coffeae, Planococcus dioscoreae* Hemiptera Pseudococcidae): Mealy bug feeds on the root and tubers of yam. Adult female is small, segmented and whitish grey. Wingless insects are covered in a mealy wax secretion. Mealy bug may present with harvested yam and can multiply to considerable number causing shrivelling on stored yam. Affected yam are not able to sprout. Only clean healthy tubers should be used for planting. Chemical treatment recommended for scale insects is also effective against mealy bug.

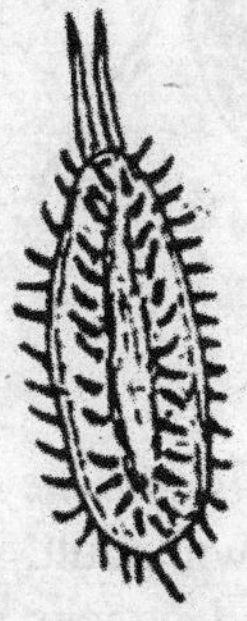

Sweet potato weevil (*Cylas formicarius* Faabr. Coleoptera: Curculionidae): The grub of this weevil bore into stem and female lays egg in tunnels made on tuber. The tunnels may be partially filled with frass. Weevils are ant like, with glassy, blue, black elytra, reddish brown legs and thorax and a black head. Weevils may be found on the leaves or in tunnel.Production of seed should be taken in the field free from weevil infestation. Planting material should be free from weevil. After harvesting field should be cleaned and trashes should be burnt. The seed production plot infected with weevil should be sprayed with 0.2% carbaxyl followed by 0.1% malathian at 10 days interval.

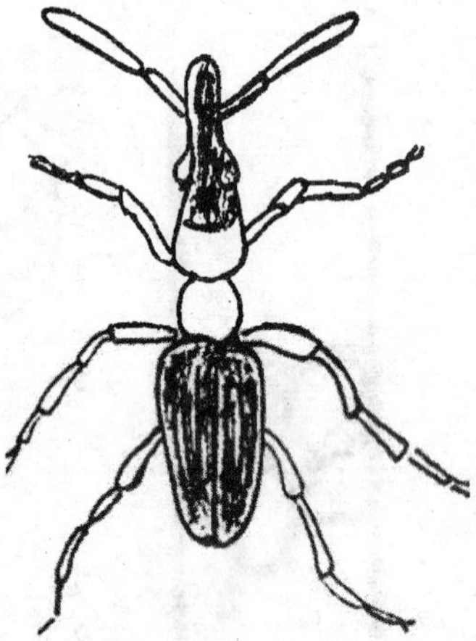

Wire worm (*Agriotes* spp.) : This is the larvae of click beetle. It lives in soil and is the damaging stage. It is about 2.5 cm long, wiry, hard shelled, shiny, smooth, and golden yellow in colour with six legs and shield like head. Infested plants show wilting of leaves and round holes in root.

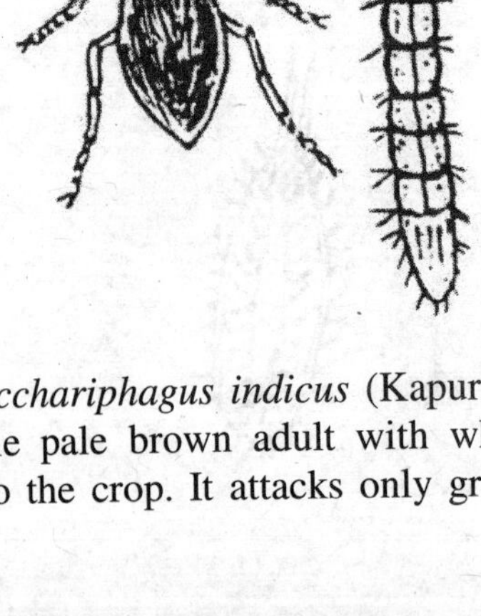

Internode borer (*Chilo sacchariphagus indicus* (Kapur)): Four violet or pink stripes and light brown head of larva while pale brown adult with white hind wings. The pest causes enormous economical damage to the crop. It attacks only grown up crop (from 90th day after planting or ratooning) till harvest and the loss caused by this pest alone ranged from 10 to 35 per cent. Caterpillar bore holes in the soft internodes of the cane with dead-heart symptoms where leaf sheath is still soft and attached tightly to the cane. Canes with dead hearts are shorter and thinner than healthy canes. Affected tissues are red in colour. Hardening of the cane after infestation force the larvae to search new cane. After coming out of the old cane it makes tunnel in the soft internode of the same cane or the adjacent cane. The bored canes show sick appearance with a constriction of the canes and shortening of the internode. Sometimes aerial roots started emerging from the bored canes.

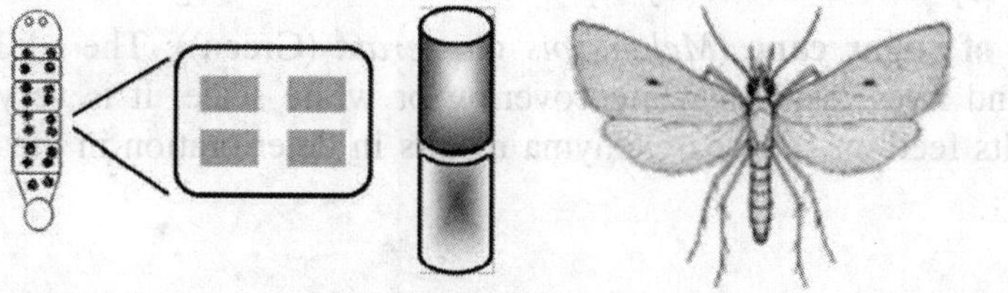

Stalk borer (*Chilo auricilius* (Ddgn.): The moth is of staw colour, Female has more wing expanse (22-30 mm) than male (16-25 mm). Golden strips are present on the forewing while the colour of hind wings is silvery white. Fully developed larva is 25-30 mm long with bluish pink body; dark brown head with five violet dorsal stripes. Larva bores in the intrenode and move from plant to plant. It causes 16% reduction in cane yield.

Plassey borer *(Chilo tumidicostalis* (Hmpsn.): Deterioration in quantity and quality of juice due to damage by caterpillar.

Gurdaspur borer (*Acigona steniellus*): Grayish brown moth with dark spots on forewings, whereas, hind wings are grayish white in colour. The full grown larvae (30-35 mm) is of pinkish brown in color with four violet stripes on the back. Caterpillars bores in the internode, and makes spiral tunnel in the upward direction up to the bud. It makes the cane weak and dry and prone to lodging.

Mealy bug of sugar cane *(Sacchariphagus sacchari*): These are curved insects with sac-like segmented body covered with white mealy powder found beneath leaf sheath on the nodes. The damage is done by both nymphs and adults by sucking the sap. Quantity and quality of juice is deteriorated due to infestation. A sooty mould develops on the secretions at the sight of puncture giving a blackish appearance to the canes.

Top borer *(Scirpophaga excerptalis)*: The adult moth is silvery white with slender and pointed abdomen of male and stout and covered with crimson or orange hair in female. The antennae are filliform and tarsai segmented. The hairy forewings are narrow. The moths are nocturnal in habit. Smooth, whitish cream coloured caterpillar with a red coloured mid dorsal line penetrates into the midrib from the underside of leaf and It goes to the spindle and moves towards the growing points and destroys the top portion of the plants resulting in the formation of dead heart. A branched top known as bunchy top is formed due to damage of central shoot.

Scale insect of sugar cane (*Melanaspis glomerata* (Green)): The adult female is without legs and wings and lives inside a white covering or white scale. It is very small in size. The Nymphs and adults feed on stem parenchyma results in deterioration in the quality of juice and drying of cane.

■■■■

Chapter 32

IDENTIFICATION OF OBJECTIONABLE PARASITIC WEEDS

Some crops have designated objectionable parasitic weeds (Table 1) at seed and field levels, in the event of more number than the prescribed standard the sample is rejected.

Table 32. 1. Objectionable parasitic weed at field and seed level

Crop	Parasitic weeds
Field level	
Sunflower	*Orobanche cumna*
Lucerne	Dodder (*Cuscutta* spp.)
Napier Grass	Dodder (*Cuscutta* spp.)
Seed level	
Lucerne	Dodder (*Cuscutta* spp.)
Napier Grass	Dodder (*Cuscutta* spp.)

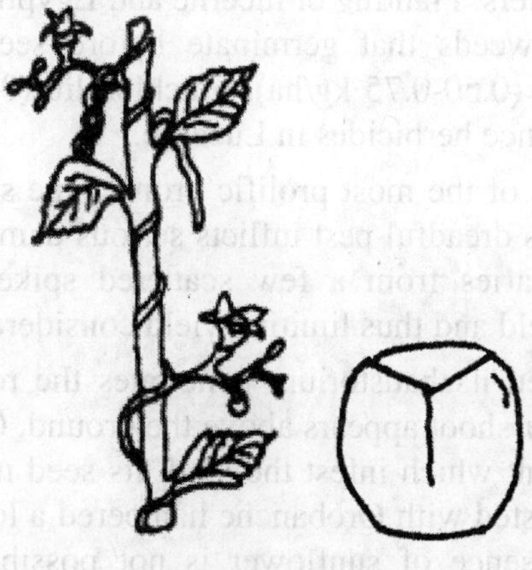

Cuscutta

***Dodder Cuscutta* spp.:** It is an objectionable parasite of Lucerne and Napier grass at field and seed level. Dodder belongs to family Convolvulaceae. Thread like, twining, slender and yellow coloured stem; leaflets reduced to a few very minute scales; flowers small, yellowish-white and present in cluster. It is a dicot seed without cotyledon. Seeds are shed when the embryos are extremely small and contain only a few cells, and completion of development occurs afterward. Seeds are small, oval to oblong, finely granular, scurfy-rough surface, triangular, seed coat colour light brown, hilum short oblong.

Orobanche

It can be controlled effectively with application of herbicide like Pendemethiline. Crop rotation with cultivation of wheat or chickpea in winter is the most effective method to eliminate *Cuscuta.* Seed of the crop is dipped in 5-10% solution of table salt for five minutes to remove seed of Cuscuta as it floats, while the crop seeds settle down at the bottom. The crop seed is washed thoroughly with fresh water to remove traces of salt. The dodder seeds are separated from Lucerne with two felt rollers as *Cuscuta* seeds stick to the felt, whereas the Lucerne seeds being smooth fall away from the rollers. Planting of lucerne and Egyptian clover in stale seedbeds using 1.0 % Paraquat to destroy weeds that germinate before seeding of the crops is effective. Application of Pendimethalin (0.50-0.75 kg/ha), Fluchloralin (0.75 kg/ha), and Oxadiazon (0.50 kg/ha) is effective pre-emergence herbicides in Lucerne.

***Orobanche* spp** It is one of the most prolific brome rape species in the world and has host preference for sunflower. This dreadful pest inflicts serious damage on sunflower and is difficult to exterminate. Infestation varies from a few scattered spikes to a large population almost attacking every plant in the field and thus limiting yield considerably.

After germination of seed its haustorium penetrates the root of host and its light yellow, reddish brown coloured flower shoot appears above the ground. One plant of parasite can produce 0.5 to 1 million seeds per plant which infest the land. Its seed may remain viable upto 18 years. The growth of crop plant infested with Orobanche hampered a lot. Control of orobanche with the help of herbicide in the presence of sunflower is not possible. Low dose of Glyphosate or Propyzamide or Pronamide is effective against Orobanche. Soil fumigation (methyl bromide or metham sodium) is recommended to kill seeds of orobanche. In the infested field green-manuring with susceptible host crop is recommended so that after germination and attachment with green manure plant, the orobanche will be destroyed by mixing of green manure in soil.

Chapter 33

IDENTIFICATION OF OBJECTIONABLE NEMATODES

Some crops have designated objectionable nematodes at seed and field levels, in the event of more number than the prescribed standard the sample is rejected (Table 1).

Table 1. Objectionable nematodes in seed certification

Crop	Nematode	Disease
Seed level		
Wheat	*Anguina tritici*	Ear cockle
Sweet potato	*Meloidogyne incognita*	
Ginger and Turmeric	*Meloidogyne incognita*	
Field level		
Banana	Nematode	
Grape	Nematode	

EAR-COCKLE OF WHEAT

Causal pathogen: *Anguina tritici* nematode and in association of bacterium *Clavibacter toxicans* causes Tundu disease.

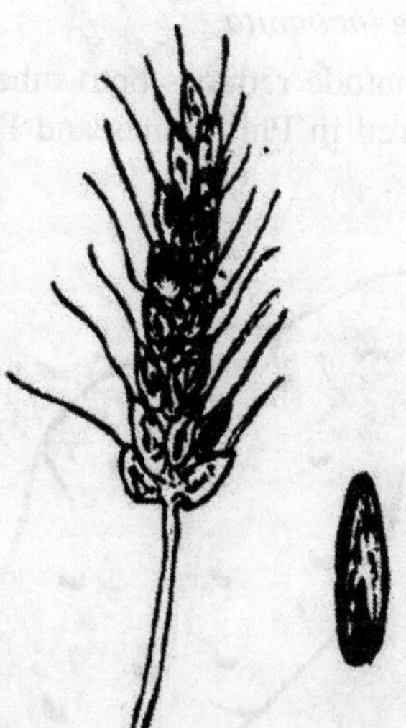

Significance: After adoption of proper seed cleaning methods the incidence in Western countries it is practically eliminated. Now disease is restricted to West Asia, China, Pakistan and Ethiopia. In India the disease has been observed in all the major wheat growing areas including parts of Haryana, Punjab, Rajasthan, Uttar Pradesh, Madhya Pradesh and Bihar. Recently in Madhya Pradesh losses up to 90 per cent have been observed in Panna district at farmer's field.

Symptoms: In diseased plants, earheads emerge 30-40 days earlier with spreading glumes. The infected earhead remains green and short. Grains are replaced with loosely attached seed gall. Galls are normally hard and dark brown in colour. Infested seedlings have twisted and crinkled leaves with swollen enlarged basal portion, more conspicuous at 20 days old seedling stage. Seed galls are black-brown which contain nematodes. Seed galls (ear cockle) are hard and contain second stage, juvenile (J_2) in dormant stage and nematodes may range from 800 to 35000 per gall. The nematode adult male measuring 3-5 mm in length and adult female 2.6 -4.4 mm long and wider than males.

NEMATODE OF GINGER AND TURMERIC

Causal pathogen: *Meloidogyne incognita*

Significance: The polyphagous nematode is a wide spread problem throughout the country. Infection results in poor yield with low quality of marketable produce. The infected rhizomes are under sized and lesser in number as compared to the rhizomes from normal healthy plants. As per the Indian Seed Act, planting should be avoided in infected fields. Specifications for Foundation and Certified seed crop are not prescribed for nematode of ginger and turmeric crop.

Symptoms: The typical root galls are observed in the infected plants. The plants become pale, look as undernourished and water deceit. The infected plants are also attacked by secondary pathogens because of the easy active passage entry available made by nematode.

NEMATODE OF SWEET POTATO

Causal pathogen: *Meloidogyne incognita*

Significance: The root rot nematode reduces both tuber yield and quality. In infested field 18 to 50% losses have been measured in Philippines and Papua New Guinea. The nematode is tuber and soil borne in nature.

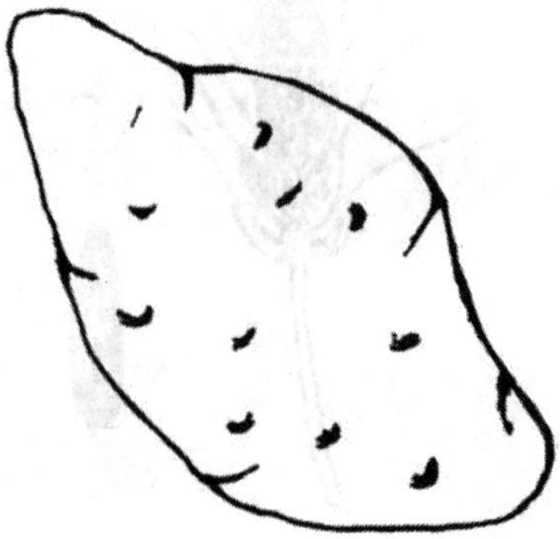

Symptoms: Development of longitudinal cracks and blisters like bumps on the tuber. Nematode is responsible for roughing and cracking of tubers and generalized decay of the root system. The nematode survives on roots in plant debris and spread through planting material.

■■■■

Chapter 34

SEED QUALITY ENHANCEMENT TECHNIQUES

Seed is treated to enhance seed longevity, establish optimum and uniform plant population of healthy plants with reduced seed rate; incorporate tolerance against biotic and abiotic stresses at the time of seedling establishment; facilitate in precision mechanical planting.

SEED INVIGORATION

Objective

To maintain longevity of seed under ambient condition by physio-chemical treatment of seed.

Deterioration of seed starts form physiological maturity and continued till sowing of the seed. The age of the seed is minimum at the time of physiological maturity and maximum at sowing. The seed is deteriorated with age (ageing) due to influence of many internal and external factors. The ageing of seed in store is always fast under humid and hot climate. The basic internal cause of seed deterioration is loss of membrane function (Tappel, 1973) and production of free radicals (Basu, 1976). The process of ageing cannot be stopped, but may be delayed by reducing its rate. The production of free radicals may be scavenging by hydration-dehydration (H-DH) treatment.

An improvement in seed performance by any post-harvest physio-chemical treatment resulting in better storability, improved germination and field performance over a wide range of edapho-climatic conditions than the corresponding untreated seed is considered as seed invigoration.

Pre storage treatments for seed invigoration

Seed invigoration treatments are given to the properly cleaned and dried seeds before storage to protect from deterioration during storage, enhance shelf life and improve performance of deteriorated seed. It is of following type:

- Halogenation
- Antioxidant treatment

Halogenation

Elements of Group XVII of periodic table are known as Halogen. Halogens are non-metals and become less reactive as move down to the group therefore Fluorine is the most reactive.

Chlorine, bromine and iodine have similar chemical properties as all of them are non-metal. All these elements react with water to form acid and all of them have one valence. Halogens show similar properties because of electrons of similar valence *i.e.,* seven. It accepts one electron easily to complete their octet and form negative ions like F^-, Cl^-, Br^- and I^- with one unit of negative charge therefore halogens are electro-negative elements.

Replacement of one hydrogen atom with halogen is known as halogenation. In seed technology pre-storage seed treatment by Halogens at room temperature is termed as halogenation. Halogens have been used in may form *i.e.,* vapour, dry and slurry.

The halogen is absorbed by unsaturated fatty acid of the seed and reduces the physiological deterioration by stabilizing the unsaturated fatty acid components of lipoprotein moiety of cellular membrane and reduction in lipid peroxidation and free radical reactions. It might reacts with C=C double bonds of polyunsaturated fatty acids by reducing their oxidation.

The Halogen protects seed from pathogen and pest during storage due to antimicrobial property.

Dry dressing: It is application of halogens like chlorine, bromine and iodine to seeds before storage as dry dusting to stabilize the lipoprotein biomembrane.

Procedure

Calcium carbonate, fresh chalk, talc charcoal or activated clay are first exposed to vapours of halogens and alcohols @ 2-5 g /Kg seed. An alternative method is to add the chemicals directly to the carrier (50 – 500 mg of chemical added to 2-5 g carrier for treating 1 Kg seed) in a closed container. Calcium carbonate has proved to be a good carrier. After equilibrating the carrier-chemical mix, the seed is thoroughly dressed with the same and kept for 24-72 hr in the closed container and thereafter stored in the usual way. Bleaching powder (calcium oxychloride) may be directly mixed with the seed at the rate of 1-4 g / Kg seed.

Vapour treatment: Seed is exposed to halogen vapour in a closed container for 16-72 hr at very low concentrations of halogens. The concentration of chemicals and the time of exposure depend on the nature of the crop.

Antioxidant treatment

Wet or dry treatments of seed with anti-oxidant substances viz., vitamin A, C, E and butylated hydroxytoluene before storage to delay the physiological ageing is known as anti-oxidant treatment. The seeds are treated by soaking in the solution of antioxidant at prescribed concentration and time. Antioxidant provides hydrogen atom for extinguish the lipid radical thereby putting an end to free radical propagation and subsequent damage to cell membrane and ultra cellular structures.

Mid storage treatments for invigoration

Treatment of seed during mid storage i.e., seed of *Rabi* crop before start of monsoon and *Kharif* crop after summer is known as mid storage treatment. In Indian conditions, it is an essential component of *Rabi* crop as it has to pass high humidity season when seed is in store.

During senescence and post harvest handling of seed especially during threshing cell membranes of seed coat and other parts of the seed are damaged. In storage, damage to cell membranes is accumulated that deteriorates the seed at faster rate e.g., soybean, onion etc. To reduce the deterioration caused by accumulated damage, the treatment is given to the seed at mid stage of storage. It is capable of reducing the age induced damages, enhancing the longevity and restoring the seed vigour to a certain extent. The seed is treated by hydration-dehydration process as mid storage treatment.

Reduction in ageing by hydration and dehydration (wet treatment) of seed

The longevity of viable seed with reduced vigour is increased successfully by hydration of seed that raises the moisture content of seed on wet basis upto 25-30% for certain preventive and restorative intracellular reactions prior to dehydration before re-storage. It is suitable for pre-sowing and mid-storage invigoration of seed.

In store the seed is tested periodically for germination and vigour. The day on which vigour of seed has gone down but not viability, the treatment of hydration and dehydration is given. Many times the treatment is given at mid way of the storage *i.e.,* if seed has to be stored for nine months the treatment is provided after four and half months.

Hydration: The seed is soaked in water or in dilute solution of certain anti-toxicant chemicals. The duration of soaking varies from 1-6 hrs depending on nature of the crop. The advancement of germination process that is not compatible with drying back should be avoided. Imbibition may be achieved by

- Immersing the seed in water.
- Adding fixed quantity of water to the seed and then the wetted seed is covered for 2-6 hrs.
- To prevent the seed (pulses) from injury of soaking, very slow hydration is provided in pre-moistened sand with or without osmo-conditioning (Poly Ethylene Glycol, -5 bar; potassium phosphate 0.3M) at 1:4 seed send ratio incubated in a covered container for 16-48 hrs. The quantity of water in the sand should be such that the moisture content of seed may not reach beyond 30% on weight basis. In place of sand vermiculture may also be used. In general, seed absorbs all the water from the sand and may be separate out by sieving.

Dehydration: The seed is properly dried upto 10% moisture level or the original moisture content by natural air or hot air before re-store. Drying under direct hot sunlight may result in seed coat bursting (soybean).

Hydration – Dehydration

Low and medium vigorous, non- leguminous seeds are soaked in water with or without added chemicals like sodium and potassium phosphate, potassium iodide, p-hydro-oxy-benzoic acid and p=amino-benzoic acid in 10^3 -10^5M for short durations to raise the seed moisture content up to 25 – 30% with no sign of germination, followed by drying back of the seed in shade to safe limits for dry storage. The hydration - dehydration treatment improves the vigour by controlling free radical reactions and consequent peroxidative damage to lipoprotein cell membranes. Type of hydration-dehydration process to be adopted as mid-storage treatment depends upon the characteristics of seed and initial vigour status of the seeds.

Soaking – Drying: Stored seed is soaked in sufficient water to cover it and kept at room temperature for 2-6 hour depending on the crop with occasional stirring. The soaked seed is surface dried in the shade and then to the original moisture content for safe storage. In place of water dilute solution of chemicals such as sodium or potassium phosphate (di and mono basic), sodium chloride, p-hydroxy benzoic acid, p-amino benzoic acid, oxalic acid, potassium iodide, etc can also be used at 10-4 to 10-3 M concentrations. As per requirement fungicidal and insecticidal formulations can also be incorporated in the solution.

Dipping – Drying : In place of soaking, seeds are dipped in water or solutions for 2-5 minutes, then the seeds taken out from the solution are kept covered for 2 – 6 hours for absorption followed by drying for storage.

Spraying – Drying: Quantity of water equal to one fourth weight of seed to be treated is sprayed in two equal installments on the thin layer of evenly spread seed. The treated seed is kept covered by a polythene sheet for 2-4 hours before drying back.

Moisture equilibration – Drying: Seeds of soaking injury prone crops viz., soybean are placed for 24-48hrs in thin layers on trays kept in a closed moisture saturated chamber with nearly 100% relative humidity at ambient temperature. The treated seed is dried back in shade.

Moist sand conditioning – Drying: The sand (three times of the weight of seed to be treated) is pre-moistened with water. Seed is thoroughly mixed in it for slow and progressive moisture and kept at room temperature for 16 – 36 hours depending on the crop. After treatment, the hydrated seed is separated from sand by sieving and dried back to the original weight.

The wet process is effective but due to following problems it is not popular

- At field level it requires enormous physical facilities for soaking and drying of large quantity of seed for both seed industry and farmer.
- The treatment has to be given at mid way of storage which is practically very difficult.
- The drying process may be affected by weather.
- Lack of mechanization for effective treatment.

Pre-sowing treatments

Germination and vigour augmenting treatments

Seed fortification/infusion: Process of enriching the seed with simple water or bioactive chemicals viz., micronutrients, growth regulators, vitamins and seed protectants to improve germination and seedling vigour is known as seed fortification (*Syn* strengthening, enrichment).

Soaking of rice seed in 1% KCl for 16h; soybean seed in 0.25M Calcium chloride or 0.2% Molybdenum or Sodium dihydrogen phosphate (NaH_2PO_4) (5%), α-tocopherol (1%), ascorbic acid (2%), thiouria (1%), Potasium dihydrogen phosphate (KH_2PO_4) (2%) and fungicides (bavistin and thiram 1g/l) significantly improve the germination and seedling vgour. Similar results have been reported in barley, tomato, onion, cotton, groundnut, wheat etc.

Procedure

Dry permeation or seed infusion: Soaking of seeds in organic solvents to improve the germination and vigour of the seed by infusion of bioactive chemicals into the seed without altering seed moisture content is known as dry permeation (*Syn* penetration, infiltration). It avoids seed damage caused by soaking of water and provides protective, regulatory and selective functions of the chemicals to improve the performance of seed.

The seed is soaked in organic solvent using ascorbic acid as a bioactive chemical in the form of acetone, petroleum ether, toluene, hexane, ethanol and dichloromethane with desired hormonal and non hormonal chemicals for 2-3 hr followed by evaporation of the solvent in air.

Seed hardening: Hydration of seed to initiate the pre-germinative metabolism followed by dehydration to fix the biochemical events is known as seed hardening. It has specialized effect to withstand abiotic stresses at the time of germination and field emergence.

Hydration and dehydration of seed reorganize the physio-chemical changes by initiating the physiological activity in germ (embryo) and associated structure (mainly cotyledon). It results in enhanced elasticity in cell wall and development of a stronger and efficient root system.

Seeds are allowed to take up certain amount of water, and then they are kept moist at 10° - 25°C for several hours before drying in air. In many crops two - three cycles of hydration and dehydration are recommended. Different amount of water, concentration of chemicals used, duration of hydration and seed to solute ratio have been standardized for different crops.

Seed priming: Controlled hydration of seeds through a carrier or a high molecular solute to a level that starts pre-germinative metabolic activity followed by dehydration to check emergence of the radicle is known as seed priming.

It is the treatment in which seeds are hydrated sufficiently to allow the preparative events for germination to take place but insufficiently hydrated to allow the radicles to emerge followed by drying before actual sowing. Primed seed exhibited uniform germination alongwith enhanced rate and speed of germination resulting in fast and uniform seedling growth. The optimal priming effect is often obtained at the least negative water potential that prevents radical emergence. The degree of enhancement from priming depends upon following factors:

Seed priming arrests the seed germination after phase-II (preparation of seed for radicle emergence) of the triphasic pattern of water uptake. It brings all the seeds to a uniform physiological and metabolic state that enables maximum uniformity in the seedling growth. Major metabolic events occur at this time too. Osmotic potential and duration of the priming contribute to the improved germination performance.

For priming, seeds are soaked at 10^0-15^0C temperature in a chemically inert, high molecular weight compound solute that does not penetrate the cell walls viz., inorganic salts, sugars and polyethylene glycol (PEG). The duration of priming varies with the crop. Seed priming is of following type

Hydro-priming: It is hardening of seed in water under anaerobic condition. Movement of water in to the seed is not controlled but germination is prevented due to anaerobic condition. It is of following types

Osmo-priming: Soaking of seeds in osmotic solutions viz., PEG, Glycerol, Mannitol, $KN0_3$. KH_2P0_4 and other salt solutions.

Solid matrix priming: Seeds are mixed with solid carrier material viz., organic or inorganic carrier such as sand and water in a sealed container for a definite period to bring the moisture content of the matrix just below the level required for radicle protrusion. Carrier holds the water during seed priming and the water is imbibed by the seed upto equilibrium. The seed water potential is regulated by the matrix potential of the seed. The mixture is dried and seeds are separated from the media and cleaned. Materials such as vermiculite, synthetic calcium silicate, calcite clay, sphagnum moss, shale and bituminous charcoal have been widely used.

Drum- priming: Seeds are hydrated to a pre determined water potential for 24hr by placing them in a horizontal rotating drum. The seed is rotated in a drum under water mist without allowing protrusion (growth) of radicle. After treatment the seeds are redried to their original moisture content.

Halo-priming: Soaking of seeds in salt solutions.

Bio-priming: Seed coating with biological agents like bacteria.

The degree of enhancement from priming depends upon following factors:

- Initial quality of the seed
- Species being treated
- Treatment conditions such as temperature and water potential
- Duration and other conditions specific to the priming medium

Effect of priming on seed

- Developmental advancement
- Membrane leakage
- Effects on germination temperature
- Metabolic repair processes
- Changes in seed morphology
- Protein and DNA synthesis
- Changes in seed water relations

Effects of priming on germination

- Increased rate of germination
- Greater germination uniformity
- Increased germination percentage
- Wider temperature range for germination
- Advancement of maturity
- Weakening of barriers to embryo growth
- Increased protein synthesis

Irradiation: Irradiation of air dried seeds by lower dose of gamma rays for short period before sowing enhances germination and vigour. Germination, field emergence and growth of plants developed from irradiated seeds are significantly enhanced due to the effect of distant irradiation action that results in acceleration of the cell multiplication, enhancement of growth and development and change of organogenesis. Higher dose or time of radiation may cause mutation.

Magnetic seed treatment: Exposure of seeds to a magnetic field for specified duration improves germination and vigour. Seed treated under magnetic field respire slowly with higher respiration quotient values that increase activity of enzymes viz., α -amylase and nitrate reductase. Germination and vigour of seed are enhanced due to increased activity of enzymes.

Seed infusion: It is impregnation of seed with bioactive chemicals through organic solvents viz., acetone, petroleum ether, toluene, hexane, and ethanol instead of water. It avoids the damage caused by soaking of water. It provides protective, regulatory and selective functions of the chemical to improve the performance of seed with the help of organic solvent.

Seed coating treatment

Seed pelleting: Enclosing or encapsulation of small seed with inert (foreign) material to produce a globular unit of standard size is known as seed pelleting. More or less spherical units developed for precision sowing, usually incorporating a single seed with the size and shape of the seed no longer readily evident. The pellet, in addition to the pelleting material, may contain pesticides, dyes or other additives. Proper shape and size of the pelleted seed helps in

- **Swing of seed**
  - Mechanical planting
  - Easy handling of small and irregular shaped seeds
  - Reduced seed rate due to precision sowing
  - Maintain plant to plant distance
  - Increase in size of seed improves scatter pattern in aerial seedling
  - Uniform field emergence
- **Protection**
  - Pellelting with fungicide provides protection against soil and seed borne pathogens
  - Pellelting with insecticide provides protection against diseases transmitted though vectors
  - Prevent damage of insect pest at the time of seedling establishment
  - Prevent seed against store grain pest
  - Incorporation of unpalatable substances provide protection against rodent
- **Vigour**
  - Enhanced seedling vigour when pelleted with fertilizer by nourishing the young seedling
  - Assured germination under abiotic stresses when pelleted with growth regulators and bio-fertilizers by promoting rooting and seedling growth

- **Environmental friendly**
  - Reduced environmental pollution
  - Precise and accurate use of chemicals
  - Improved ballistic property

Material and equipment

Pelleting material

Amalgam of pelleting materials: Clay, limestone, calcium carbonate, talc, vermiculite

Binding agents: *Gum arabic*, gelatin, methylcellulose, polyvinyl alcohol, polyoxylethylene glycol-based waxes

Seed protectants: Insecticide fungicide and antibiotics, herbicide and chemicals to enhance resistance against abiotic stresses

Nutrients: e.g., Diammonium phosphate to stimulate prolific root growth

Inoculants (bio-fertilizer): *Rhizobium, Mycorhiza* etc.

Coating drum: In this seed can be mixed with other ingredients as cement mixer plant

Procedure: Seeds are placed in a coating drum and rotated after spraying of water. In this, prescribed quantity of amalgam of pelleting materials, seed portectants, nutrients and inoculants are added. By rotation the wet seed is coated with the dry pelleting material. The size of pellet gradually increases with each turn of coating drum. With the increase in number of rotation the size of pellet is enhanced with improvement in roundness. At the end of the pelleting process, a binder is added to harden the outer layer of the pellet. The pelleted seed is dried and handled as unpelleted seed. The ingredients of the seed pellets exert respective positive influence on seed performance.

Encrusted seed Units more or less retaining the shape of the seed with the size and weight changed to a greater or lesser extent. The encrusting material may contain pesticides, fungicides, dyes or other additives.

Seed granules Units more or less cylindrical, including types with more than one seed joined together. The granule, in addition to the granulating material, may contain pesticides, dyes or other additives.

Seed tapes Narrow bands of material, such as paper or other degradable material, with seeds spaced randomly, in groups or in a single row.

Seed mats Broad sheets of material, such as paper or other degradable material, with seeds placed in rows, groups or at random throughout the sheets.

Film coating: A continuous layer of film or chemical is coated on the surface of the seed without altering the shape or size. The advantage of film coating over other coating process is the minimum emission of dust particles, minimum loss of coated material, high retention capacity of chemical on seed, low quantity of chemical requirement, accurate dose of chemical with low coefficient of variation, proper availability of nutrient to the seedling, bio-pesticide can be enclosed successfully, less hazardous to operator and farmer and improvement of seed flow in planting equipment.

Seed is coated with a thin layer of durable water permeable coating usually polymer binder in the slurry coating equipment by spraying or dipping the seed in the dissolved additives followed by immediate drying. Polymer film coating involves application of 1% composition of vinyl acetate + vinyl chloride + ethylene + acrylate with 30% cellulose ether as a stabilizing substance followed by immediately drying. eg., coating of seed with Bavistin protects the seedling from wilt disease. The rate of water vapour movement for equilibrium into the stored seed is controlled effectively by polymer coating. Polymer coating alongwith fungicide has added advantage on storage capacity of seed eg., film of Polyvinylidene chloride on the seed of maize and soybean controls the invasion of storage fungi.

Coating of seed with hydrogel polymer helps in obtaining optimum plant population under excessive moisture and drought at the time of seedling establishment due to erratic behaviuor of monsoon. It absorbs 100-200% moisture of its weight and assures supply of moisture at the time of seedling establishment under drought. Coating of seed with temperature responsive polymer ensures germination only at optimum temperature.

Seed colouring: Precise amount of natural or artificial dyes or pigments is applied directly on the surface of the seed. The colouring agent should not be absorbed in the seed. The purpose of seed coluring are

- Prevent accidental use of treated seeds for food or feed purposes
- Identification of the seeds of parental lines 'A' line, 'B' line, 'R' line
- Brand identification
- Distinct and attractive look
- Improvement in storability
- Enhanced germinability and vigour

Fluid drilling or gel seeding: Mixing of germinated seed (only radicle is emerged out) in a viscous gel, which is sown with an appropriate seed drill to maintain the seed moisture of the pre geminated seeds and to prevent injury to emerged radicle.

Seeds are placed in an aerated water unit upto emergence of radicle. These seeds with the initiation of germination are mixed with a viscous gel e.g., alginate gel to provide a cushion to the emerged radicle thus preventing any mechanical damage during drilling and protection of seed against any moisture loss.

Seed sanitation

Treatment of seed with pesticides i.e., fungicides, insecticides or a combination of both for disinfection or disinfestation from various seed borne pests and storage insect pests is known as seed sanitation. For sanitation the seed is dipped or soaked in the chemical solutions and the fungicides applied as dust, slurry or liquid in recommended concentrations before storage.

Seed disinfection: Treatment of seed with the prescribed fungicide for eradication of pathogen established within the coat or in more deep-seated tissues of seed is known as seed disinfection.

Seed disinfestation: Destruction of surface borne organisms that have contaminated the seed surface but have not infected the seed surface is known as disinfestations.

Seed treatment against abiotic stresses

The application of exogenous salicylic acid provides protection against high or low temperature, heavy metals etc. Salicylic acid may cause oxidative stress to plants by accumulation of hydrogen peroxide. Antioxidative capacity is enhanced by low concentration of H_2O_2 tolerance. *Halopriming* of sets with 100 mM NaCl improves the percent and rate of germination of varieties of sugarcane sensitive to salinity and drought tolerance. Priming with *Plant Growth Regulators*, gibberellic acid and kinetin increases percentage germination and seedling growth of sugar beet, wheat, maize, soybean and sunflower under salt stress conditions.

Treatment of maize seed with a 2,4-diacetylphloroglucinol (DAPG)-producing *Pseudomonas* strain ameliorated abiotic stress disorder in a low-pH soil.

Designer seed: Coating/pelleting of all the possible useful active ingredients (fungicide + insecticide + micronutrients + colouring agents + microorganisms etc) on the seed/fortified seed/ primed seed with an adhesive is known as designer seed. The sequence of the treatment is

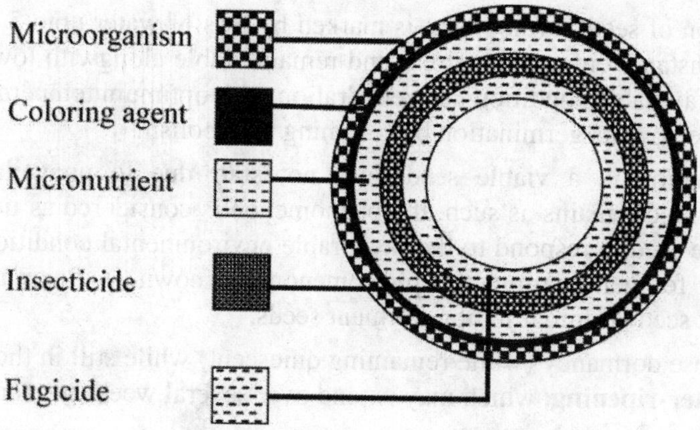

■■■■

Chapter 35

SEED DORMANCY AND ITS MANAGEMENT

Object

Overcome the problem through breaking of seed dormancy to facilitate germination.

Definition

The completion of seed development is marked by loss of water upto 5 - 15 percent. At dry state seed may withstand temperature stress and remain viable alongwith low metabolic activity. In the presence of favorable condition *i.e.*, rehydration with optimum temperature and availability of oxygen, viable seed starts germination by resuming metabolism.

When germination in a viable seed does not start due to unavailability of favorable conditions and the seed remains as such, the phenomenon is considered as **quiescence**. Whereas, inability of a viable seed to respond to the favourable environmental conditions *i.e.*, temperature, water and oxygen for germination, the phenomenon is known as **dormancy**. The metabolic activity of dormant seed is similar to non-dormant seeds.

Many seeds lose dormancy (while remaining quiescent) while still in the mature dry state, in a process called **after-ripening**, which may extend over several weeks to many years.

Dormancy is a mechanism to ensure the seed germination after optimum time. For example, seed of *Kharif* crops has to germinate at the start of *Kharif* and of *Rabi* at the start of *Rabi* to complete the life cycle successfully. Therefore crop of *Rabi* season requires cold temperature to break dormancy while of *Kharif* season high temperature. This ensures that the seedling stage shall not encounter unfavourable environmental conditions for establishment. This is important for small seeded species with small food reserve.

Lack of temporary dormancy is detrimental as mature seeds of mung bean, urid bean, wheat etc. start pre-harvest-sprouting due to favourable condition at maturity prior to harvest.

TYPES OF DORMANCY

Primary dormancy/ Innate dormancy: It is the state of the seed as the seed gets dispersed from the mother plant, the dormancy may be induced before maturity, during maturity and after maturity but before seed is dispersed from mother plant. The viable seed that does not germinate immediately after maturity under favourable conditions due to exogenous, endogenous or combination is considered as dormant. It starts germination under favourable condition after resting period. It is known as primary dormancy. It is of following types:

Exogenous dormancy: Dormancy because of unavailability of water, oxygen or mechanical restriction due to physical barrier is considered as exogenous dormancy.

Hard seededness: Dormancy due to unavailability of water or oxygen or both to embryo due to impermeable seed coat is considered as hardseededness. Hard seeds are also formed due to forced maturity. It is of following types:

Impermeabilty of seed coat to water: It is the most abundant type of seed coat dormancy. The impermeablity of seed coat to water due to genetic or environmental factors prevents a viable seed from germination under favourable conditions. Seed coat is responsible for impermeability to water in family Leguminoceae and Malvaceae, whereas nucellar membrane in Cucurbitaceae. No attempts are made to break the dormancy in the crops of given families and in seed testing dormant seeds are considered as germinated. These seeds start germination after resting period. Structure of testa, testa thickness, no pore and wrinkling on testa surface, closed hilum fissure, closed micropyle (strophiole-epistate) and deposition of phenol, tannin, calcium (Calcification), wax, cutin (watermelon), suberin (Subernization-grasses), lignin (Lignification), pectic substances (Pectinization) etc., in testa, pericarp or nucellar membrane are responsible for impmermeability to water.

Impermeabilty of seed coat to oxygen: In many crop species the hard seed coat allows water to enter but not oxygen. Oxygen is essential for germination and viability. In the initial stage these seeds are viable but in later stage due to unavailability of oxygen, embryo is not able to remain viable. Therefore, hard seeds observed in crops from families other than Leguminoseae are not considered as germinated during seed testing. The endocarp in coffee and pericarp in *Xanthium* restricts the entry of oxygen. In white mustard mucilage layer around the seed coat is produced under high moisture condition that induces dormancy by restricting gaseous exchange. In seed of crops from family Compositeae tegmen, layer of endosperm and nucellus remain active even after separation of seed from mother plant. These layers induce dormancy by restricting gaseous exchange.

Mechanical dormancy: In many plant species very hard and strong seed coat offers mechanical resistance to the growth of germinating embryo inside and does not allow it to emerge out. Example plantain, Amaranths, peach, walnut and cherry.

Endogenous dormancy: It is the inherent property of seed. Failure of viable mature embryo to germinate even when it is isolated from the seed coat is considered as endogenous dormancy or embryo dormancy. This phenomenon is common in the woody members of family Rosaceae. It is of following types

Immature embryo: Dormancy is expressed by seed because the embryo is not fully matured (rudimentary / immature) and developed. It needs after ripening period for complete growth that leads to germination. e.g. oil palm, anona and hog weed.

Dormant embryo: Dormancy is expressed by seed with fully matured and developed embryo. It may be due to metabolic blockage in the path of food availability or presence of inhibitory substances.

Chemical dormancy: chemicals present within the seed coat inhibit germination and cause endogenous dormancy (Table 1). It is of following types:

- Presence of inhibitory substances: Some crops develop inhibitory substance on the surface of seed at the time of physiological maturity. These inhibitory substances have ability to check the germination of viable seeds. Removal of these substances from the seed results in normal germination.

- Growth regulator induced dormancy: Presence of growth regulator *viz.*, abscisic acid (ABA), that comes in the seed from the female parent during seed development may induce seed dormancy.

Table 1. Chemical induced dormancy in field crops

Species	Chemical	Location of chemical
Avena sativa	Unknown	Hull
Beta vulgaris	Phenolic acids, ABA	Pericarp
Corriandrum sativum	Coumarin	Pericarp
Elaegnus angustfolia	coumarin	Pericarp, testa
Gossypium spp.	Absicic acid	Pericarp, testa
Helianthus annus	Hydorcyanic acid	Pericarp, testa
Hordeum vulgare	Coumarin, Phenolic acids	Hull
Oryza sativa	Probably ABA	Hull
Triticum spp.	Catechin, tannins	Pericarp, testa

Physiological or intermediate seed dormancy

Physiological dormancy is a result of seed requiring some type of physiological condition to be met in order to germinate. Seed of some crop species has to be dried down/dehydrated after maturity to make them capable of germination. Other seeds need exposure to light to initiate the process of germination. Brief exposure to light, dehydration, or slight chilling treatments in the absence of moisture can help to overcome physiological or intermediate dormancy.

Osmotic inhibition: High osmotic potential substances can induce dormancy. High concentration of sugar or salt prevents optimum imbibition. Example sugar beet.

Metabolic inhibition: Few chemicals present in the seed inhibit the metabolic process/ pathway. Example cyanide in apple and peach, phenolic compound in barley and coumaine in coriander.

Secondary dormancy: **It means** a viable seed removed from the plant is not able to germinate. Seed with secondary dormancy imbibe water as a normal seed without any sign of germination. The activation process of germination is started in seed but does not progress. Causes of secondary dormancy include unfavourable environmental conditions for germination viz., thermo (temperature), photo (light) or skoto (darkness) extremes. Water stress, or oxygen extremes--anoxia or hypoxia (too much or too little oxygen) may also induce secondary dormancy. In wheat secondary dormancy can be induced one and half months after physiological maturity by storage of seed at high moisture content in airtight containers at 50ºC for 24 hrs or placement of seed under water in dark for one to 3 days at 2ºC.

Other types of dormancy

Deep Dormancy: Seed of temperate plants remain dormant until conditions are favorable for germination it is known as deep dormancy. Seed germinates in worm weather may be killed by repeat change in temperature especially freezing in early spring. Therefore deep dormant seed does not germinate at the initiation of warmth environment. It requires 2-3 months of chilling temperature to break the dormancy. This chilling process is termed stratification. Artificial stratification treatments may be applied to break the deep dormancy.

Light dormancy: Phytochrome of seed has light-perception system which is activated by the wavelengths rich in red. In forest, the red wavelength of light is absorbed by the leaves of trees. Therefore seed in forest has to remain dormant or quiescent upto fall of leaves. e.g., *Chenapodium album.*

Relative dormancy: This type of dormant seed exhibits dormancy at one particular temperature, while not exhibiting dormancy at another. Wheat seed just after harvest does not germinate at 20^0C temperature while the same seed germinates at the temperature lower than 15^0C. After one and half month of harvest all the seeds germinate at both the temperatures.

Combination of dormancy : Combination of more than one type of dormancy posed by the same seed needs more than one treatment to break the dormancy.

Methods to overcome dormancy

A seed requires a long resting period to breakdown the dormancy naturally by completion of certain metabolic changes.

After ripening: Many seeds remain dormant in the mature dry state to complete metabolic and enzymatic activities the process is called after-ripening, it may extend over several weeks to many years. Dormancy of hydrated seeds in the soil may be broken by one or more environmental factors viz., light, (usually for a short duration, with sunlight being the most effective), low temperature (around 1 to 5°C for several to many weeks) and; fluctuating day-night temperatures (5 to 10°C) and chemicals (eg., presence of nitrate in the soil).

Dry storage: Seeds of the crops with physiological dormancy are stored under ambient condition for required period of time to break the dormancy. Example: sunflower.

Soaking in water: To break physical or chemical dormancy the seed may be soaked in cold or hot water for few hours to several days. It is of following types

Cold water treatment: Some inhibitors associated with dormancy are present in the seed viz., coumarin in coriander and hydrocyanic acid in sunflower. To remove inhibitors the seed is soaked in water for 3 days with change of water after every 12hrs to avoid fermentation. Soaking of seed in running water for a day may leach out the inhibitors.

Hot water treatments: Seed is soaked in boiled water for one to 80 minutes to break physical dormancy caused by hard seed coat or chemical dormancy due to inhibitors.

Pre-washing: Inhibitory substances from the surface of the seed coat are removed by washing the seed in running water at 25°C. Seeds are then dried at room temperature prior to germination test. Example: *Beta, Xanthium* and *Fraxinus.*

Scarification

Seeds of some crops (fababean, rice bean etc.) are scarified mechanically to break the hard seededness by following methods

Piercing: Seed coat is pierced with the help of a sharp needle without any injury to other parts of seed.

Chipping: Seed coat is chipped with a sharp razor without any injury to other parts of seed.

Filing: Seed coat is scratched at the suitable site with a file or sand-paper for entry of water.

Acid scarification: The seed is treated with concentrated sulphuric or nitric acid to make the seed coat pitted for a pre-standardized duration with often stirring to overcome physical dormancy caused by hard seed coat. The treated seed is washed 5-10 minutes in running water to remove all traces of acid before shade drying. Example: rice

Bio-scarification: Use of animals or microorganisms to make the hard seed coat permeable is known as bio-scarification. To break the dormancy of *Acacia nilotica,* the pods are fed to sheep or goats and the scarified seeds are collected from the droppings for sowing purpose. The combination of moisture, warmth and chemical action of the digestive juices in the stomach of animal make the hard seed coat permeable. Similarly, termites break the seed coat dormancy by feeding the hard cover on seed known as exocarp. In nature, impervious seed coats or structure inducing mechanical dormancy in pits of stone fruits are softened by action of microorganisms in soil.

Removal of structure around the seed: Inhibitory structures like involucre of bristle, lemma and palea, hard shell etc. are removed with the help of forceps prior to germination test.

Stratification

Exposure of imbibed seed to cool or warm temperature prior to germination in order to break the dormancy is known as stratification.

Hot stratification: Exposure of imbibed seeds to high temperature for a prescribed period of time to overcome mechanical or morphological dormancy is known as hot stratification. Seeds are immersed in water at approximately 30-35°C. After 48hrs of soaking the water is drained off. The imbibed seeds are mixed with moistened sand, sand peat mixture or vermiculite and stored at a constant temperature of 20-25°C or alternating temperature of 20°C and 30°C in a container for 2 to 16 weeks depending on species. Seeds are mixed weekly and the water is added as per requirement.

Preheating treatment: Seeds are arranged on a paper towel for germination test. These rolled towels are heated by free air circulation with maximum 30-35°C temperature for 7 days before placement for germinabiltiy test. Example: groundnut (40°C), wheat, oat, barley, sunflower and Egyptian clover (30-35°C) and rice (50°C).

Cold stratification: Exposure of imbibed seeds to low temperature for a prescribed period of time to overcome physiological dormancy is known as cold stratification. Seeds are immersed in cold water at approximately 3-5°C. After 48hrs of soaking the water is drained off. The imbibed seeds are mixed with moistened sand, sand peat mixture or vermiculite and stored at a constant temperature of 3-5°C in a container for prescribed period of time depending on species. Seeds are mixed periodically and the water is added as per requirement (eg., *Abeis* spp, *Eucalyptus delegatensis*) etc.

Pre-chilling treatment: Seeds are arranged on a paper towel for germination test. These moistened paper towels containing seeds are kept at a low temperature (5-10°C) upto 7 days before germinability test. Example: Lentil, sunflower, onion, oat, cauliflower, cabbage, barley, linseed, lucern, radish, sorghum, wheat, fababean, lettuce, opium, spinach, onion, cole crops and turnip.

Light: Many seeds require light for germination. Light with 750-1250 lux intensity from cool white lamp is illuminated for 8 hrs in every 24 hours during the high temperature period to break the dormancy of grass seeds with alternate temperature. In lettuce seeds germination is stimulated by red light and inhibited by far-red. The species with germination under the control of phytochrome are not able to germinate when sown deep in the soil. For germination seed has to be exposed to light as Infrared system influences the growth rate, whereas ultra violet light receptor influences direction of seedling. Example: *Jussiae, Pavlonia tomentosa* and *Kalanchoe blossfeldiana.*

Potassium nitrate: A solution of KNO_3 (0.2%) is prepared by dissolving 2 g KNO_3 in 1 litre of water. The substratum selected for germination test is moistened with this solution instead of water. Seeds are arranged on it and kept for germination.

Example: Brassica, tomato, cole crops and chilli.

Treatment with growth regulators

Endogenous type of dormancy due to presence of inhibitors can be broken down by application of low concentration of growth regulators.

Gibberelic acid: During seed development dormancy is associated with an increase in abscisic acid. This dormancy can often be broken artificially by treatment with gibberelic acid. GA_3 solution of 0.05% is prepared by dissolving 500 g GA_3 in 1 litre water. Seeds are arranged on a substratum moistened with this solution and placed in germinator. Example: Oat, barley and wheat.

ABA: The seed of cotton shows dormancy due to presence of ABA in pericarp that can be broaken down by application of Ethrel.

Treatment with chemicals

Nitrogenous compounds like thiourea, hydroxylamine, Nitric acid and nitrate can be applied to break dormancy. 10^{-2} to 10^{-3} molar Thiourea is applied to break dormancy of both light and chilling treatment requiring seeds e.g. lettuce. Dormancy is also broken down by application of sulphidral compounds like 2 mercapto ethanol and 2,3 dimercapto ehtanol and root exudation containing strigol from the plant infested by striga parasite. Soaking of groundnut seed in 50ppm solution of GA_3 + 100ppm KNO_3 for 1 hr breaks the dormancy

Sealed polythene envelope: The fresh ungerminated seeds of Egyptian clover obtained after germination test are arranged on a paper towel. It is placed in a sealed polythene envelope for the prescribed period and conditions recommended for germination to induce germinability.

Fire: Many shrubs and trees of sub-tropical and semi-arid regions have extremely hard seeds with seed coat very impervious to water e.g., *Albizia lophantha*. Germination in these seeds is promoted by ethylene. It is formed as an ecological response to fire which occurs naturally in forest. Burning vegetation generates trace amounts of ethylene and clears the ground for germination of seed e.g., *Calluna vulgaris*.

Pressure treatment: Dormant seeds are placed in an autoclave at required pressure to break the dormancy.

Magnetic seed treatment: Seeds are placed in the magnetic field for about 1 to 10 days to break dormancy of many forest trees.

Chapter 36

HYBRID SEED PRODUCTION

1. OBJECTIVE

Production of hybrid seed.

Hybrids are F_1 seed produced by hybridization of selected inbred lines. Performance of hybrid is better than the parents and other varieties because of heterosis.

2. SELECTION OF PARENTS

Parents for hybrid seed production programme are selected on the basis of specific combining ability and heterosis. Parent used as female is known as seed parent, whereas, male parent is designated as pollinator in hybrid seed production programme.

3. METHODS OF HYBRID SEED PRODUCTION

3.1. Hand emasculation and pollination

The male part of seed parent is removed by hand before anthesis to prevent self-pollination from the crop plants with bisexual flower having fertile male and female part. This process is known as emasculation. These emasculated flowers then pollinated by the pollens collected from pollinator parent to produce hybrid seed.

3.2. Monoecious nature of the plant

The male inflorescence (maize) or male or bisexual flowers (castor) are removed from the seed parent to prevent self pollination. Natural cross pollination between pollinator (male parent) and seed parent (female) is allowed in order to harvest hybrid seed from seed parent.

3.3. Male sterility

Some genotypes of crop plants have bisexual flowers with sterile male part due to genetic, cytoplasmic or cytoplasmic genetic interaction. This condition is exploited to produce hybrid seed by using male sterile line as seed parent.

3.4. Self-incompatibility

Inability of a functional gamete to perform fertilization and/or to induce seed set under self-pollination is known as self-incompatibility. This condition is exploited to produce hybrid seed.

3.5. Gametocide

In order to overcome the hand emasculation and pollination some chemicals are spread on seed parent to make them male sterile by killing their male gametes. These chemicals are known as gametocide or chemical hybridizing agent (CHA).

3.6. Barnase-barstar system

The gene introduced from the bacterium *Bacillus amyloliquefaciens* produces enzyme BARNASE in plant system that has ability to kill the cell by cleaving the RNA molecules of the cell. The other gene from *B. amyloliquefaciens* forms an active substance BARSTAR with the ability to neutralize the action of barnase.

4.1. Nucleus and Breeder's seed production

Seed of both seed (female) and pollinator (male) parents are multiplied at Nucleus and Breeder's seed stage by self pollination. Female parents of hybrids involving male sterility are maintained with the help of maintainer lines, whereas other lines are produced by self- pollination under isolation to avoid out-crossing (Fig. 1).

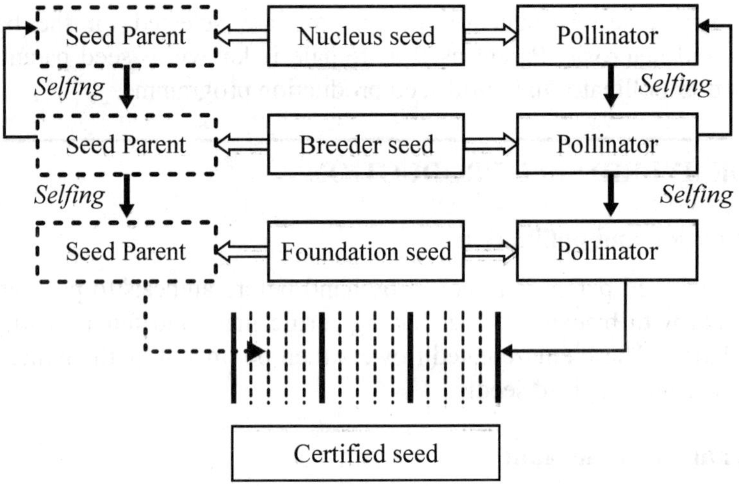

Fig. 1. Seed multiplication chain for hybrid

4.2. Foundation seed production

Seed of seed (female) and pollinator (male) parents are required for hybrid seed production programme. It is produced under the supervision of Seed Certification Agency and considered as Foundation seed after its certification at field and seed levels.

4.3. Certified seed production

Hybrid seed produced by registered growers after certification by Seed Certification Agency is known as Certified seed.

5. IMPORTANT POINTS TO BE CONSIDERED FOR HYBRID SEED PRODUCTION PROGRAMME

5.1. Selection of field

For seed production programme of hybrid and parent selected field should be free from volunteer plants.

5.2. Isolation

In order to maintain the genetic purity of the parent or hybrid seed, cross contamination should be avoided by maintaining the isolation distance from field of other varieties, seed production of other hybrid and seed production programme of same hybrid not conform the seed certification standards. Modification of isolation distance is not allowed for hybrid seed production programme except maize.

5.3. Planting ratio

Planting ratio refers to the number of rows of the male parent of that of the female parent, in the hybrid seed production plot. Appropriate line ratio of seed and pollinator parents in a unit area is the key factor for seed production of hybrid, male sterile line or monoecious nature of the crop plant. Planting ratio depends on the nature of pollination, pollen load, mode of pollination, pollen viability, stigma receptivity, wind velocity, density and movement of vector; height of the pollinator and agronomical management. Ideal planting ratio for hybrid seed production programme differs from crop to crop, genotype to genotype and place to place. First and last row of the hybrid or male sterile line seed production programme should preferably be of the pollinator. To facilitate the movement of pollen and vector from pollinator to seed parent, row should be across the wind direction.

Determination of planting ratio: In an isolated field, pollinator parent is planted across the wind at the side from where wind is normally coming. Seed parent is planted in line towards the direction of windblown, at fixed row to row distance. Plants of seed parent are harvested row wise and the data of seed yield/row is recorded. Number of rows, just before the row showing significant difference in seed yield/row in comparison to first row are selected as rows of seed parent for planting ratio.

5.4. Nicking

Flowering in seed and pollinator parents at the same time is termed as nicking. Nicking is essential for enhancing producibility of hybrid and male sterile seed. Flowering time of parents is mainly influenced by genetic constitution of parents, climatic conditions, and agronomical practices. Difference in flowering time of parents varies from hybrid to hybrid and place to place. Flowering in all the plants of a population/variety at the same time is termed as synchronous whereas between populations it is termed as nicking.

Sorghum: Four week old plants of both the parents are critically examined to check the flower primordia and the apex of male and female plants by stripping the leaves of stem. The difference in the time of initiation and size of the panicle bud (primordial) indicates the difference in the time to 50% flowering in between male and female lines. The growth and flowering in lagging parent can be enhanced upto one week by application of additional dose of 50kg N/ha, followed by foliar spray of 1-2% urea and spray of GA_3 at primordial initiation stage. Alternatively, selective irrigation of one parent and delayed irrigation of the other also promotes nicking. Flowering may be delayed by irrigation under cool climate. Growth rate of male parent at early stage can be reduced by cutting alternate plants to allow vegetative growth by promotion of tillers. It may be boost up with additional dose of nitrogen.

Sunflower: Foliar spry of 2% urea/DAP at button formation stage may advance the flowering in late parent by 2-3 days.

Pearl millet: Foliar application of 3% urea, 4, 5 and 6 weeks after sowing preponed the flowering of late plant by 2-6 weeks. Early flowering in any line can be delayed by jerking of plants.

5.5. Staggering

Nicking can be achieved by altering the date of planting of male and female parents. The difference in planting time for nicking is known as staggering and it is normally achieved by early sowing of late flowering parent.

5.6. Supplementary pollination

Natural pollination is supplemented by hand or other device to increase the seed setting during hybrid and female line seed production programme. It is known as supplementary pollination.

5.7. Pollen shedder

Plants of seed parent in monoecious or male sterile system showing male fertility in complete plant/inflorescence or partially in some part of the inflorescence / plant are considered as pollen shedder and partial pollen shedders, respectively. Pollen shedders are rouged out as and when observed.

5.8. Rouging

Off-types and rogue plants are removed from both pollen and seed parent whereas, pollen shedders should be rouged out from the line of seed parent. For rouging of pollen shedders, field should be visited every alternate day during flowering period. Rouging of plants infected with seed borne diseases should be done regularly upto physiological maturity.

5.9. Field inspection

Field inspector inspects the Foundation and Certified seed production programme atleast four times. Out of these four inspections first is made before flowering, 2nd and 3rd during flowering, and fourth at physiological maturity to verify isolation distance, planting ratio, planting pattern, offtypes, pollen shedder, objectionable weed plants, incidence of objectionable seed borne diseases and true nature of the parents.

5.10 Seed testing

The submitted sample is tested for verification of seed standard of physical purity, number of weed, other crop and objectionable weed seed, incidence of objectionable seed borne diseases, germination (%) and moisture content.

5.11 Testing of genetic purity

Genetic purity of Foundation and Certified seed is verified by ODV and GOT test.

6. HYBRID SEED PRODUCTION BY HAND EMASCULATION AND POLLINATION

Crops with commercial hybrids

Field crops: Maize, sorghum, pearl millet, rice, sunflower, safflower, pigeonpea, cotton, potato, sugarcane, castor etc.

Vegetable crops: Brinjal, watermelon, onion, tomato etc.

6.1. Cotton

Nucleus and Breeder's seed production programmes: Seed of male and female lines are produced as per the method described for a normal variety by selfing under isolation by avoiding out-crossing. Meticulous rouging with adoption of recommended package of practice is essential.

Foundation seed production programme: It is similar to the method described for seed production of a normal variety with same isolation distance, field standard, seed standard, genetic purity and field inspection. Seed production programme should be isolated from contaminator to avoid genetic contamination (Table 1). For Foundation seed production programme a minimum of three field inspections are required. Out of these three inspections, two are made during flowering and one at maturity to verify the isolation distance and offtype.

Certified seed production programme: Seed production plot is divided into two blocks. Female parent is sown in the first block consists of 80% area whereas male parent is sown in second block consist of 20% area with isolation distance of 5 meter between the blocks.

Table 1. Minimum isolation distance (m) required for Foundation and Certified seed production programme of hybrid cotton

Factor	Foundation	Certified
Field of other varieties, same commercial hybrid seed production programme of same hybrid not confirming to varietal purity requirement.	50	30
Field of other spp. and between parental lines of the hybrid (in hybrid seed production programme)	5	5

Approximately, 1200 hills of female and 300 hills of male parent for seed production programme of one hectare are required. Row to row and plant to plant distance should be maintained with optimum seed rate (Table 2). Seeds of male parent should be sown in three different dates with 7-10 days interval to get sufficient pollen load throughout the blooming period of seed parent. Offtypes should be rogued out at seedling, square initiation and flowering stage from both the parents.

Table 2. Recommended spacing and seed rate for Foundation and Certified seed production of cotton.

Particulars	Seed parent	male parent
Spacing (cm)		
Row to row	150	100
Plant to plant	150	100
Seed rate (kg/ha)	0.5	0.25

All the buds, appeared to open next morning in seed parent are emasculated in the evening between 2-6 pm with the help of sharp thumb nail or blade in such a way that the corolla tube and all the anthers get detached without any injury to ovary. This method is known as *Doak* method. The emasculated bud may be covered with a red coloured paper bag. Next day morning, male flowers opened on the same day are collected for pollination and dehiscence of pollen from anther is tested. For proper anthesis male flower may be kept under light source for 15-30 minutes. From seed parent all the selfed (red coloured) flowers are removed. Anthers from male flower are rubbed on all the lobes of stigma in between 7-10 am. Pollen released by one male flower is sufficient to pollinate 3-5 emasculated flower. After pollination a soda straw may be placed on the stigma and is tied up with square by thread. A thread is tied on the pedicel of pollinated flower as mark of identification. After completion of pollination no red colour i.e., of paper bag or selfed flower should be present on the plant of seed parent.

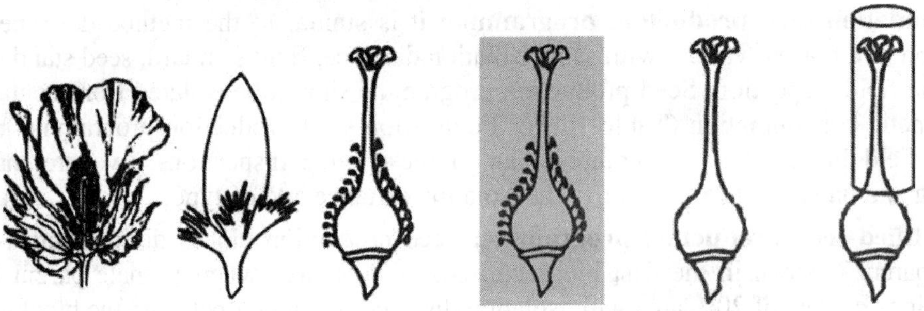

For Certified seed production programme atleast four field inspections are required as per detail given in Table 3.

Table 3. Stage of crop during field inspections and factors to be verified for Certified seed production programme

Inspection	Stage of crop	Factors to be verified
First	Before flowering	Isolation, genetic purity
Second and third	During flowering	Offtypes
Fourth	During picking of cotton in female parent	Elimination of selfed bolls

Field standard: Maximum permitted limit for offtype in Foundation and Certified seed production programme at any one inspection at and after flowering is 0.1% and 0.5%, respectively. During Certified seed production programme same standard should be met separately by both the parents.

Completely open bolls having thread in its pedicel should be sorted out and dried separately. Mixing of selfed boll or boll of different varieties should be avoided in every step. Bolls from seed parent containing hybrid seed should be stored in gunny bag until supplied to processing plant. Ginning of these cotton bolls should be done in the gin approved by Seed Certifying Agency.

Seed standard: Foundation and Certified seed should have pass as per seed standard for certification in Seed Testing Laboratory.

Table 4. Certification standard for seed testing

Factor			**Foundation**	**Certified**
Pure seed	Minimum	(%)	98	98
Inert matter	Maximum	(%)	2	2
Other crop seeds	Maximum	Number/kg	5	10
Weed seed	Maximum	Number/kg	5	10
Germination	Minimum	(%)	65	65
Moisture	Maximum	(%)	10	10
Vapour proof container	Maximum	(%)	6	6

Genetic purity of the hybrid seed is verified by seed certification agency in off season. Grow out test is performed with minimum 400 seeds in two replications of 200 each. Genetic purity of every individual plant is verified on the basis of expression of marker characteristics. Hybrid seed should have 90% genetic purity. The percentage of other variety plant segregates or other offtype excluding selfed seed of parents should not be more than 1.5%, whereas, it should not be more than 10% when selfed seeds of the parents are included. The reject number of grow out test is 44 offtypes out of 400 observed plants.

7. HYBRID SEED PRODUCTION OF CROPS WITH MONOECIOUS NATURE

7.1. Maize

It is monoecious in nature with male flower (tassel) present on the top of the plant and female flowers (spadix) in the middle part of the plant.

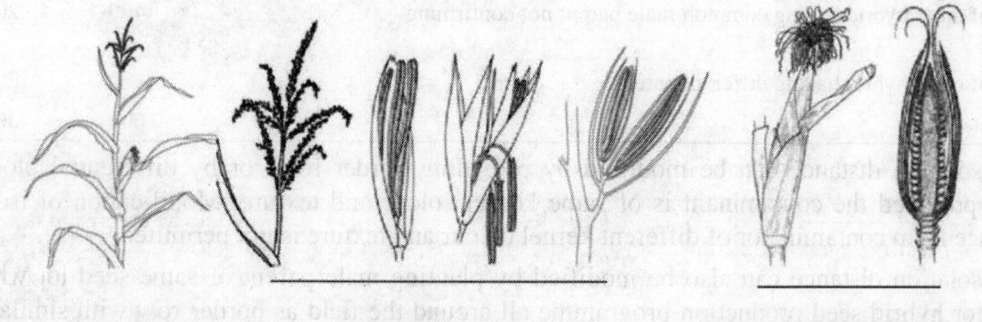

Plant **Male inflorescence (Tassel)** **Female inflorescence (Spadix)**

Seed Production: Hybrid seed production in maize involved two parents to produce following types of hybrids

Single cross: Hybrid seed produced by controlled crossing between two selected inbreds (A x B).

Double cross: Hybrid seed produced by crossing between two certified single crosses [(A x B) x (C x D)].

Three way cross: Hybrid seed produced by crossing between an inbred used as male and a certified single cross hybrid [(A x B) x C] as female parent.

Top cross: Hybrid seed produced by crossing of inbred line with a certified open pollinated variety.

Double top cross: Hybrid seed produced by the crossing between a certified single cross and a certified open pollinated variety.

Nucleus and Breeder's seed production programmes: Nucleus and Breeder's seed of male and female parents are produced by selfing under isolation of 400 meter from the field of other varieties.

Foundation seed production programme: Inbred lines and single cross hybrids are produced as Foundation seed (details are given with Certified seed).

Certified seed production programme: Production of single, double, three way, top and double top cross hybrid is known as Certified seed. Foundation and Certified seed production programme are taken at recommended isolation distance (Table 5).

Table 5. Recommended isolation distance for seed production programme of hybrid maize and its parents

Contaminants	Isolation distance (m)		
	Foundation	Certified	
	Inbred	Single	Hybrid
Maize with same kernel colour and texture	400	400	200
Maize with different kernel colour and texture	600	600	300
Field of same inbred/hybrid seed production not confirming MSCS	400	400	200
Field of other hybrid having common male parent confirming MSCS	-	5	5
Field of other hybrid having common male parent not confirming MSCS	-	400	200
Field of other hybrid having different parent	-	400	200
Teosinte	600	600	300

Isolation distance can be modified by providing border rows or by differential blooming dates provided the contaminant is of same kernel colour and texture. Modification of isolation distance from contaminator of different kernel colour and texture is not permitted.

Isolation distance can also be modified by planting male parent of same seed lot which is used for hybrid seed production programme all around the field as border row with similar seed rate, spacing and time of sowing as recommended for seed production programme. Distance between border row and seed parent should not be more than 5 meter and the population in border row should be uniform. Isolation distance between two plots of hybrid seed production programme, which involves different pollinators, can be modified by planting border rows in both the fields. Planting of one border row can reduce 12-13 m isolation distance. Modification of isolation distance by other barriers is not permitted.

Isolation distance can also be modified by creating difference in blooming date of seed parent and contaminator provided 5% or more of the seed parent plants do not have receptive silk when more than 0.05% plants of field at isolation have shedding pollen.

Optimum planting ratio of seed and pollinator parent for production of hybrid seed and seed of male sterile parent is 2:6 (female). 12 kg seed of seed parent and 5 kg seed of pollinator is sown in one hectare field with 70 cm row to row and 25 cm plant to plant spacing. Normally flowering in both the parents occurs at the same time but in some hybrids staggering period of 7 days is required for synchronization of flowering.

Removal of male inflorescence from the female plant is known as detaselling. With the emergence of tassel (before anthesis) the stalk of female plant is hold just below the tassel with one hand and the entire tassel is removed with other hand by upward pull without leaving any part of tassel and any injury to plant. This tassel is immediately thrown on the ground. Detasseling is done daily till all the seed parents get detaselled.

Offtype plants from male parent should be rogued out before flowering. Whereas, offtype and pollen shedder plants in seed parents should be removed as and when observed. Seed production programme is inspected for atleast 4 times (Table 6). One field inspection is made before flowering, whereas, remaining three during flowering to certify Foundation and Certified seed production programme. Sucker tassels, portion of tassels and main tassel of offtype or seed parent with more than 5 cm spike having exerted anthers from glumes are considered as offtype and pollen shedder, respectively.

Table 6. Seed certification standard at field level for production of Foundation and Certified seed of hybrid maize

Factors	**Foundation**		**Certified**
	Inbred	**Single cross**	
Offtype plants that have shed or shedding pollen when 5% or more of the plants in the seed parent have receptive stigma			
(a) Offtype plants	0.2	-	-
(b) Offtype plants in male parent	-	0.2	0.5
(c) Offtype plants in female parent	-	0.5	1.0
Total of offtype pollen shedding tassels including offtype tassels observed in previous inspections	-	1.0	2.0
Offtype plants in seed parent in final inspection	-	0.2	0.5

Harvesting: Cobs from female lines are harvested when the husk turned yellow and grains become hard. These cobs are stored as such after removal of their husk.

Seed standard: Cobs are inspected after harvesting and dehusking and the lot containing more than 0.2% offtype cobs including cobs with off coloured seed is rejected.

Shelling is started after approval from the seed certification agency. Samples submitted to seed testing laboratory are tested for verification of seed characters (Table 7).

Table 7. Certification standard for Foundation and Certified seed of hybrid maize

Factors	**Foundation**		**Certified**
	Inbred	**Single cross hybrid**	
Pure seed (%) maximum	98	98	98
Other crop seed (no./kg) maximum	5	5	10
Weed seed (no./kg) maximum	None	None	None
ODV (no./kg) maximum	5	5	10
Germination (%) minimum	80	80	90
Moisture (%)maximum			
(a) Sample not in vapour proof container	12	12	12
(b) Sample in vapour proof container	8	8	8

8. USE OF MALE STERILITY FOR HYBRID SEED PRODUCTION

In all three types of male sterility system are used for commercial production of hybrid seed Cytoplasmic. Genetic and Cytoplasmic and genetic

8.1. Cytoplasmic male sterility (CMS)

Principle : Male sterility under the control of cytoplasmic factors without any influence of nuclear gene is considered as cytoplasmic male sterility. The hybrid produced by this system is sterile due to maternal inheritance of the cytoplasm (Fig. 2).

The cytoplasmic gene responsible for male sterility

S = Sterile cytoplasm

F = Fertile cytoplasm

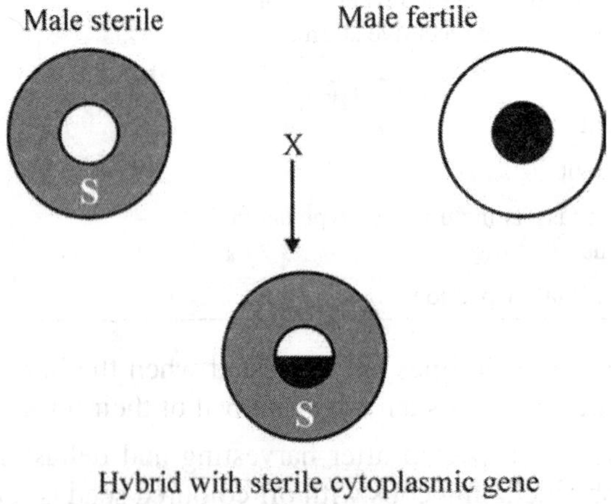

Fig. 2. Use of CMS system for production of hybrid seed

Seed production: This system requires three lines for hybrid seed production programme (Table 8).

Table 8. Cytoplasmic constitution of the parents and behaviour of male parent

Cytoplasm	Nuclear gene	Behaviour of male part	Name
Fertile	different than seed parent	Fertile	Male parent
Sterile	different than male parent	Sterile	Seed parent
Fertile	same as seed parent	Fertile	Maintainer
Sterile	50% male parent 50% seed parent	Sterile	Hybrid

Nucleus and Breeder's seed production programme: Seed of male parent and maintainer lines are multiplied by selfing whereas, seed of male sterile line is multiplied by crossing between male sterile line and maintainer in an isolated field.

Foundation seed production programme: Seed production programme of seed (male sterile) and male parent are organized on government farms under the supervision of seed certification agency as per method described for Nucleus and Breeder's seed production programme (Fig. 3).

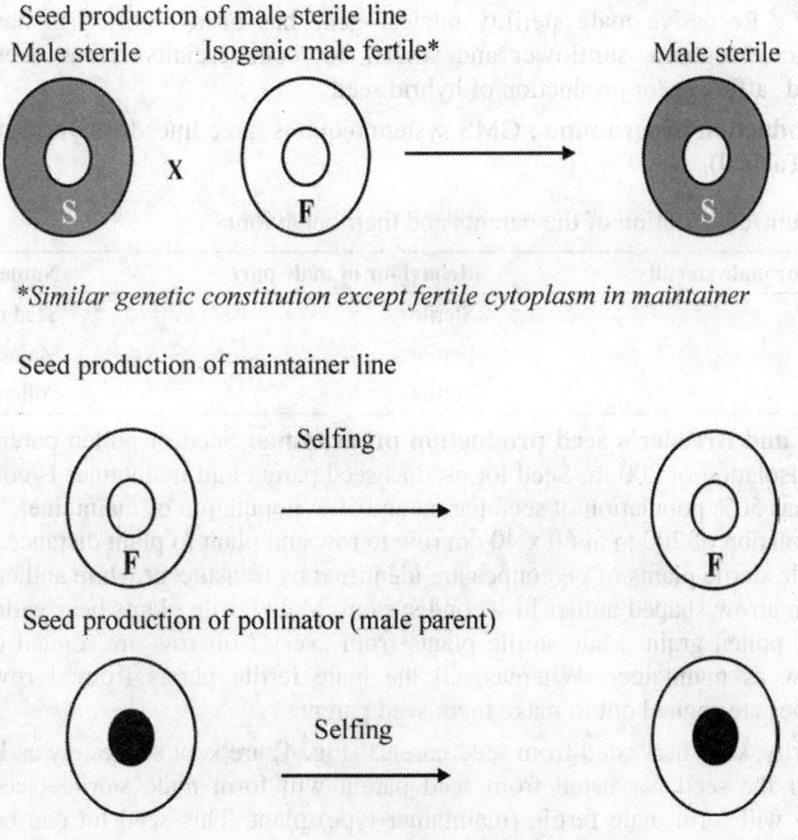

Fig. 3. Foundation seed production in cytoplasmic male sterile system

Certified seed production programme: Hybrid seed produced by the farmer with the use of Foundation seed of seed (male sterile) and male parent under the supervision of Seed Certification Agency is known as Certified seed (Fig 4).

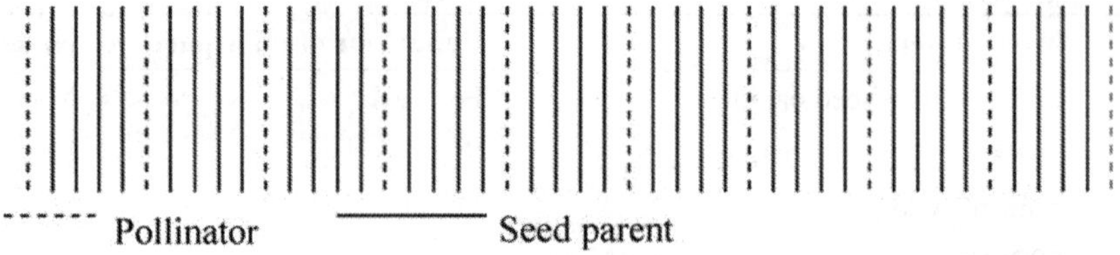

Fig. 4. Certified seed production programme of hybrid seed using CMS system

Example : This male sterility system is exploited in crops where seed is not used as food like onion or ornamental plants like *Petunia*.

C.2. Genetic male sterility (GMS)

Principle : Male sterility controlled by nuclear gene without any influence of cytoplasm is known as GMS.

Example : Recessive male sterility nuclear gene has been reported in barley, *Brassica*, pigeonpea, rice, sorghum, sunflower and wheat, but commercially it has been exploited in pigeonpea and safflower for production of hybrid seed.

Seed production programme : GMS system requires three lines for hybrid seed production programme (Table 9).

Table 9. Genetic constitution of the parents and their behaviour

Nuclear gene for male sterility	Behaviour of male part	Name
msms	Sterile	Seed parent
Msms	Fertile	Maintainer
MsMs	Fertile	Pollen parent

Nucleus and Breeder's seed production programme: Seed of pollen parent is multiplied by selfing in isolation of 200 m. Seed lot used as seed parent and maintainer is common because this seed lot has 50% population of seed parent and 50% population of maintainer. This seed lot is sown in an isolation of 200 m at 60 x 40 cm row to row and plant to plant distance. At the time of flowering male sterile plants of pigeonpea are identified by translucent white anther in one system or dark brown arrow shaped anther in second system. Male fertile plants bear yellow anther with abundance of pollen grain. Male sterile plants from every fifth row are rogued out in order to make the row as maintainer. Whereas, all the male fertile plants from 4 rows in between maintainer lines are rogued out to make them seed parent.

At maturity, seed harvested from seed parents (Fig. 4) are kept separately as Breeder's seed. 50% seeds of the seed harvested from seed parent will form male sterile (seed parent type) whereas 50% will form male fertile (maintainer type) plant. This seed lot can be used as seed parent for production of hybrid seed or again as source seed for multiplication of seed and maintainer parent.

Foundation seed production programme: Seed production of seed and pollinator parent by the method described for Breeder's seed production programme at the government farm under the supervision of seed certification agency is known as Foundation seed (Fig 5).

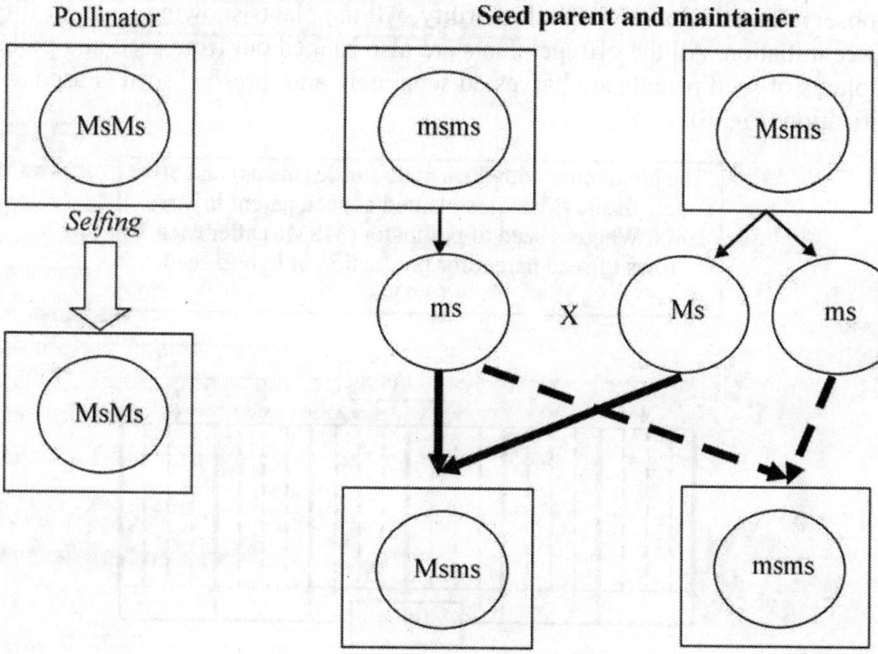

50% population will be male fertile (Msms) and 50% male sterile (msms) without any mark of identification on seed

The population with 50% male sterile (msms) and 50% male fertile (Msms) is used as sowing material to produce seed for of both Seed and Maintainer parent

In the filed male fertile (Msms) and male sterile (msms) plants are tagged with thread of different colours

The seed (Msms/msms) from male sterile (msms) plants are harvested to use as seed parent / maintainer

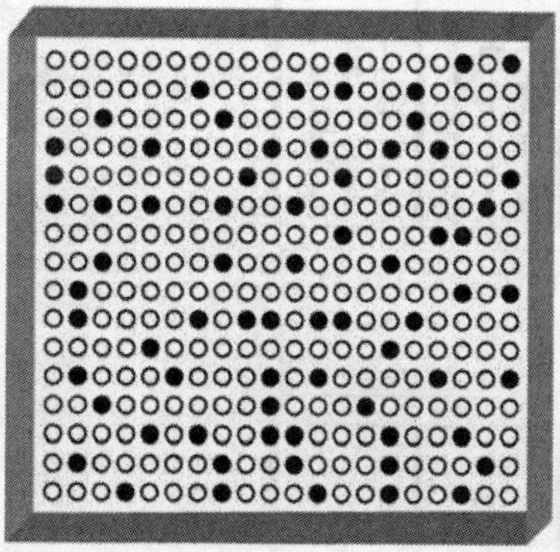

*** Heterozygous line for fertility with all the remaining characters as seed parent*
**Homozygous line for sterility*

Fig. 5. Seed production of Seed parent, Pollinator and Maintainer in Genetic Male Sterility System

Certified seed production programme: Seed of pollen and seed parent are sown at 1:5 (female) planting ratio with 60 x 40 cm of row to row and plant to plant spacing in the field maintaining 100 m isolation distance. At the time of flowering each and every plant of seed parent is observed meticulously for male sterility. All the plants showing male fertility are rouged out at flower initiation. All the offtype plants are also rouged out from seed and pollen parent. At maturity, plants of seed parents are harvested separately and threshed seed is used as hybrid seed after certification (Fig. 6).

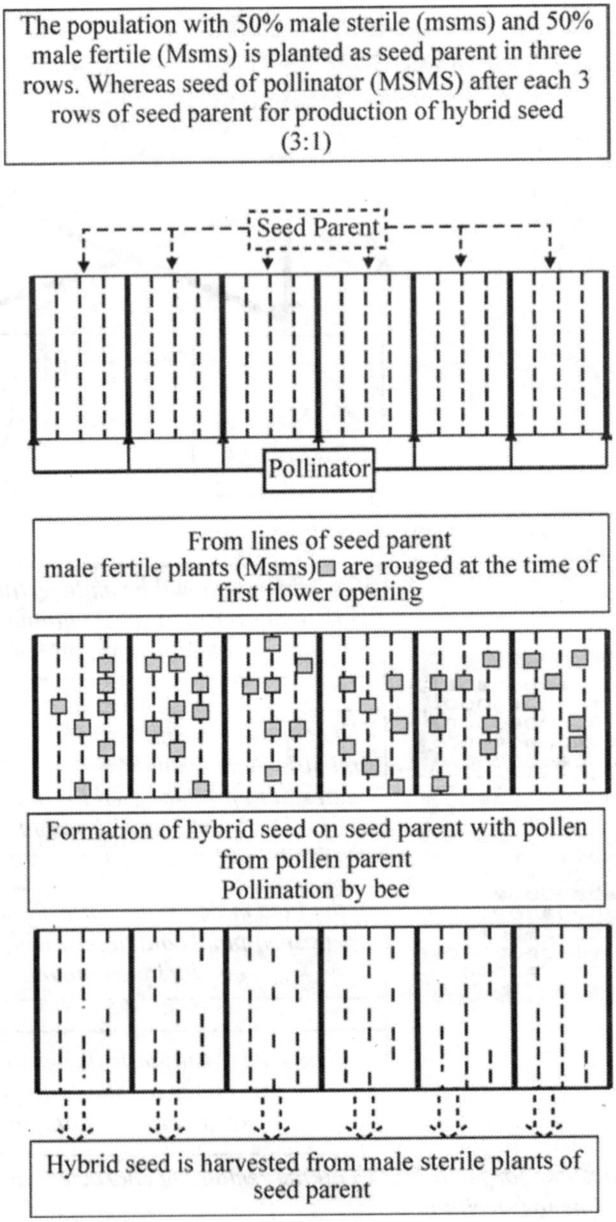

Fig. 6. Hybrid seed production in Genetic Male Sterility System

C.3. Cytoplasmic genetic male sterility (CGMS)

Principle : In this system male sterility is controlled by the interaction of cytoplasm with nuclear gene. Male sterility is expressed by homozygous recessive gene (msms) only in the presence of sterile(s) cytoplasm. The corresponding dominant allele (MS) expresses male fertility in both types of cytoplasm (Fig. 7).

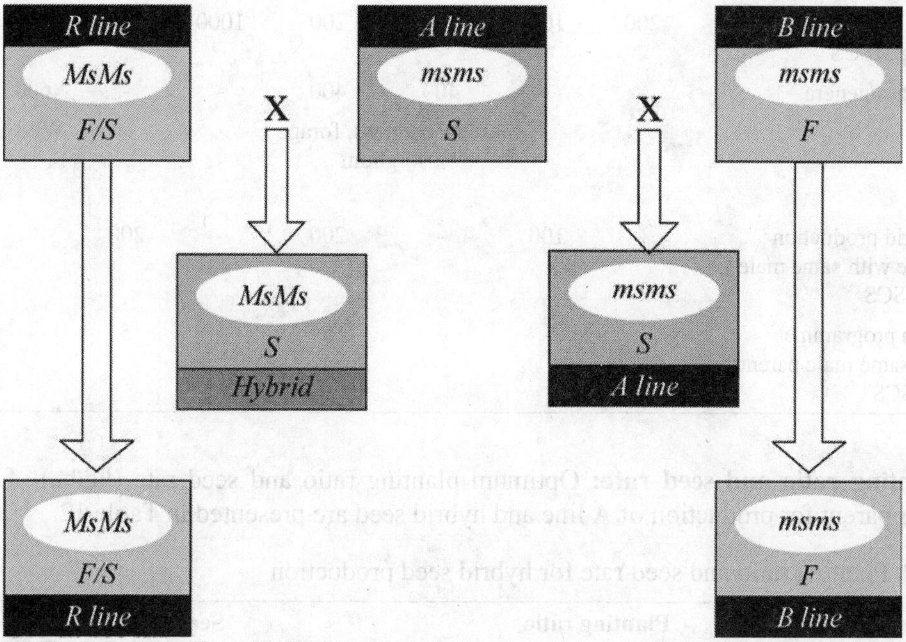

Fig. 7. Production of hybrid seed by CGMS system

Example: CGMS system has been commercially exploited for hybrid seed production of rice, sorghum, pearl millet, sunflower and pigeonpea.

Seed production: It requires three lines for production of hybrid seed (Table 10).

Table 10. Genetic constitution of parents and their behaviour in CGMS system

Nuclear gene	Cytoplasm for sterility	Behaviour of male part	Name
msms	Sterile	Sterile	Male sterile line (A)
msms	Fertile	Fertile	Maintainer (B)
MsMs	Sterile/Fertile	Fertile	Restorer (R)
Msms	Sterile	Fertile	Hybrid

Table 11. Requirement of isolation distance for Foundation and Certified seed production programme

Source of contamination	Isolation distance (m)							
	Rice		Sorghum		Pearlmillet		Sunflower	
	F	C	F	C	F	C	F	C
Other varieties / parents	200	100	300	200	1000	200	600	400
Same seed production not confirming MSCS	200	100	300	200	1000	200	600	400
Other species/genera	-	-	400	400	-	-	600	400
			Johnson grass, forage sorghum				Wild sunflower	
Other hybrid production programme with same male without MSCS	-	100	-	200	-	200	-	400
Production programme involving same male parent having MSCS	-	-	-	5	-	5	-	400

Planting ratio and seed rate: Optimum planting ratio and seed rate (kg/ha) of seed and pollinator parent for production of A line and hybrid seed are presented in Table 12.

Table 12. Planting ratio and seed rate for hybrid seed production

Crop	Planting ratio		Seed rate (kg/ha)	
	Pollinator	Seed	Pollinator	Seed
Rice	1	06	10	25
	2	08	15	20
Sorghum	2	04	02	08
	2	06	1.5	8.52
Pearl millet	2	04	0.75	1.5
	2	14	0.5	2.0
Sunflower	1	04	2.5	07
Pigeonpea	2	04	1.5	10

In sunflower a block with total 1/5 to 1/4 area is planted separately with male parent whereas in the remaining area female parent is planted. The pollen grains are collected from the male parent and transfer to female parent by hand for increase in seed set percentage.

Spacing : Optimum planting distance for hybrid and A line seed production programme is presented in Table 13.

Table 13. Plant geometry (cm) for hybrid and A and R line seed production

		Pollinator x pollinator	Pollinator x seed parent	Seed x seed parent
Rice	Row	30	30	15
	Plant	15	15	15
Sorghum	Row	75	75	75
	Plant	7.5	7.5	7.5
Pearl millet	Row	75	75	75
	Plant	50	50	20
Sunflower	Row	60	60	60
	Plant	30	30	30
Pigeonpea	Row	90	90	90
	Plant	30	30	30

Synchronization of the flowering : Pre-ponement of flowering by 2-3 days in late flowering parent of all the crops (barring pigeonpea) can be achieved by foliar spray of 3% urea just before flowering. Flowering can be delayed by 4-7 days by jerking the ear heads of early flowering parents of sorghum and pearl millet.

Supplementary pollination: In most of the *indica* CMS lines of rice 25-30% spikelets of a panicle remain inside the flag sheet. To expose the flowers of rice for enhancing cross pollination 1/2 to 2/3 parts of leaves including flag leaf over lapping the panicles are cut down 2-3 days prior to panicle emergence. Spray of 50 g gibberellic acid/hectare in three split doses i.e., 10 g at 5%, 20 g at 10% and 20 g at 30% panicle emergence stage help in complete exertion of panicle. Natural pollination in rice is supplemented by rope pulling or rod driving at 30 minutes interval during anthesis in the morning hours. Pollen shedding of restorer line of sorghum can be increased by tapering of the plant at the time of anthesis. Pollination in sunflower is supplemented by rubbing the 'puff' on flower of male parent then on female parent when population of bee is low.

Rouging : Offtype, plants infected with objectionable seed borne disease and pollen shedder are rouged out as and when observed. Plants of sorghum infected by sugary/ergot disease are also rouged out in order to maintain the maximum seed certification standard at seed level.

Field inspection : Foundation and Certified seed production programmes are inspected minimum of 4 times for verification of certification factors at field level (Table 14).

Table 14. Stages of field inspection and factors to be verified

Crop	Number of field inspection/stage of crop		
	I	II and III	IV
	Before flowering	During flowering	At maturity
Rice	Isolation, volunteer plants, planting ratio, error in planting	Offtypes, pollen-shedder	True nature of the parent
Sorghum	Isolation, volunteer plants, planting ratio, error in planting	Offtypes, pollen-shedder	True nature of the parent, grain and head smut in both the parents, separately

Pearl millet	(within 30 days after flowering) Isolation, volunteer plants, planting ratio, error in planting, Incidence of downy mildew and offtypes	Offtypes, pollen shedder and downy mildew/ green ear disease	True nature of the parent, green ear and ergot infected plants
Sunflower	(at 6-7 pairs leaf stage) Same as pearl millet	Pollen shedder, offtypes, infection of downy mildew and wild Sunflower	True nature of the parents, downy mildew, orobanche and wild sunflower
Pigeonpea	Isolation, volunteer plants, planting ratio, error in planting	Offtypes, pollen-shedder	Offtypes, True nature of the parent

Offtype : Field is certified on the basis of seed certification standard prescribed for offtypes in pollinator and seed parent and pollen shedder in seed parent (Table 15). Sterile plants present in pollinator (R or B line) are not considered as offtype.

Table 15. Maximum certification standard for offtypes and pollen shedder

Crops	Seed parent				Pollinator	
	Offtype(%)		Pollen shedder (%)		Offtype (%)	
	F	C	F	C	C	F
Rice	0.05	0.20	0.05	0.1	0.05	0.20
Sorghum	0.05	0.10	0.05	0.1	0.05	0.10
Pearl millet	0.05	0.10	0.05	0.1	0.05	0.10
Sunflower	0.20	0.50	0.05	1.0	0.20	0.50
Pigeonpea	0.10	0.20	0.10	0.20	0.10	0.20

Objectionable weed plants: Objectionable weed plants are verified at and after flowering. In the event of more number of objectionable weed plants than the MSCS (Table 16) the seed production programme is rejected.

Table 16. Maximum seed certification standard for objectionable weed plants (%) at field level

Crop	Objectionable weeds	F	C
Rice	*Oryza sativa* var. *fatua*	0.01	0.02
Sunflower	Wild *Helianthus*	Nil	Nil

Objectionable diseases: Objectionable seed borne diseases are verified at prescribed stage of field inspection (Table 17) on the basis of described symptoms. In the event of more number of grain and head smut infected earhead in sorghum (Table 18) than the seed certification standard the request can be made only once for reinspection after removal and burning of the diseased earhead. Complete earhead of pearl millet is considered as one infection unit of green ear disease. Infection of ergot in pearl millet even within seed certification standard should be reported in order to recommend the flotation treatment with brine before seed analysis. Pearl millet seed with infection of grain smut more than the seed certification standard can be certified with recommendation of seed treatment by any mercurial fungicide within one month prior to sowing.

Table 17. Maximum seed certification standard for objectionable seed borne disease

Crop	Disease	A line		R/B line		Stage of field inspection
		F	C	F	C	
Sorghum	Grain and head smut	0.05	0.1	0.05	0.1	Final
Pearl millet	Downy mildew/green ear	0.05	0.1	0.05	0.1	Any
	Ergot	0.02	0.04	-	-	Final
	Grain smut	0.05	0.1	-	-	Final
Sunflower	Downy mildew	0.05	0.05	-	-	Each
	Orobanche	Nil	Nil	-	-	Final

Harvesting : The pollinator (R or B) should be harvested first and the produce should be kept in isolation from hybrid or A line seed.

Seed testing : Sample submitted in seed testing laboratory is verified for seed certificates standard as described in Table 18.

Table 18. Seed certification standard at seed level

Factors	Rice		Sorghum		Pearl millet		Sunflower		Pigeonpea	
	F	C	F	C	F	C	F	C	F	C
Pure seed (%)	98	98	98	98	98	98	98	98	98	98
Husk less seed (%)	2	2	-	-	-	-	-			
Other crop seed (number/kg)	10	20	5	10	10	10	Nil	Nil	5	10
Total weed seed (number /kg)	10	20	5	10	10	20	5	5	5	10
Objectionable weed seed (number/kg)	2	5	-	-	-	-	Nil	Nil	-	-
	wild rice						wild sunflower		-	-
Disease infection	0.1	0.5	0.02	0.04	0.02	0.04	Nil	Nil	-	-
(% by number)	bunt		ergot		ergot		*Orobanche*		-	-
Germination (%)	80	80	75	75	75	75	70	70	75	75
Moisture (%)										
Not in vapour proof container	13	13	12	12	12	12	9	9	9	9
In vapour proof container	8	8	8	8	8	8	7	7	8	8

D. Chemical hybridizing agents (CHA)

These are the chemicals with the ability to produce male sterility by killing the male gametes. Since late 1970's efforts have been made to use CHA as male gametocide for abolition of male gamete.

It has following advantages (Mahajan and Nagrajan, 1998)

- No need to develop and maintain sterile and restorer parents.
- Any genotype may be used as female parent
- Sufficiently large quantity of seed can be produced.
- An extensive testing can be made that avoid losses by ensuring poor cross-pollination
- It saves time needed for transferring male sterility to agronomically useful variety through back crossing.

Quality of gametocide

- It should not affect the female fertility
- It's effect should be uniform during whole blooming period
- It's effectiveness should not be influenced by environment and genotype
- It should have minimum side effect on plant growth
- It should be systemic in nature
- It should be cost effective
- No harm to human health

Time of application: By nature gametocides are crop specific (Table 19) and applied just before the initiation of flowering in ppm.

Table 19. Crops and effective gametocides

Crop	Gametocide	
Rice	Zinc methyl arsenate, sodium methyl arsenate, ethrel, Gibberellin, TIBA, Natrium arsenale	
Wheat	Ethrel, malic hydrazide	
Maize	Gibberellins	
Sunflower	Gibberellins	
Groundnut, cotton	FW 450	
Tomato	FW 450, malic hydrazide	
Onion	Malic hydrazide, gibberllins	
Sugarbeet	Ethrel FE 450, (2 chloroethyl phosphoric acid)	
Lettuce	Gibberellin	
New CHA		
Phenyl cinnoline carbonylates SC 1058 and Sc 1271		Azetidine-3-carboylic acid
Hybrex		Diketones
Gametan		SD 84811
MON 8164		Chlorogenic acid

Major problems associated with CHA

- It expresses partial to complete male sterility.

- It may affect female fertility at higher dose.
- Study on dose and duration of treatment requires more specificity as at low dose male sterility remain low and at higher dose female fertility is affected.
- Performance of chemicals is different on different crops/genotypes. Broad-spectrum chemical may have more advantage.
- Many chemicals have different performance on different tissues as some chemical may inhibit spore formation and some bring changes in the tapetum cell whereas other deteriorates vascular system.
- The effectiveness of the chemical depends on the developmental stage of the plant organ receiving chemical treatment. In wheat, the gametocidal effect is efficient when applied during the premeiotic DNA synthesis, which is not synchronous in all the tillers and spikelets, thus resulting in inconsistencies in expression of male sterility.
- The performance of chemical is modified by the physiological state of the plant and prevalent environmental condition. Therefore it is very difficult to recommend specific concentration and time of spray to achieve optimal result.
- The spray has to be done regularly to get the pollen sterility for longer period of time.
- Increase female sterility due to damage in ovule is the major draw aback.
- Chemicals may induce morphological, phenological and genetic variations.
- The chemicals may have phytotoxic effect. Chemical deposit on the seed may be health hazardous.
- The chemical may significantly alter the nutritional quality of the produce.

30 E. Use of self-incompatibility in hybrid seed production

Many bisexual flowering plant species possess a reproductive strategy called self-incompatibility (SI) that enables the female tissue (the pistil) to reject self but accept non-self pollen for fertilization. Three different SI mechanisms have been reported, each controlled by two separate, highly polymorphic genes at the *S*-locus. Inhibition of self pollination due to inability of viable pollen to germinate on the stigma of the same flower or fertilize the ovule present in the same flower is considered as self-incompatibility. Self-incompatibility was first reported by Köelreuter (1761-1766) in *Verbascum phoenicuem*. It has been reported in more than 300 species representing 71 families of angiosperm (de Nettancourt, 1977). It is of following types

- Heteromorphic
- Homomorphic
- Gametophytic
- Sporophytic

Out of these only homomorphic self incompatible system is discussed in reference to hybrid seed production. In this system, self-incompatibility is under the genetic control rather than morphological difference in gynoecium and androecium.

The reasons of self-incompatibility are following pre-fertilization barriers

- Non germination of viable pollen grain

- Pollen tube is not able to enter in the stigma
- Pollen tube is not able to reach upto embryo sac due to short of length
- Slow growth of pollen tube
- Degeneration of embryo
- Genetics

Many SI systems are phenotypically plastic considered as pseudo- self compatibility (Darwin 1876) that may show some seed formation on selfing.

The first use of SI in hybrid seed production was reported in Chinese cabbage in the year 1949. The phenomenon is used to produce hybrid seed in many horticultural (cabbage, cauliflower and broccoli), forage crops (rape) and ornamental (*Ageratum haustonianum* and *Petunia hybrida*) crops (Table 20).

The advantages of self incompatible system over male sterility system are

- Presence of active nectar in the flower.
- Lower breakdown with environmental fluctuations.
- Higher hybrid seed reproducibility / unit area due to hybrid seed is harvested from both the parents.

Table 20. Hybrid seed production in vegetables

System	Crops
Male sterility	
Cytoplasmic	Carrot (*Daucus carota*); Leek (*Allium porrum*); Onion (*Allium cepa*)
Genetic	Celery (*Apium graveolens*)
Diocious nature	Asparagus (*A. officinalis*); Beet and chard; (*Beta vulgaris*); Spinach (*Spinacia oleracea*)
Hand pollination	Bitter gourd (*Momordica charantia*); Cucumber (*Cucumis sativa*); Eggplant (*Solanum melongena*); Gourd (*Benincasa hispida*); Luffa (*Luffa angulata, L. cylindrica*); Melons (*Cucumis melo*); Okra (*Abelmoschus esculantus*); Peppers (*Capsicum annuum*); Pumpkin (*Cucurbita moschata*): Tomato (*Lycopersicum esculentum*); Watermelon (*Citrullus lanatus*); Zucchini (*Cucurbita pepo*)
Self-incompatibility	Broccoli (*Brassica oleracea*); Cabbage (B. *oleracea*); Cauliflower (*B. oleracea*); Chinese cabbage (*B. rapa*); Chinese mustard (*B. juncea*); Pakchoi and Petsai (*B. rapa*); Radish (*Raphanus sativus*); Turnip (*B. rapa*)

In SI system based hybrid seed production programme, hybrid seed is harvested from both the parents. Therefore the parents are designated as parent 1 and 2.

Hybrid seed production with the help of self-incompatibility system requires cycling of S alleles for retrieval of specific genotypes for S allele combination.

Six types of crosses involving S alleles are possible in homomorphic self-incompatibility system (Fig. 8). Both the self-incompatibility systems are divergent in cycling of S allele.

First parent		Second parent
S_1 S_3 and S_2S_3	X	S_1S_3 and S_1S_4

First parent	Gamete	Second parent			
		S_1S_3		S_1S_4	
		S_1	S_3	S_1	S_4
S_1 S_2	S_1	X	S_1S_3	X	S_1S_4
	S_2	X	S_2S_3	X	S_2S_4
S_2S_3	S_2	S_1 S_2	X	S_1S_2	S_2S_4
	S_3	S_1S_3	X	S_1S_3	S_3S_4

Fig. 8. Possible crosses involving S alleles for self-incompatibility

It shows that with the availability of atleast four alleles of gene S, hybrid seed production and maintenance of male and female line is possible in gametophytic system of self-incompatibility.

30 E 1. Gametophytic system

When growth in pollen grain on female reproductive part after pollination depends on genetic constitution of male gamete and stigma of female than it is known as gametophytic self-incompatibility.

- At genetic level it is under the control of S gene with multiple alleleism i.e., S_1, S_2, S_3, S_4----S_n .
- None of the allele is dominant over other.
- Pollen grain is haploid in nature and phenotype of pollen grain is same as of its genotype.
- Stigma is diploid in nature with both the alleles have expression.. i.e., no dominance effect.
- It is an interaction between genetic constitution of stigma and genotype of pollen grain.
- A self incompatible plant with S_1S_2 genetic constitution forms two types of male gamete S_1 or S_2, whereas the genetic constitution of stigma will be S_1S_2. Due to presence of S_1 allele in stigma, pollen with S_1 allele will not germinate, similarly, due to presence of allele S_2 in stigma, the pollen with allele S_2 will not germinate. Therefore, self pollination is not possible. Pollen from any other plant with the same allele i.e., either S_1 or S_2 will also not be able to fertilize the stigma with S_1S_2 allele. For fertilization it is essential that the pollen should have other than S_1 and S_2 allele for stigma with S_1S_2 allele (Table 21).

Table 21. Behavoiur of gametophytic self-incompatibility system

Pollination	Male gamete	Stigma	Fertilization
Self	S_1	S_1S_2	No
	S_2	S_1S_2	No
Cross	S_1	S_1S_2	No
	S_2	S_1S_2	No
Cross	Other than S_1 and S_2	S_1S_2	YES

- It shows that
  - If the genetic constitution of male gamete is same as any one allele of stigma than the reaction will be incompatible.
  - If the genetic constitution of male gamete is different than both the alleles of stigma than the reaction will be compatible.
- Three types of reaction may arise in gametophytic system of SI
  - Complete incompatibility
  - Partial incompatibility
  - Complete compatibility

I	Female	Male		
	S_1S_2	S_1S_2		
Reproductive part	Stigma	Pollen grain		
Chromosome number	Diploid	Haploid		
Interaction	S_1S_2	S_1	S_2	
	S_1S_2	X	X	Complete incompatibility
II	**Female**	**Male**		
	S_1S_2	S_2S_3		
Reproductive part	Stigma	Pollen grian		
Chromosome number	Diploid	Haploid		
Interaction	S_1S_2	S_2	S_3	
	S_1S_2	X	S_1S_3 S_2S_3	Partial incompatible
III	**Female**	**Male**		
	S_1S_2	S_3S_4		
Reproductive part	Stigma	Pollen grain		
Chromosome number	Diploid	Haploid		
Interaction	S_1S_2	S_3	S_4	
	S_1S_2	S_1S_3 S_2S_3	S_1S_4 S_2S_4	Complete compatibility

- In gametophytic type of self-incompatibility, genetic constitution of male parent (S_1S_2 XS_1S_3) can be recovered but genetic constitution of female type cannot.
- Homozygote cannot be formed.
- Plant species with gametopthytic self-incompatibility system have binucleate pollen grains.
- Gametophytic system is mainly expressed in the plants with wet stigma.

Gametophytic self-incompatibility in plants involves rejection of pollen when pistil and pollen share the same allele at the S locus. This locus is highly multi-allelic, but the mechanism by which new functional S alleles are generated in nature has not been determined and remains one of the most interesting conceptual barriers to a full understanding of self- incompatibility.

Examples

Family	**Crop**
Leguminoceae	Red clover (*Trifolium pratense),* White clover (*Trifolium repens),* Alsike *(Trifolium hybridium),* Lucerne (*Medicago sativa)*
Solanaceae	Tobacco (*Nicotiana sanderae),* Tomato (*Lycopersicon peruvianum),* Potato (*Solanum araccpapa)*
Gramineae	Pearl millet (*Pennisetum glaucum),* Rye (*Secale cereale),* Wild barley (*H. bulbosum),* Festuca (*Festuca pratensis),* Phalaris (*P. coerulescens)*
Chenopodiaceae	Sugar beet (*Beta vulgris)*
Liliaceae	*Gasteria*
Commelineaceae	*Tradescantia*
Malvaceae	
Rosaceae	
Scrophulariaceae	

30 E 2. Sporophytic system

- At genetic level it is under the control of S gene with multiple alleleism *i.e.,* S_1, S_2, S_3, S_4----S_n. It has different behavior in male and female parents.

Male parent: In sporophytic system dominance effect is observed in the microspore i.e., pollen grains

S_1 is dominant over all the alleles

S_2 is dominant over all the alleles except S_1

S_3 is dominant over all the alleles except S_2 and S_1

- It means all the pollen grains formed in the flowers of one plant have phenotypic expression of dominant allele present in the male parent regardless of the genotype of pollen.

Example: If the genotype of male parent is S_1S_2. At genotypic level it will form two types of pollen grains i.e., S_1 and S_2. But, phenotypic expression of both the pollens will be as of S_1 because S_1 is dominant over S_2. The allele transferred by S_2 will be S_2 but its phenotypic expression will be as of S_1.

Pollen parent	S_1S_2		S_1S_3		S_2S_3		S_1S_4		S_3S_4	
	Pollen									
Genotype	S_1	S_2	S_1	S_3	S_2	S_3	S_1	S_4	S_3	S_4
Phenotype	S_1		S_1		S_2		S_1		S_3	

Female parent: There is no dominance expression in the stigmatic tissues of female parent. It behaves exactly same as in gametophyte system i.e., both the alleles will express. It means on

S_2S_3 stigma, pollen with phenotypic expression of S_2 or S_3 will not be able start process of fertilization.

Table 22. Behavoiur of sporophytic self-incompatibility system

Female	Male	Pollen			Seed formation	Product
Stigma		Genotype		Phenotype		
S_1S_2	S_1S_2	S_1	S_2	S_1	No	
S_1S_2	S_1S_3	S_1	S_3	S_1	No	
S_1S_3	S_3S_4	S_3	S_4	S_3	No	
S_1S_3	S_2S_3	S_2	S_3	S_2	Yes	S_1S_2, S_1S_3 S_2S_3, S_3S_3
S_1S_2	S_2S_3	S_2	S_3	S_2	No	
S_2S_3	S_1S_2	S_1	S_2	S_1	Yes	S_1S_2, S_1S_3, S_2S_2, S_2S_3
S_2S_3	S_3S_4	S_3	S_4	S_3	No	
S_3S_4	S_2S_3	S_2	S_3	S_2	Yes	S_2S_3, S_2S_4, S_3S_2, S_3S_4
S_1S_2	S_3S_4	S_3	S_4	S_3	Yes	S_1S_3, S_1S_4 S_2S_3, S_2S_4

1. Homozygous product may be formed.
2. Parental type both as male and female parent may appear.
3. Expression is either no seed formation or seed formation. There is no class as partial.
4. Reciprocal crosses may behave differently.
5. Sporophytic system of self-incompatibility is expressed by the plant species with dry stigma.
6. Plant species with sporopthytic self-incompatibility system have trinucleate pollen grains.

Examples

Family	Crop	
Cruciferae	Radish	*Raphanus sativus*
	Cabbage	*Brassica oleracia*
	Cauliflower	*Brassica oleracia*
		Eruca sativa
Compositae	Sunflower	*Helianthus annus*
	Niger	*Guizotia abyssinica*
		Cosmos bipinnatus

30 E.3. Maintenance of self-incompatible lines

For maintenance and reproduction of SI lines on a large scale for hybrid seed production, it is necessary to overcome the SI reaction after self-pollination by means of a chemical treatment.

Self-incompatibility lines are multiplied by selfing achieved with following methods for commercial seed production

- Bud pollination
- An increased concentration of CO_2
- Spraying inflorescences with NaCl solution

Bud pollination

Pollination of immature non receptive stigma with mature pollen with same genetic constitution forms seed in self incompatible crop. It appears that the chemical which inhibits the pollen tube growth on self pollination is not present in young stylar tissues. Bud pollination works for some crops but not for others, and not all flowers are easily manipulated for bud pollination.

Increased concentration of CO_2

Hybrid seed of vegetables from species of Brassica are generally produced with the exploitation of self-incompatibility system. To maintain inbred lines of self incompatible nature the plant, CO_2 gas treatment has been routinely used as a common method to overcome the self-incompatibility in seed companies. It is recommended to obtain a smaller amount of seeds. The procedure of the treatment is

- The plant is grown in a pot.
- At the time of flowering the plant is covered with a bag to avoid out-crossing and promote self-pollination.
- After pollination the pot is shifted into a closed chamber containing approximately 4% of CO_2 gas.
- The pollens of the same flower successfully germinate and penetrate the cuticle layer of the pistil bypassing the self-incompatibility in the atmosphere with 4% CO2 concentration.

CO_2 is effective when supplied to a self-pollinated flower while hundreds of pollen grains are germinated on the stigma. The effective time coincides with the attachment of pollen tubes to papilla cells. CO_2 produces a metabolic change in these cells during attachment.

Spraying inflorescences with NaCl solution

In oilseed rape, self-incompatibility is overcome by spraying 2–10% solution of NaCl (Table salt) on inflorescence. In the field conditions, 5% NaCl solution applied every 3–5 days is enough for the practical reproduction of SI lines. After spraying 5% NaCl solution at intervals of every 3 to 5 days, the compatibility index after self-pollination is almost equal to that of bud-pollination.

For research purpose self-incompatibility can be overcome by

Bud pollination: Pollination of buds 2-7days prior to anthesis eg., *Brassica, Raphanus, Petunia, Nicotiana* etc.

Delayed pollination: Pollination after 3-9 days of anthesis eg., *Brassica, Lilium* etc.

Irradiation of pistil/pollen: Irradiation of pistil or pollen with X or UV rays before pollination eg., *Petunia, Lilium, Oenothera, Rubus, Ribes, Dendrobium* etc.

High temperature treatment: Exposing plant to high temperature eg., Brassica, Raphanus, Nemesia, Lilium, Trifolium, Secale, Lycopersicon, Tradescantia etc.

Hot water treatment: Treatment of style and stigma with hot water (50ºC) for 5-6 minutes before pollination eg., *Oenothera, Lilium.*

Application of hormones: Spraying of hormones viz., Indole Butyric acid, Naphthalene acetic acid, Benzyl adenine, Gibberellic acid etc. on stigma before pollination eg., *Lilium, Petunia, Tagetes, Trifoulim, Brassica, Lycopersicon* etc.

Use of mentor pollen: Ineffective compatible pollens are known as mentor pollen. These are made ineffective by ionization/treatment with organic solvent/repeated freezing and thawing. Pollination with the mixture of incompatible viable and mentor pollen is effective in breaking incompatibility.

In few case stigma is first fertilized by mentor pollen and after few hrs by incompatible pollen to break self-incompatibility. In this case incompatible pollens are termed as pioneer pollen. Eg., *Apple, pineapple, Brassica, Nicotiana, Pyrus, Lilium* etc.

Treatment of stigma with organic solvent: washing of stigma with organic solvent viz., Hexane before pollination is effective in breaking self-incompatibility in *Brassica.*

Exposing plant to high humidity: Under high humidity incompatibility of *Brassica* is broken sown.

Exposing plant/flower to high concentration of CO_2: Incompatibility of *Brassica* is broken down under high concentration of CO_2 (4%).

Surgical treatment of the stigma: removal of upper surface of the stigma or stigma before pollination may break the incompatibility in *Brassica* and *Raphanus.*

Electric aided pollination: Application of electric current of 100v at the time of pollination on pollen and stigma for 2-3 seconds may break the incompatibility in *Brassica.*

Use of specific sugar and lectins: Treatment of pollen with glucose or lactose sugar and stigma with lectin before pollination may break the incompatibility in *Petunia* and *Eruca.*

■■■■

Chapter 37

SEED PRODUCTION OF SEEDLESS VARIETIES

1. PRINCIPLE

A plant with triploid (3x) ploidy level gives rise to seedless fruits.

Example : Watermelon

A tetraploid parent is pollinated by diploid watermelon to produce triploid watermelon. During sexual reproduction, the new seed inherits half of each parent's genetic material. As a result, the new watermelon receives one set of chromosome from the diploid parent, and two sets from the tetraploid, ultimately making it triploid. The triploid hybrid is virtually seedless. It produces very few seeds, and these can be planted to grow new watermelon. But the new plant must be pollinated by standard diploid watermelon in order to produce fruit. Triploid seedless watermelon produces little or no pollen. This is a result of chromosomal imbalance produced when triploid germ line cells attempt to undergo meiosis. Seeds arrest post fertilization in early development (after pollination from normal diploid plants) again as a result of the chromosomal imbalances generated during meiosis in the egg mother cell. The lack of viable gametes makes the process inactive with regards to gene containment.

2. REQUIREMENT

Diploid seed of female and pollinator parent and chemical colchicine

3. NUCLEUS AND BREEDER SEED PRODUCTION PROGRAMMES

Seeds of diploid seed of both female and pollinator parents are multiplied by selfing under space isolation.

4. FOUNDATION SEED PRODUCTION PROGRAMME

Seed of parents are multiplied under the supervision of seed certification agency in an isolated field (Table 1) by selfing (Fig 1). The field and seed are tested for certification standard (Table 2 and 3) by field and seed inspectors.

Table 1. Space requirement to keep the Foundation and Certified seed production programme of seedless watermelon in isolation to avoid out-crossing

Contaminator	Foundation	Certified
Field of other varieties, same seed production programme without seed certification standards and wild water melon (*Citrallus colocynthis*)	1500m	1000m

Commercial seed grower obtains seed or seedlings from the developer and planting is done in an isolated field. The supplied seedling may be of tertapolid parent after colchicine treatment. In the event of seed of diploid parent is provided then induction of polyploidy is essential. The field for production of tetraploid seed should be 1500m away from any other watermelon field, so that bees (pollen vector) do not pollinate the seedless plants with other varieties. The grower carefully maintains the plants, by treating them with a growth hormone. Bees pollinate the plants, to produce ample quantity of stock seed.

Induction of polyploidy: Diploid seed is sown for production of tetraploid seedling. One drop of 0.2 to 0.4% aqueous solution of colchicine is applied on the stem apex growing points of young seedlings at 1 to 2 true leaf stage for consecutive days to induce tetraploidy (44 chromosomes). The treatment is given under controlled conditions avoiding direct sunlight. The tetraploid plants are characterized by broad thick leaves, bigger pollen grain, broad and bigger seed.

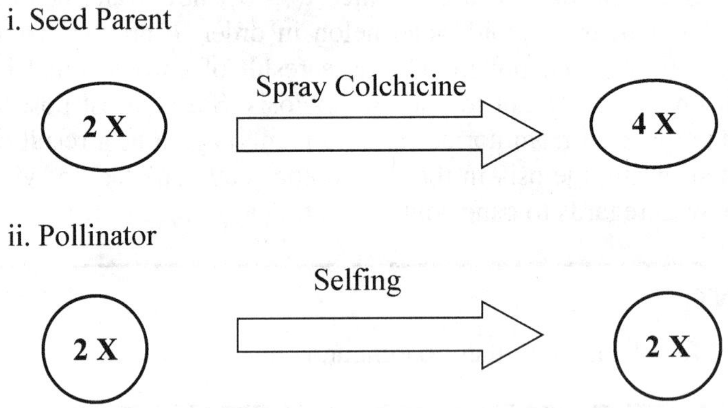

Fig. 1. Foundation seed production programme (schematic representation)

For extraction of **tetraploid seed** fermentation method should be avoided and before sowing tetraploid seed should be soaked overnight. The maintenance of tetraploid lines at stable level is important and continuous selection for improvement of quality and vigour in tetrapolid lines has to be done for atleast 4-5 generations. The seed of diploid line is produced in isolation.

Certified seed production programme: Foundation seed of parents (Fig 2) are planted in two different blocks with 5 m distance between them in a field located at safe isolation distance (Table 1).

The stock seed (tetraploid and diploid) is planted in a seed production field along with a variety of regular diploid watermelons. Pollen from the male flower of diploid plant is transferred to the female flower of tetraploid plant by hand. The fruit thus developed contains numerous triploid seeds. These seeds are marketed to produce seedless fruit.

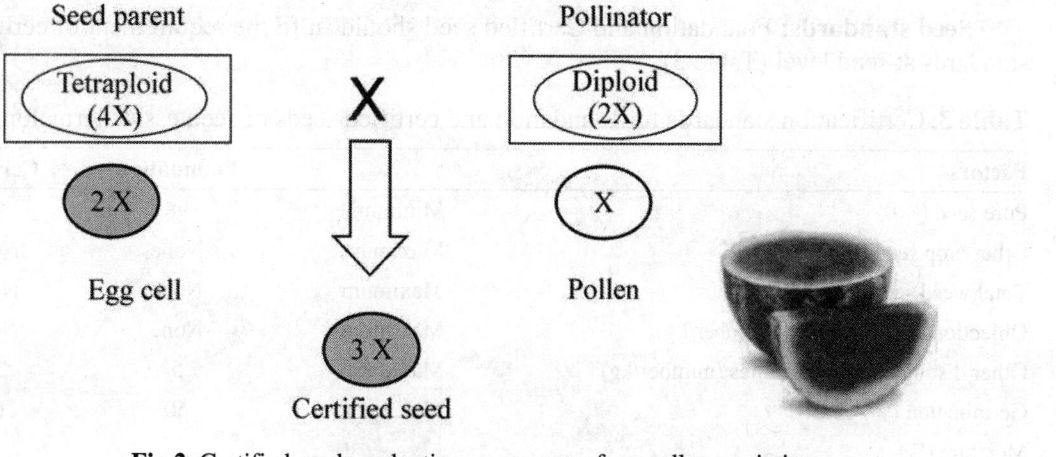

Fig 2. Certified seed production programme for seedless varieties

Seed Production of Tripliod (3x)

Watermelon is a monoecious type of species with presence of both male and female flowers on the same plant at different places. The tetraploid and diploid plants are planted in 1:1 ratio. Triploid seed of watermelon is produced by

(a) Hand pollination in female tetraploid with pollen from diploid male and removal of male flower from tetrapoloid before anthesis.

(b) Removal of male flower from tetraploid line and allow to open or natural cross pollination with diploid.

From one hectare land nearly 15 kg seed/ha is produced under open pollination. The cost of triploid seed production is very high.

Rouging : Offtype plants and plants of wild watermelon should be removed before flowering from seed and pollinator parents.

Emasculation : Male flowers from seed parent should be plucked-off with care prior to anthesis.

Field inspections: For Foundation and Certified seed production programme a minimum of four inspections should be made to verify isolation distance and volunteer plants before flowering; plucking of male flowers from seed parent (certified seed production programme) and offtypes during flowering and; true characteristics of fruits at maturity as per prescribed seed certification standards (Table 2).

Table 2. Maximum certification standards for offtype and objectionable weeds at field level for production of seed of seedless watermelon

Factors	**Foundation**	**Certified**
Offtype in seed parent (Tetraploid)	0.01%	0.05%
Offtype in pollinator	0.01%	0.05%
Pollen shedding flowers in seed parent	0.10%	None
Wild water melon (*Citrallus colocynthis*)	None	None

Seed standards: Foundation and Certified seed should fulfil the requirement of certification standards at seed level (Table 3).

Table 3. Certification standards for foundation and certified seeds of seedless watermelon

Factors		Foundation	Certified
Pure seed (%)	Minimum	98	98
Other crop seed (number/kg)	Maximum	None	None
Total weed seed (number/kg)	Maximum	None	None
Objectionable weed seed (number/kg)	Maximum	None	None
Other distinguishable varieties (number/kg)	Maximum	5	10
Germination (%)	Minimum	60	60
Moisture (%)			
(i) Sample in vapour proof container	Maximum	6	6
ii) Sample not in vapour proof container	Maximum	7	7
Genetic purity for triploid nature (%)	Minimum	95	95

Testing of genetic purity

Certified seeds are verified for seedlessness by triploid ploidy nature under laboratory conditions.

- In the off-season, seeds are planted in the polyhouse on sand in plastic portray.
- The properly washed root tips of 7-8 days old seedlings are plucked under water and placed in the Farmer's fixative for 48 hr for fixation.
- The Farmer's fixative is prepared by mixing 3 part of ethyl alcohol with 1 part of acetic acid (95-100%+absolute alcohol). It is prepared just before use by adding the acid to the alcohol.
- After 48hr the fixed root tips are transferred to ethyl alcohol of 70% strength.
- Roots that have been set to sprout for two to five days seem to give the highest likelihood of finding actively dividing cells.
- The root tip is placed on a clean micro slide with the help of mounted needle.
- About 2.0 mm part with growing root tip is cut down with discard of remaining part.
- A drop of acetocarmine stain is added and leave for 2 minutes.
- The tissue is crushed by hammering with the back side of the needle.
- It is covered by a cover slip. It is pressed on a flat surface with a strong vertical pressure of thumb by placement of the slide and cover slip between double layers of paper towel/ blotter paper.
- The cover slip may be tapped 20 times by backside of the needle or a pencil.
- The root tips are viewed under a microscope (x 400 magnification) to count the number of chromosome in the cell at metaphase stage.

- Seedlings with 33 number of chromosome are considered as triploid.
- The seeds are certified for genetic purity as per prescribed standard .

Cultivation of seedless certified seed

- Triploid plants are usually grown in a greenhouse due to the fragile nature of seedling which is easily killed by fluctuation of temperatures.
- Due to hard thick seed coat germination of triploid seed is very poor. Seed is germinated under controlled conditions and then transplanted in the field to produce seedless fruit. To speed up the germination of triploid seeds the rounded end of the seed with thick coating is removed by scratching or cutting before planting without any damage to other part of the seed. Mechanical weakening of the seed coat structure such as scarification, seed coat nicking, and seed coat lateral splitting has been reported to successfully enhanced germination performance of triploid watermelon seed.
- The seed is planted in polythene bag containing light soil with mixture of sand and compost. The temperature of the polyhouse is maintained at 30°C. After germination, the temperature is maintained at 25±2°C.
- Three to four weeks old seedlings with two or three leaves are transplanted in the field.
- To stimulate formation of fruit in the flower on triploid plant pollination is essential therefore a diploid cultivar is also planted in the field with triploid plant with transfer of pollination by bee.
- In the field after every three rows of triploid, one row of diploid watermelon of same flowering duration is planted. The seed of Diploid line may be planted directly in the field earlier or at the same time. The flowers of triploid lines are pollinated by bees with the use of pollen from diploid line. After fertilization the triploid line will produce seedless fruit.

Watermelons bear separate male and female flowers on the same plant (monoecious). Only the female flowers set fruit. Bees transfer pollen from male flowers to female flowers, making fruit set possible. For successful seedless watermelon production, an adequate bee population is important to transfer the pollen from the pollenizer variety to the seedless variety (seedless watermelons do not produce pollen). It is recommended that at least three honey bee colonies be introduced for every hectare during the blooming period since native bee populations may not be adequate, or may not coincide properly with the blooming period. Research indicates that the harvest period was advanced a week or more and shortened by one week, reducing the pickings necessary by 33%. At least eight visits of bees to an individual flower of the seedless variety are necessary for adequate hormonal stimulation for normal fruit development.

- When first flowers occur under periods of temperature or moisture stress, first fruits are often seedy, rough, small and of poor quality. To reduce this risk, growers often remove first fruit by hand when it is small or delay placement of bees for one or two weeks or until the first female flowers have dropped off in the seedless variety.
- To avoid confusion at harvest time, use a distinctly different variety for the normally seeded diploid pollinator in order to easily distinguish the seedless fruit from the pollinator for marketing purposes eg., different rind color.

- Three spray of Gibberellic acid @ 5g/ha in 500 litre of water is recommended for stimulation of fruit set first before flowering bloom and two after 10-14 day interval.
- Next year farmer has to purchase fresh seed to produce seedless fruit.

Tissue culture: With watermelons, the sprouting tip of a seedless plant is placed in a Petridish filled with growth regulators and nutrients, and one tip will sprout in as many as 15 clone plants. This technique has also been used to produce seedless tomatoes.

Triploid seeds are difficult to produce in quantity often requiring close to decade to produce sufficient amounts for commercial use. This has led to the development of potentially faster molecular strategies. The basic molecular strategy to generate seedless fruits centers on the temporal seed specific expression of a cytotoxic gene, or cytotoxic combination of genes, resulting in the prevention of seed maturation at a stage that does not affect fruit development and ripening.

Parthenocarpy can be induced by the application of the plant hormone auxin on developing flowers have been suggested. Over production of Indole Acetic Acid (IAA), results in localized toxicity that arrest seed development in a fruit. The ectopic overproduction of IAA is controlled by two gene cytotoxic system (IaaM and IaaH) that convert tryptophan to IAA in embryonic cells. Expression of these genes is under the control of seed specific promoters. A hybrid with both the genes of cytotoxic system developed by involving parents contributing one of the genes individually arrest seed development. The timing of expression of the genes should be such that seeds abort too early as in later abortion it may interfere with fruit development and ripening.

RolB gene of *Agrobacterium* linked to a promoter specifically in the ovary between anthesis and early fruit development is sensitive for auxin formation to induce parthenocarpy.

Colchicine ($C_{22}H_{26}NO_6$) is a natural product extracted from *Colchicum autumnale*, known as meadow saffron. Colchicine inhibits microtubule polymerization by binding to tubulin, one of the main constituents of microtubules. Availability of tubulin is essential to mitosis, and therefore colchicine effectively inhibits the formation of spindle.

Movement of chromosome during cell division at anaphase stage towards both the pole is due to microtubules form spindle. Application of colchicine inhibits the formation of spindle thus out of two cells formed by cell division one will contain no chromosome while other will contain double the number of chromosome of the original cell *i.e.* tetraploid instead of diploid). It is fatal in animal cells, but in plants it normally tolerates it with double number of chromosome resulting in larger, hardier and fast growing plant. Colchicine is applied on the growing tip of the seedling such as apical tip, shoot or sucker to create polyploidy.

■■■■

Chapter 38

POTATO MICRO TUBERS

1. OBJECTIVE

Production of small sized tuber of potato for sowing purpose.

Microtubers are the small tubers of 100-150 mg weight produced *in vitro* (Wang and Hu, 1982). In India, potato microtubers are successfully used in basic seed production programme because it results in large number of disease free, robust and long survival (in dark) microtubers that can be produced from a single individual in short duration involving less space. These microtubers can easily be handled, transported, stored and planted for production of commercial crop, multiplication of seed and exchange of germplam.

2. CHEMICALS AND EQUIPMENT

Solid propagation medium: MS medium (Murashige and Skoog, 1962) containing gibberellic acid (0.1 mg/liter), calcium D-pantothenate (2 mg/liter), 1-naphthalene acetic acid (NAA) (0.01 mg/liter), sucrose (30 g/liter) and agar (0.6%).

Tuber induction medium: MS medium (Murashige and Skoog, 1962), containing benzylamino purine (10 mg/liter) and sucrose (8%).

Equipment: Incubator, laminar flow, autoclave and refrigerator

Chemicals: Sodium hypochlorite, distilled water

Glassware: Test tubes (150 x 25 mm), 250 ml Erlenmeyer flask, Petriplates, beakers (250 and 500 ml)

Instruments: Forceps, biological needle, scalpel

Other: Paraffin, cotton plug

3. PROCEDURE

Plant part use to initiate tissue culture is known as *explant*. It is obtained from meristem tip or axillary segment. These explants are thoroughly surface sterilized by dipping in 2% solution (v/v) of sodium hypochlorite for 5-30 minutes. After surface sterilization explants are thoroughly washed for at least 3 times with distilled water. 2-3 explants are transferred under aseptic condition for initiation of culture in pre sterilized test tube (150 x 25 mm) containing 10-15 ml of solid propagation medium. Forceps, scalpel and needles are disinfected by dipping in 95% alcohol and flaming. Tubes are closed with polypropylene plugs and sealed with paraffin.

These tubes are kept in an incubator at 25 ± 1°C under 16 hr photoperiod of 150 w/m^2 fluorescent light. The apical/axillary buds of the explants develop into shoots (plantlets) of 6-7 cm height within 15-20 days.

These plantlets are cut into pieces each containing 3-4 nodes, 15-20 such cuttings is transferred under aseptic conditions in 250 ml Erlenmeyer flasks containing 40 ml tuber induction medium. The flasks are incubated at 18 ± 1°C under complete dark condition. Tuber formation initiates within 10-20 day and it attains harvestable size in 60-80 days.

Harvested microtubers are dried at 20°C in dark and stored in sealed Petriplates at 4 ± 1°C after treatment with fungicide. These microtubers are sown directly in field to raise commercial crop. Whereas, minitubers can also, be produced by planting them in nursery bed @ 100 microtubers/m^2.

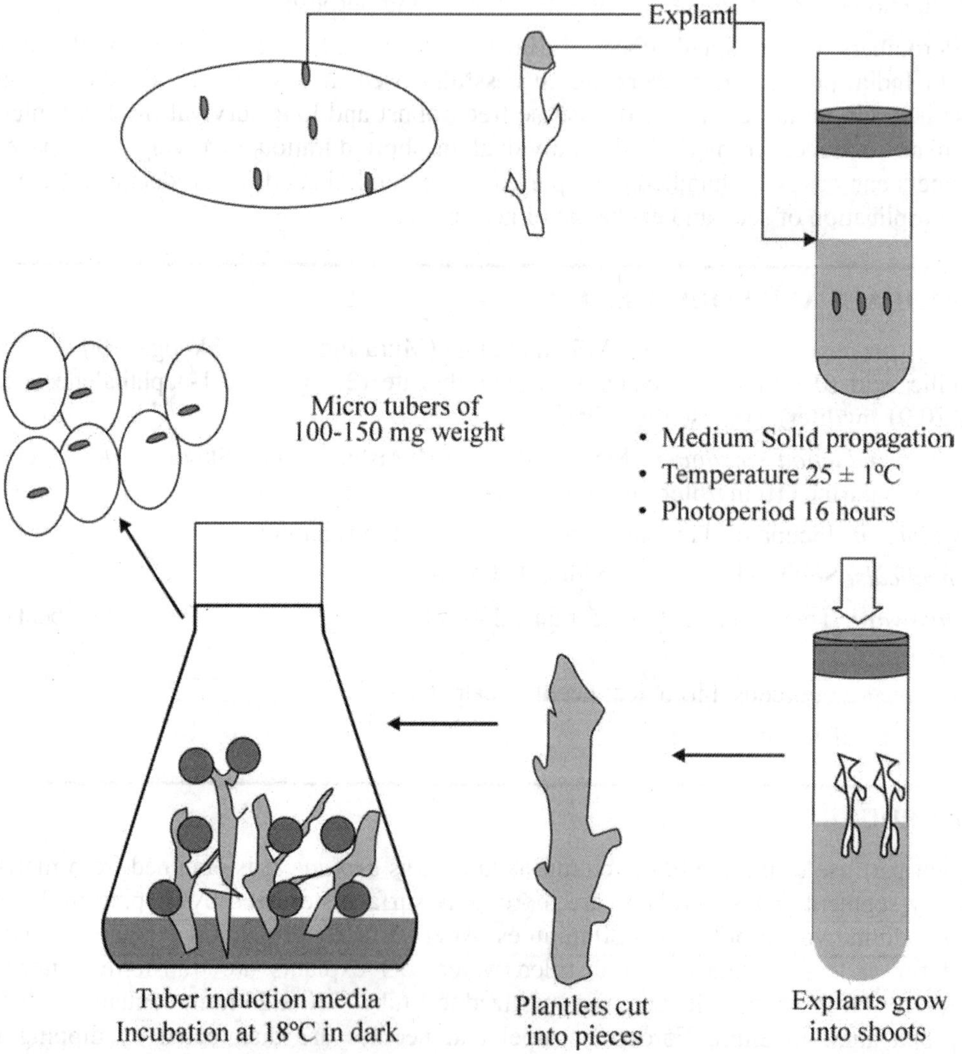

Fig. 1. Production of micro tuber

4 REFERENCES

Murashige, T. and F. Skoog (1962) A revised medium for rapid growth and bioassays with tobacco tissue culture. Physiol. Plant, 15: 473-497.

Wang, P.J. and C.Y. Hu (1982) *In vitro* mass tuberization and virus free seed production in Taiwan. Amer. Potato J. 59: 33-39.

■■■■

Chapter 39

TRUE POTATO SEEDS AND ITS PRODUCTION

1. OBJECTIVE

Production of true (botanical) potato seed

2. PROCEDURE

Potato (*Solanum tuberosum* L.) is conventionally raised from seed-tuber with 2.5 t/ha seed rate. Production of disease free tubers in such a large quantity, its transportation and storage are often encountered as the major constraints. Since, 1978, work on production of potato hybrid from the botanical seed i.e., true-potato-seed (TPS) initiated which concluded that only 100g seed is sufficient for one hectare. TPS is practically pathogen free and quite easy for transportation and storage.

The potato flower consists of 5 sepals and 5 petals. The petals are united and form a tubular structure. The stamen (5) attached to the tube have erect anthers, which form a cone around the style. Pistil consists of bilobed stigma, single style and two celled ovary (Fig. 1).

The fruit of potato is bilocular polyspermous berry. The seeds are small and flat with 0.4 to 0.6 g test weight and resemble to those of tomato and chilli (Fig. 1).

Potato plant requires long photoperiodic conditions for flowering. In plains flowering can be achieved by providing 5 hr of extra illumination during flowering i.e. December-January from 4 to 9 pm with sodium vapour lamp of 150 watt. One sodium vapour bulb is sufficient to cover an area of 100^2 m.

Selection of parents is based on the quantity of hybrid TPS, which can be produced from per unit area apart from heterosis. Usually, clones of *S. tuberosum* are used as female and *S. andigena* as male that results in heterosis with added advantage of high pollen load.

Cut pieces of tuber obtained from male and female plants are planted in separate blocks with required staggering period i.e., 7 days early planting of male than the female to facilitate pollen availability. Recommended planting geometry for male is 60 x 20 cm and 50 x 15 cm for female. The block with tuber of female parent (crossing block) is kept one meter wide with a walking space of 80 cm between blocks. 120 kg nitrogen and 80 kg phosphorus is the recommended nutritional requirement of the crossing block.

To minimize duration of pollination between first and last flower of a branch and synchronous bud and berry development flower bunches are trimmed to retain 6 buds/branch.

The male flowers with anthers ready for dehiscence are collected in the evening and stored over night at room temperature for higher recovery of pollens. Manual extraction of pollen by shaking of anthers with the help of forceps require 20 minutes, by electric buzzer 15 minutes and by motorized mini mixer 16-20 minutes/100 flowers. Motorized mini mixer gets cent per cent pollen from the anther.

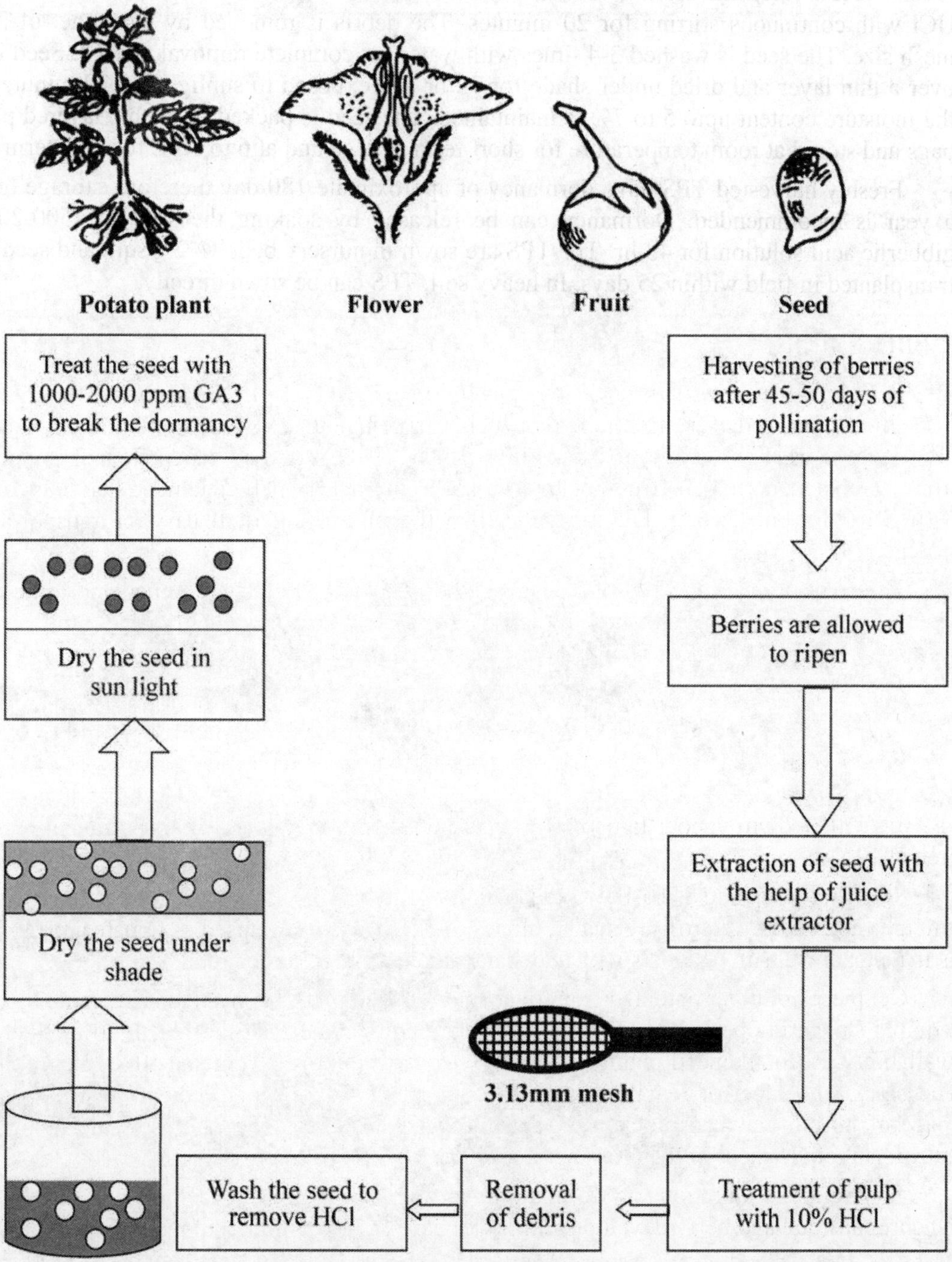

Fig 1. Flower, fruit and true seed of potato

In commercial seed production programme pollination on one female flower is done 2-3 times at the interval of 8 hr without emasculation to save the time. It allows 1-5% selfing.

Berries are picked after 45-50 day of pollination and allowed to ripen. A hand operated reverse screw juice extractor is used to extract the seed. The seed and pulp is treated with 10% HCl with continuous stirring for 20 minutes. The debris is removed by nylon net of 3.13 mm mesh size. The seed is washed 3-4 times with water for complete removal of HCl. Seed is spread over a thin layer and dried under shade for 72 hr and exposed to sunlight for 30 minutes so that the moisture content upto 5 to 7% is maintained. The seed is packed in double layered polythene bags and stored at room temperature for short-term storage and at 6 to 10ºC for long term storage.

Freshly harvested TPS have dormancy of approximate 180 day therefore, storage for atleast a year is recommended. Dormancy can be released by soaking the seed in 1500-2000 ppm gibberlic acid solution for 48 hr. The TPS are sown in nursery beds @ 2 g/sqm and seedlings are transplanted in field within 25 days. In heavy soil, TPS can be sown directly.

Chapter 40

PRODUCTION OF SYNTHETIC SEED

1. OBJECTIVE

Encapsulation of somatic embryo to serve as seed for commercial cultivation.

In some crop species propagation through seed is not successful or economically viable because of heterozygosity, minute seed size, very small food store, requirement of mycorrhyzal association for germination (orchids), vegetative propagation, production of hybrid seed is not economical, seedless varieties, very low seed multiplication ratio, peculiar climatic condition for flowering and fertilization, poor longevity, poor transportability etc. Synthetic seed is an effective and efficient alternate method of seed multiplication and availability that has developed new vistas in agriculture.

Encapsulation of tissue culture derived somatic embryo, shoot buds, cell aggregates or any other tissue that can form a plant on sowing as seed in the field and can be stored is termed as synthetic seed (not synthetic variety). The concept of seed was first conceived by Murashige (1978) and first synthetic seed was prepared by Kitto and Janaick (1982) in carrot.

Somatic embryos are bipolar structures with both apical and basal meristematic regions, capable of forming shoot and root, respectively. Somatic embryos are morphologically same as zygotic embryos but at genetic level it contains genetic material from one parent only. Therefore, all the plants formed by somatic embryo from one mother plant have same genetic constitution. This characteristic of somatic embryos allows clonal propagation with introduction of specific and directed changes in the desirable individual by inserting isolated gene sequences.

2. MATERIALS AND EQUIPMENT

Equipment: Laminar flow, autoclave, incubator, analytical balance, hot plate and pH meter.

Glassware: Culture tubes (150 x 25 mm), Petriplates (90 mm), Erlenmeyer conical flasks (100 ml) and beakers (all the glassware should be used after autoclaving).

Instrument: Forceps and scalpel, they should be kept in 90% ethanol and flamed thoroughly before use. Spirit lamp/burner, paraffin, glass marking pen and cotton.

Chemicals: Distilled, sterilized and deionized water, sodium hypochlorite, ethanol, MS medium, crop specific chemicals for inducing callus formation and somatic embryogenesis i.e., growth regulators, sodium alginate, saccharose, calcium nitrate/chloride and water soluble plastic resin.

3. PROCEDURE

Preparation of nutrient stock medium: Required quantity of MS (Murashige and Skoog) medium is dissolved in sterilized water and supplemented with growth regulators and 6-8 g agar/litre of the medium.

This medium (1000 ml) is autoclaved at 121°C (1.0 kg/cm^2) for 30 minutes. The autoclaved medium is then poured in sterile plates under aseptic condition of laminar flow.

Isolation of explant: Healthy organ made up of diploid cell is selected as explant to initiate tissue culture. These explants are submerged for 30 minutes in 2% solution of sodium hypochlorite for sterilization. After sterilization these explants are rinsed with distilled water for at least 2-3 times and kept in a sterilized container after proper drying.

Inoculation of explants: In aseptic condition slices of 1 mm thickness or diameter are cut from the sterilized tissues with the help of sharp scalpel. These pieces are transferred in test tube containing MS medium (Murashige and Skoog, 1962) with chemical inducing callus formation and somatic embryogenesis. Somatic embryogenesis is the process by which the somatic cells or tissues develop into differentiated embryos, and each fully developed embryo is capable of developing into a plantlet.

The process of embryogenesis involves various stages of differentiation and development *viz.,* proembryo, globular, torpedo and heart shaped embryos. Properly sealed test tubes are incubated at optimum required condition and time (Fig 1). The initiation of somatic embryogenesis requires auxins and/or cytokinins for initiation and proliferation of embryogenic callus and exogenous abscisic acid for synchronous maturation and low sucrose concentration in the medium.

Isolation of somatic embryos: Calli formed in incubated test tubes are observed for embryogenic callus. Somatic embryo with globular and torpedo structures are isolated. Partial dehydration of mature somatic embryos followed by rehydration enhances germination.

Encapsulation of somatic embryos: Synthetic seed is encapsulated to simulate the seed coat and endosperm. Out of several gel tested for encapsulation (agar, carboxy methyl cellulose, carrageenan, gelrite, guargum, sodium pectate, tragecanth gum etc.) alginate hydrogel was found most suitable because of

- Moderate viscosity
- Low spin-ability of solution
- Low toxicity for somatic embryos
- Quick gelation
- Low cost
- Bio-compatibility
- Long storability (The seed encapsulated by alginate has better storability than agar gel matrix).
- It enhances capsule formation
- Alginate beads provide better protection to the encased somatic embryo against mechanical injury

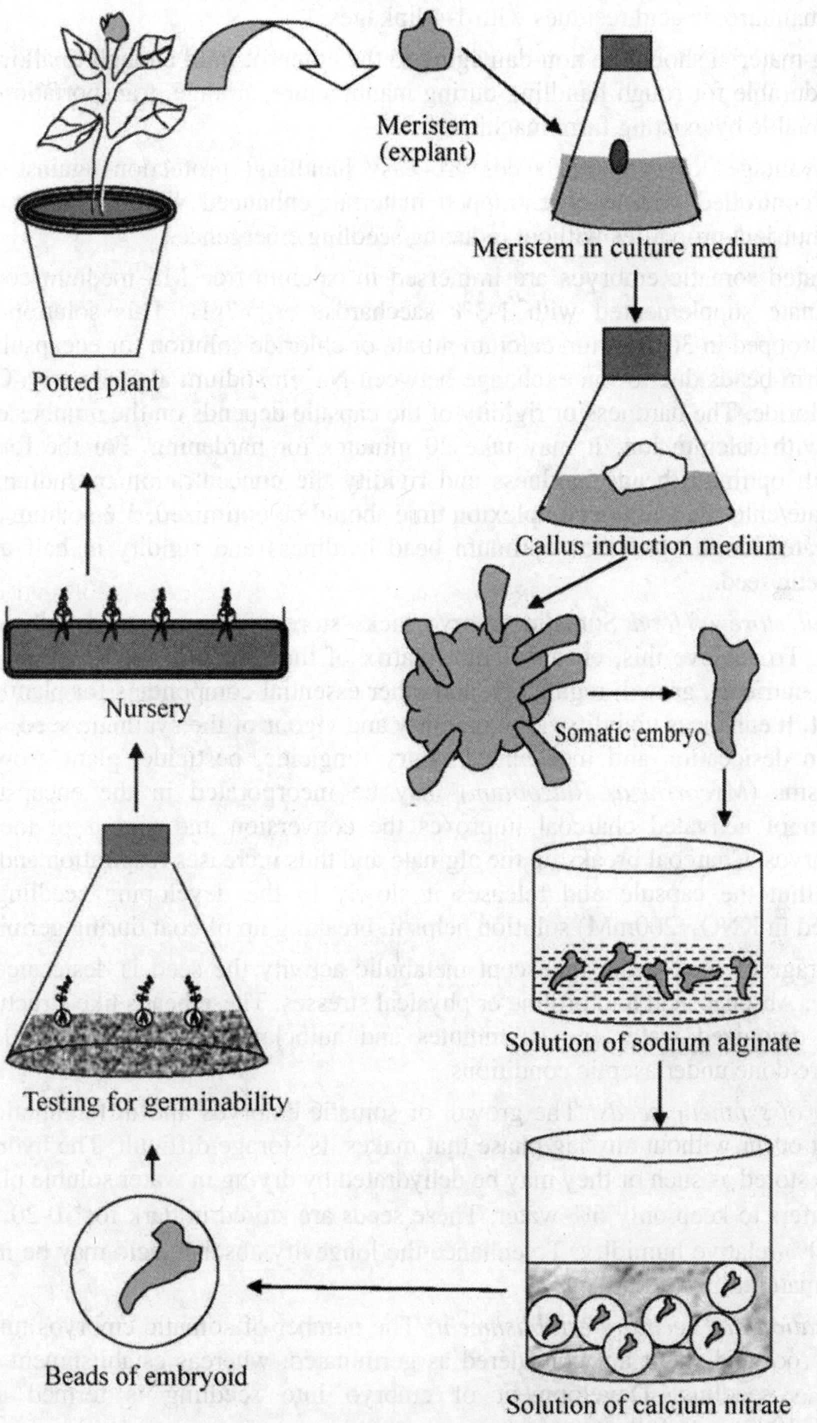

Fig. 1. Procedure of synthetic seed production

Alginate is a straight, chain, hydrophilic, colloidal polyuromic acid composed primarily of hydro β- D manuuronic acid residues with 1-4 linkages.

Coating material should be non-damaging to the embryo; mild enough to allow germination; sufficiently durable for rough handling during manufacture, storage, transportation and planting; and transplantable by existing farm machinery.

The advantages of synthetic seeds are easy handling; protection against environmental fluctuation; controlled release of entrapped material; enhanced viability; addition of bulking agents with nutrient properties without reducing seedling emergence.

Dehydrated somatic embryos are immersed in calcium free MS medium containing 2-3% sodium alginate supplemented with 1-3% saccharose at 5.7pH. This solution with somatic embryos is dropped in 30-100 mm calcium nitrate or chloride solution for encapsulation. It forms round and firm beads due to ion exchange between Na^+ in sodium alginate with Ca^+ of calcium nitrate or chloride. The hardness or rigidity of the capsule depends on the number of sodium ions exchanged with calcium ion. It may take 20 minutes for hardening. For the formation of the capsules with optimum beads hardness and rigidity the concentration of sodium alginate and calcium nitrate/chloride and the complexion time should be optimized. 3% sodium alginate in 1% calcium nitrate/chloride provides optimum bead hardiness and rigidity in half an hour to the viable synthetic seed.

Artificial store of food Somatic embryo lacks store of food that provides energy during germination. To achieve this, encapsulation matrix of the synthetic seed may be supplemented with growth nutrients, growth regulators, and other essential components for plant formation and development. It enhances viability, germinability and vigour of the synthetic seed. To prevent the embryo from desiccation and mechanical injury fungicide, pesticide, plant growth promoting microorganisms (*Mycorrhiza, Rhizobium*) may be incorporated in the encapsulated matrix. Incorporation of activated charcoal improves the conversion and vigour of the encapsulated somatic embryos. Charcoal breaks up the alginate and thus increases respiration and it also retains nutrients within the capsule and releases it slowly to the developing seedling. Soaking of synthetic seed in KNO_3 (200mM) solution helps in breaking up of coat during germination.

For storage of seed under quiescent metabolic activity the seed is desiccated by chemical treatment *viz.* Abscisic acid and proline or physical stresses. These beads like structures are gently rinsed with deionized water for 10 minutes and autoclaved at 121°C for 20 minutes. All operations are done under aseptic conditions.

Storage of synthetic seeds: The growth of somatic embryos and differentiation in seedling development occur without any lag phase that makes its storage difficult. The hydrated synthetic seeds can be stored as such or they may be dehydrated by drying in water soluble plastic resin and hardened wafers to keep only 6% water. These seeds are stored in dark for 10-20 weeks at 25 ± 1°C at 40-50% relative humidity. To enhance the longevity abscisic acid may be incorporated in the embryo maturation medium.

Germination and seedling establishment: The number of somatic embryos turns green with initiation of root and shoot are considered as germinated, whereas establishment of plantlets is considered as seedling. Development of embryo into seedling is termed as *conversion* (Redenbaugh,1993) with following criteria

- Germination of embryo

- Development of root system
- Development of shoot axis
- Formation of two true leaves
- Complete connectivity between root system and shoot axis
- Absence of hypocotyl swelling
- No/minimum callus growth on the hypocotyl
- Formation of green plant with normal phenotype

Advantage of somatic seed: Synthetic seed may help in storage marinating genetic constitution, transportation and other aspects of agriculture in following ways (Hussain *et al.*, 2000)

- It assures availability of planting material.
- Fast multiplication *i.e.,* millions of plants in a few days.
- It assures genetic purity of seed.
- Low cost, high volume propagation system in comparison to true seeds and transplants.
- An alternate viable technology for vegetatively propagating crops.
- The plant slow or difficult in propagation may be multiplied very fast.
- Easy transportation of seed in comparison to the crops multiplied vegetatively.
- Pathogens transmitted through vegetative parts are checked.
- It improves the storability in comparison to vegetative part as seed.
- Synthetic seed is an answer to maintain hybrid vigour without production of hybrid seed every year.
- It is a solution for exploitation of hybrid vigour in crops where male sterility system is not available and production of hybrid seed is not economical and feasible.
- It plays a major role in forestry and horticulture.
- It offers the opportunity to store heterozygous combination year after year that is not possible by conventional breeding.
- It may enable long-term storage of recalcitrant seeds.
- Somatic embryogenesis from the explants of inter-specific crosses allow propagation of desired genotypes and back cross.

Example of synthetic seed: rice, maize, *Brassica* Spp., Cotton, sugarcane, lucerne (*Medicago sativa*), *Apium graveolens*, carrot (*Dauchus carota*), *Lactuca sativa*, *Velerina* sp., *Santalum* sp., *Cassava* (*Manihot esculenta*), ashwagandha (*Withania somnifera*), sandalwood, banana, cardamom and mulberry.

■■■■

Chapter 41

SEED PACKAGING AND HANDLING

Storability of seed is mainly affected by ambient RH and temperature. Therefore seed is stored in container made of moisture impervious material under controlled temperature after proper drying.

PACKAGING MATERIAL

After processing, seed are packed in containers of specified net weight for safe storage and transportation. The size of package depends on the requirement of seed per unit area. In India for field crops acre is considered as unit for bagging of seed. The packaging materials should protect most physical qualities of seed and should have sufficient tensile strength, bursting strength and tearing resistance to withstand the handling stresses. It should be environment friendly with ability to reduce to debris easily. The best packaging material has ability to provide protection against climatic fluctuation and infestation of insect pests with minimum deterioration in seed during transportation.

TYPES OF PACKAGING MATERIAL

Moisture pervious container: These are made up of moisture vapour permeable material viz., gunny (jute) bag, cloth bag, paper bag, multiwall paper bag, paper, cardboard, fibre board etc. Seed of orthodox category is stored in moisture pervious container when seed store has facility to control temperature and relative humidity. Special care has to be taken to control insect infestation.

Moisture resistant container: These are made up of moisture vapour resistant material viz., jute bag laminated with thin polythene film. Initially the container is resistant to the moisture but over a long period of time, relative humidity inside and outside the container equilibrates. Polyethylene is not suitable for long-term storage of orthodox seed for genetic conservation, because there is no absolute control over moisture uptake by the seed. It is suitable for short or medium term storage of orthodox seed.

Moisture impervious container: These are made up of moisture vapour proof material viz., tin can, polythene bags, aluminimum foil pouches, glass bottles etc. Orthodox seed is stored in sealed moisture impervious containers after drying upto safe moisture level. This avoids extensive dehumidification requirement. Moisture impervious containers are effective for long term storage of properly dried orthodox seed under low temperature. This method has an added advantage of exclusion of oxygen.

PACKAGING

Jute bag: As per notification of Ministry of Textiles on 30th June, 1997, it is mandatory in India to pack food grains cereals, pulses etc., in jute bags. Environment friendly, biodegradable nature, strong seam strength, rough surface texture, convenient in loading and unloading, resistance to hooking, excellent capacity utilization, excellent stability of stack, high drop test performance etc. make jute bag an excellent packaging material. But, due to moisture vapour permeable nature it maintains equilibrium with environment. Therefore, special care has to be taken during storage of seed.

Cloth bag: Environment friendly, biodegradable nature, strong seam strength, excellent capacity utilization, high drop test performance but smooth surface texture, inconvenient in loading and unloading, not fit for hooking, non stability of stack and cost of bag make cloth bag a packaging material for small quantity of quality seed only. Due to moisture vapour permeable nature it maintains equilibrium with environment. Therefore, special care has to be taken during storage of seed.

HDPE bags : High-density polyethylene (HDPE) is defined by a density of greater or equal to 0.941 g/cm^3. HDPE has a low degree of branching and thus stronger intermolecular forces and tensile strength. HDPE can be produced by chromium/silica catalysts. The lack of branching is ensured by an appropriate choice of catalyst and reaction conditions. In general, HDPE is used in products and packaging such as milk jugs, detergent bottles, margarine tubs, garbage containers and water pipes. But due to poor seam strength, poor capacity utilization, poor stack ability, poor resistance to hooking and poor drop and end use resistance, it requires special care during handling and transportation. Cost effectiveness, durability, high bag strength, light weight, Ultra Violet Stabilized nature, minimal seepage, and moisture impervious nature make the 700 gauge HDPE bags effective for long term storage of properly dried orthodox seed under low temperature.

HDPE interwoven non-laminated bags: To enhance seam strength, capacity utilization, stack ability, resistance to hooking and drop and end use resistance interwoven HDPE bags without lamination are used to store high volume seed. These bags are cost effective with ease in transportation and handling.

Polylined gunny bag: To overcome the problem of moisture vapour permeability in gunny bag a layer of polythene sheet is fixed at the inner side of the bag. This type of packaging has all the advantages of jute bag alongwith no vapour moisture permeability. It is a cost effective packaging martial for storage of high volume seed under poor storage condition.

Bubbled poylster cover to the jute cotton bag: A 10% decrease on germination of soybean seed during road transport is reported due to 7-12% increase in mechanical damage of seed. This can be minimized significantly by using bubbled polyster cover to the jute bag as packaging material.

Aluminum foil (AF) : Small quantity of seed is stored after proper drying in pouch made up of aluminium foil. Due to moisture impervious nature the seed may be stored under poor storage condition. High cost involvement made the packaging material effective only for vegetable and ornamental seed.

Metalized Polyethylene Terephthalate (MPET): Polymer film coated with a thin layer of metalized films with reflective silver surface similar to aluminium foil reduces the permeability to light, water and oxygen. The properties of the film remain, such as higher toughness, ability to be heat sealed, and a lower density at a lower cost than an aluminium foil. This gives metalized films some advantages over aluminium foil and aluminium foil laminates. It may protect the seed for longer period of time from fluctuation of climatic condition. Aluminium foil bags are used widely in gene banks for storage of the seed of cultivated species. However, it is not possible to see from the outside that the seed of which bag has problem as 20% bags are not able to keep their tightness.

Metal bins: For storage of high volume seed as bulk, metal bins are used. Properly dried and cleaned seed are packed in pre cleaned and fumigated metal bins. The bins are placed in dry store to open in the next season. It is store grain pest resistant packaging material but not fit for transportation of seed.

Paper bag: Weak strength, weak seam strength, smooth surface texture, inconvenient in loading and unloading, poor resistance to hooking, poor capacity utilization, poor stability of stack, low drop test performance, vapour permeable nature and no resistance for entry of insect pest make paper bag a poor packaging material for storage of high volume low value seed in poor store or geographical area with high RH and temperature. Its cost effectiveness, availability, environment friendly and biodegradable nature make it a suitable packaging material for storage of low quantity of seed for short duration.

Polythene bags: Seed are stored in small quantity by packaging machine in the polythene bags. It is very popular packaging material in the private sector due to effective printing, easy availability and low cost but due to non biodegradability it is not promoted in the public sector.

Plastic container: All plastic containers allow movement of moisture inside the container. The water molecule is small enough to get through the pores of the polymers used to make the plastics and in the medium and long term storage. Polyvinyl bags are permeable to humidity. The plastic box with a lid are also not appropriate for long term storage due to use of different metals for lid and base, they expand and contract differently at differ temperature regime. It forms cracks on the body that allow movement of moisture to seed preset in the container.

Glass container: The glass containers *viz.,* bottle, jar etc. are moisture impervious but repeated use of the container with twist-on or off the lid may make it moisture pervious. The oxidation of metal lid of glass container often show oxidation stains with the age. Transparency of the containers is a key quality to monitor the seed from outside.

Flame-sealed glass vial: Small quantity of seed in genebank is stored in the glass vial after drying upto 3-4% moisture level. The tip of the vial is sealed with flame and the vial is stored at low temperature. This is the most popular method for storage of small valuable accessions of endangered species genotypes, crosses, mutants etc.

SELECTION OF PACKAGING MATERIAL

Seed should be packed in clean, hygienic bags of any material, which does not affect the seed and prevent it from absorbing moisture. It depends on

Quantity of seed to be packed: Seed of vegetable/ornamental crops are stored in small paper or aluminum pouch for kitchen garden. Whereas, seed of field crops are stored in large gunny or HDPE bags of 40-60kg capacity bags.

Kind of seed to be packed: Selection of packaging material depends on cost of seed. Better packing material is utilized for storage of Nucleus seed in comparison to other classes of seed. Packaging material of the crop with orthodox seed is different than the crops with recalcitrant seed.

Crop to be packed: Some crops are very sensitive to transportation and loss in germinability therefore seed of such crops should be stored in improved packaging material *i.e.,* soybean, groundnut and sunflower. Storage of seed in metal bins or polylined bags extends the storability of poor storer crops. Loss of vigour and viability of vegetable crops during storage is fast. Packing of low moisture vegetable seed in moisture resistant containers under ambient conditions enhances the storability of seed. Properly (5.5% moisture) dried seed of onion, cauliflower and tomato packed in sealed moisture impervious packaging material maintain the germinability for longer period in comparison to moisture pervious containers.

Duration upto which seed has to store: Packaging material for long term storage is different than short term storage. It depends on cost effectiveness, moisture resistance, distance for transport and kind of crop.

Environment of store: Under poor storage condition best packaging material has to be used with minimum moisture vapour permeable nature whereas under best storage condition i.e., no environmental fluctuation seed may be stored in gunny or paper bags. For cryo-preservation specific packaging material is required.

Suitability for transportation: Safe and easy to handle during transportation, strong seam strength, rough surface texture, convenient in loading and unloading, resistance to hooking, excellent capacity utilization, high drop test performance, sufficient aeration facilities are the main criteria for transportation.

Suitability for storage: Excellent stability of stack, resistant to climatic and environmental fluctuations and protective strength to the packed seed are the main criteria of packaging material for storage.

Moisture content of seed at the time of store: Seed of orthodox category with high moisture content should never be stored in moisture resistant or impervious packaging material. Packaging material of the crop with orthodox seed is different than the crops with recalcitrant seed.

Value of seed: Hybrid, transgenic and other category of seed of low volume high value category are stored in better package than the seed of crops with high volume low value category.

Cost of packaging material: It should be cost effective, economical and readily available.

Marketing: Packaging material should be attractive for display and convenient and suit the need of the customer.

Prevention of Food Adulteration (PFA) standards : It should conform to the requirements and standards of PFA.

Environment friendly: The packaging material should be natural, biodegradable, eco-friendly and recyclable.

The geographical area: Proper packing material is required to store the seed in area near to costal region with high humidity and temperature which is not suitable for long term storage. Similarly, at high altitude seed may be stored in moisture and temperature pervious packaging materials. Storage of seed in metal bins or polylined bags extends the storability at the location with high RH and temperature range.

Season of store: To store the seed of *Rabi* crops better packaging material in comparison to *Kharif* crop is required as *Rabi* crop has to remain in store under high humid condition.

STEPS FOR PACKING OF SEED

Seed filling: Required quantity of seed is filled in the bags with the help of seed filling machine. The empty bag is fixed in the outlet spout of the machine by a bag clamp. The required quantity of seed is poured in the bag by operating trip lever. It automatically controls the quantity of seed to be poured in the bags. Manual, semi-automatic or automatic operating seed filling machines are available.

Weighing: Bag is weighed to confirm required quantity of seed. Weighing machine is a portable platform, with digital display of weight. The bags are placed manually on the platform and then 'even weighed' with a hand scoop. Automatic bagging and weighing machines are used to pack the seed of high value low volume crops mainly vegetable and ornamental crops.

Literature: Leaflet or pamphlet prepared in simple local language is placed in the bag to provide information of variety and improved cultivation practices.

Tagging and labeling: Proper tag and certificates are attached on the bag.

Closing of the container: The top of the jute or inter woven plastic bags is sewed with a precision, high speed bag sewing machine equipped with thread of proper size. It should be operated and maintained properly to prevent frequent break-downs and a short operating life. The sewing machine is suspended by a cord on the roof. Bags are moved upto the sewing station so that operator can sew the bag with ease and accuracy. HDPE bag and aluminum foil pouch are sealed with heat, whereas paper bag are closed with glue.

The container should be perfectly tight, preventing water vapour intake. If the container is not perfectly vapour tight, the seed will tend to get balanced with the external air humidity. In this way, any potential benefit of low temperature will be offset by an increase in seed moisture. It must be noticed that relative humidity inside any uncontrolled cold room is usually very high. Further to this, it should be emphasized that dried seed are highly hygroscopic and – even if often very slowly - there is time enough ahead for moisture to enter the container.

Handling

Several types of conveyors are available for movement of unpacked or packed seed into, through or away from processing plant in vertical, horizontal or inclined directions. Different types of conveyors are used in processing plants, e.g., bucket elevators, belt conveyors, vibrating conveyors, pneumatic conveyors, screw conveyors, chain conveyors and lift trucks. Selection of the proper type of conveyors, for each operation receiving seed in the plant, moving seed from dryers, shellers and from one processing machine to another and finally moving packaged seed into storage, has an important bearing on the efficiency of processing operations. Conveyors should be selected on the basis of the kinds of seed handled, direction and length of conveying, capacity of equipment from which or to which seed are conveyed and should be properly cleaned.

■■■■

Chapter 42

STORAGE OF SEED

OBJECTIVE

To maintain planting value of harvested seed from harvest till planting.

The planting value of seed is the ability to achieve optimum population of uniform, healthy and vigorous seedling of the selected variety to produce high yields. The aim of the Seed Act is to ensure availability of required quantity of quality seed at the time of sowing to harvest maximum yield by exploiting the potential of the variety under optimum management.

Crop has to pass unfavorable environment for plant as seed. Seed has ability to remain viable until the condition become favorable for sowing to translate genetic information into plant. The period from physiological maturity till sowing is termed as storage period. Viability and vigour of seed are highest at the time of physiological maturity, which usually occurs before harvest. Ability of seed to remain viable during storage differs from crop to crop. In all the crops seed start deterioration during storage. The rate of deterioration can be reduced by management but seed cannot remain viable indefinitely as finally it has to deteriorate and die. Management of seed in storage enhances its longevity by minimizing its activity of life. Optimum storage conditions slow down the aging process by minimizing the activities of life that enhance its longevity. Same storage conditions are provided for Foundation and Certified seed.

Storage conditions are determined by duration upto which seed has to be stored. Seed can be stored under normal storage conditions for short period, whereas for longer storage period special care is required to enhance longevity. The two major factors required for germination of seed viz., high moisture and high temperature are responsible for deterioration of seed during storage. High moisture and temperature that initiate process of germination result in heating, molding, rotting and damage of the seed in storage.

STAGES OF SEED STORAGE

The seeds are considered to be in storage from the moment they reach physiological maturity until sowing for germination. Therefore storage of seed can be divided as

Storage in mature fruit on mature/ immature plant in field: Seed is stored from physiological maturity to physical (harvestable) maturity on mature plant (determinate habit) and or on mature fruit (indeterminate nature). After physiological maturity, fruit and seed have no association with plant due to formation of abscission layer. Matured seed reaches at quiescence by evaporation with start of deterioration by weathering in field.

Storage on plant after harvesting till threshing: Mature plant is harvested with the help of sickle and stored in bundle. For threshing the harvested crop is stacked near the threshing floor in such a way that rain water does not affect the stack and runs down the stack by virtue of the slope without affecting the quality of seed. At this stage loss of water from seed and plant is the main physical process. It takes nearly 5-10 days depending on the crop. In *Kharif* crops the duration of storage on harvested plant is enhanced because threshing is delayed upto sowing of *Rabi* crops.

Storage during drying: After threshing seed is stored as heap on the threshing floor for sun drying. The seed harvested by combine is stored as a layer to minimize deterioration due to very high temperature and humidity. Seed is stored before processing as bulk/heap or in gunny bags in shed/store.

Storage for processing: Seed lots produced by individual farmers are stored under isolation based on crop, variety and class in gunny bags.

Storage in seed store: After processing and certification at seed level seed is stored in properly labeled and tagged containers with other information on the container.

Storage during transportation: The Certified seed has to be stored properly during transportation as it may take few hrs to several days to reach upto indenter/farmer. This is usually a critical phase of storage as much damage to seed occurs during transport. Transport is a mobile storage with many physical and abiotic stresses on the stored seed that leads to fast deterioration. Seeds of few crops are very sensitive for deterioration during transportation viz., soybean.

Storage in retail store: Seed is stored in the godown and shop of retailers for few days to several months for sale.

Storage in producer's field before sowing: The purchased seed has to be stored for few days in the producer's field before sowing.

Crops can be divided into two following categories depending on the moisture content required for storage of seed

Orthodox seeds: These are the long-lived seeds that can be successfully dried to moisture contents as low as 5% without injury and are able to tolerate freezing temperatures.

Recalcitrant seeds: They are short -lived seeds, which cannot be dried to moisture contents below 30% without injury and are susceptible to freezing viz., coconut, coffee, cacao, citrus etc. They are difficult to store successfully because of their high moisture content that encourages microbial contamination and results in more rapid seed deterioration. Storage of these seeds at subzero temperatures causes the formation of ice crystals, which disrupts cell membranes and causes freezing injury. Mature seeds of these crops are covered with fleshy or juicy fruit cover and impermeable seed coat. The more moisture content (50-70%) is present in recalcitrant seed in comparison to orthodox seed at physiological maturity. Recalcitrant seeds never show dormancy.

Formation of ice in the recalcitrant seed under low temperature condition of store is avoided by using endogenous seed inhibitors such as abcisic acid or replacing the water content with sugar or ethylene glycol that avoid seed damage due to no ice-crystal formation.

FACTORS INFLUENCING THE LIFE SPAN OF SEEDS

Genetic factors: Seeds of crop species have genetic control on longevity viz., onion, soybeans, ground nut etc are poor storer. Seed longevity has positive association with seed coat attributes viz., its hardseededness, impermeability etc. The oilseed crops have poor storability followed by protein rich (pulses) then carbohydrate rich (cereals) crops. Crops with similarity in nutrient level may differ in storability due to differences in genetic potential viz., *Paspalum scrbiculatum* has better storability than other crops of family Gramineae. These genetic factors affect seed storability and have lead to classify the seeds based on their relative storability. Variation in seed longevity exists among cultivars of one crop. Some cultivars have better longevity than others e.g., in soybean black seeded and yellow small seeded genotypes with low oil content have better storability in comparison to yellow bold seeded oil rich genotypes. Environment strongly alters the genetic potential of a cultivar for seed longevity. Storability of many hybrids differs from their parents.

Initial seed quality: Storability of seed is greatly influenced by the physical and physiological state of seed to be stored. Physically damaged seeds viz., broken and cracked seed and wrinkled, burst or cracked seed coat deteriorate more rapidly than undamaged seeds. Environmental stresses during seed development and prior to harvesting (field weathering) reduce the longevity of seeds. Immature, shriveled and small seeds of a seed lot deteriorate faster than healthy well developed seed. Seed longevity extends with increase in hard seededness. Therefore, quality seed with high germination and vigour which has been properly processed and treated, should be stored after proper cleaning and drying to enhance storability as seed lot with undeteriorated seeds have better storability than the lot with deteriorated seed. For safe storage seed should be free from storage pathogen and insect pest and should have not encountered adverse conditions i.e., abiotic and biotic stresses during development and maturation.

Seed moisture: Moisture content of the seed is an important factor influencing the viability of seed during storage. The rate of seed deterioration increases with increase in moisture. High moisture stimulates physiological activity in storage, just as it is necessary for germination in the field. Storage is basically a period of suspended growth, and the seeds must be kept at minimal levels of physiological activity at which the life can be maintained. Since the life span of seeds largely depends on the moisture content it is necessary to dry it to safe moisture limits before storage. However the safe moisture content again depends on length of storage, type of storage structure and kind of the seeds to be stored.

Temperature: The temperature is one of the most important environmental factor which influences seed viability and vigour during storage. The lower the temperature, the longer will be the maintenance of germination capacity. Thus, temperature should be controlled either by ventilation, insulation or by refrigeration.

Relative humidity (RH): Relative humidity is the amount of water present in the air at a given temperature in proportion to its maximum water holding capacity. Due to hygroscopic nature seed attains specific moisture content when exposed to atmospheric humidity. This characteristic moisture content is known as equilibrium. At equilibrium moisture there is no net gain or loss in seed moisture content. The rate of deterioration of crop seed in storage increases as seed moisture content increases. Mature seed are hygroscopic in nature. Their moisture content varies with the relative humidity of the atmosphere. Moisture absorption or evaporation from seed is a slow process. Equilibrium moisture content varies among crops. In general, the equilibrium moisture content of oilseed is lower than that of starchy seed at the same RH and temperature.

Relative humidity and temperature: The most important factors that influence the lifespan of seed is combination of relative humidity (RH) and temperature. The seed starts losing viability at 80% RH and 25-30°C temperature. The rate of seed deterioration at high moisture content enhances both at upper and lower ranges of temperature. At low moisture content (>4%) breakdown of membrane structure due to extreme desiccation deteriorates the seed. Whereas, at high moisture content the seed is deteriorated due to enhanced biological activity of seeds, insects and moulds. Reduction in both temperature and seed moisture is an effective means of maintaining seed quality in storage. Equilibrium moisture content of seed at a given RH increases with reduction of temperature. Thus the moisture content of seed in storage is maintained in response to RH and temperature.

The thumb rules of Harrington (1973) for assessing the effect of moisture and temperature on seed storage are

- For every decrease of 1% seed moisture content at moisture content range of 4-14% the life of the seed doubles.
- For every decrease of 10°F in storage temperature at range between 32 to 122°F (0°C to 50°C) the life of the seed doubles.
- Ideal seed storage environment is achieved when the sum of relative humidity in percentage of storage environment and the storage temperature in degrees Fahrenheit is less or equal to 100. However, to maintain viability and vigour of stored seed, the RH of store should be 30% with seed moisture content in equilibrium and temperature levels should not exceed 21-26°C (70-80° F). Store should be dry and cool and relative humidity and temperature of the store should not be change with environmental fluctuation.

Provenance effect: Climatic condition of the geographical area during storage is an important consideration. The area with low relative humidity and temperature is safe for storage of seed. The sample of seed of the same crop and variety obtained from different locations may show difference in viability. The seeds harvested from regions of high relative humidity and temperature at the time of maturation or harvesting have poor viability in comparison to seed harvested from the regions of low relative humidity with moderate temperature.

Pre and post harvest conditions: Environmental variations during seed development has little effect on the viability of seeds, unless the ripening process is interrupted by premature harvesting, weathering of maturing seeds in the field, particularly in conditions of high humidity, forced maturity or freezing temperature that results in a deterioration of seed expressed as poor longevity. Fluctuating temperature during seed formation and maturation and pre-harvest rain affects the longevity of seed significantly. Mechanical damage during harvesting reduces the longevity of seed significantly *viz.,* soybean and other large seeded legumes. Crops with small seed, additional protective cover on seed and spherical seed shape escape from injury during threshing. During storage opening of seed coat and injury on seed serve as entry point for pathogen and pest. Careful harvest and threshing to prevent mechanical damage to the seeds, and harvest at the proper stage of seed maturity enhance the seed longevity. Immediate drying to safe moisture content and proper processing in the shortest possible time have significant impact on storability of seed. Prompt harvest, drying and thorough cleaning of the seed prevent entry of insects into the seed stores. Drying followed by cooling before bagging and ventilation of bulk stored seed is important for long term storage. High temperature during drying or drying too quickly or excessively can dramatically reduce viability.

Oxygen pressure during storage: Increase in pressure of oxygen gas during the storage decreases the period of longevity by enhancing the biological activity of the seed. Therefore to enhance the storability the biological activity of the seed during storage is minimized with low level of oxygen pressure achieved by enhancing the concentration of other gases viz., nitrogen and carbon dioxide in the atmosphere. Storage of seed in vacuum or in the environment of nitrogen is based on reduction of biological activity during storage.

Effect of storage conditions on the activity of organisms associated with seed: Pathogen, insect-pest and other organisms which feed on stored seed cause significant damage to the seed by eating, injuring, damaging and increasing heat or moisture that stimulates attack of mold, mites and other pathogens. The organisms associated with deterioration of seed in storage are:

Fungi: Two types of fungi viz., field fungi and storage fungi attack the seed in the storage. Primary infection of field fungi is always during development or before harvesting in the field, but never in store viz., *Alternaria, Fusarium* and *Helminthosporium* spp. Whereas, the primary infection of storage fungi viz., *Aspergillus* and *Penicillium* is always during storage of seed. Infection of fungi decreases the viability by producing mycotoxins that develop stuffiness and clumps due to enhanced temperature. Storage fungi never grow at low level of moisture percentage i.e., 13 % in cereals and 7-8% in pulses and oilseed at relative humidity below 68%. Infestation results in loss of viability, discoloration, enhancement of temperature and enhanced metabolic activity.

Insects and mites: Deterioration of seeds by insects and mites is a serious problem, particularly in warm and humid climates. Insect pest viz., weevils, flour beetles, borers etc. deteriorate the seed with increase in moisture content (>15%) and temperature (>30°C). The activity of these insect pest is checked at low moisture content (<8%) and temperature (<20°C). Activity of mites is associated only with high relative humidity (>60%), but not with extremes of temperature. Pre-harvest sanitation spray in pulses to control the insect infestation comes from field eg., bruchid.

Rodents and birds: Rodents are serious problem of storage that may result into a complete loss of seed. Entry of rodent in the store can be prevented by elevating the floor by 90 cm above the ground level with escorting of 15 cm around the building at 90 cm height. A removable step or gap between step and escorting at the entrance for loading and unloading of seeds at the store also prevents the entry of rodent. Birds are also responsible for huge loss of seed in poor storage condition. All openings including ventilators should be screened/sealed to prevent the entry of birds in the store.

Storage structure: Storage should be thermal insulated and moisture proof made up of concrete on high plinth to avoid the effect of environmental fluctuations on seed. For long-term storage, air-conditioning and dehumidification could be added to an insulated, vapour-proofed godown. Proper ventilation, sanitation, free from leaks and dampness and rodent and bird proofing are the prerequisites. To break the convection current of moisture exhaust fan can be used. The floor should be three feet above the ground level with sand and/rock/wooden plank as barrier of moisture on/under the concrete.

Sanitation and maintenance: Before arrival of new produce, storage and seed containers should be thoroughly cleaned and disinfested to kill insects. It is essential to control the infestation of insect pest during storage as they often lay down eggs on old seeds in accessible places viz., conveyor belts, screens, lift hoppers etc. Unreachable spots are disinfested by fumigation or spray of insecticide. The sweepings, spilled seeds, leftover seeds, clear trash, litter etc., should be removed from the processing plant, store and containers. Chemicals such as insecticides, fertilizers should not be stored along with seeds.

Cracks and holes present in the floors and wall should be sealed properly. Sealing of cracks and holes of the store by spray of a vinyl plastic sealing compound is known as Cocooning. Crevices and openings along the door are closed with heavy grease.

Packaging: Ideal storage conditions can be provided by combination of good storage and protective seed bag. The kind of packaging depends on quantity of seed to be packed, duration of storage, storage condition, cost of package, value of the seed, crop, kind of seed, moisture level of seed etc. Old jute or cloth bags and other containers are fumigated before packing of seed. For short-term, seed is stored in sealed, vapour-proof bags of heavy polythene or polylined gunny bags. Packaging is discussed in detail in separate chapter.

Duration of storage: Long-term storage requires more care and appropriate conditions of storage than short-term storage.

Use of desiccants: Desiccant like silica gel can maintain the moisture content in equilibrium with the relative humidity at 45%. Dried silica gel is placed @ 1 kg / 10 kg of seeds. With the absorption of moisture it turns to pink in colour. It can be reused after drying at 175°C.

Stacking: Bags in store should be stack on the wooden or steel pallets with proper matting/tarpaulin to avoid absorption of moisture in the seed from floor. Stacks should always be kept away from the walls to minimize the damage by latent heat. Sufficient space should be provided between stacks for free movement and accessibility. Number of bags in one stack should never be more then 8-10 to avoid the stack pressure on the seed of lower most bag. Four tier arrangement of bags per stack is recommended for stack pressure sensitive crop soybean to maintain germination capacity of seed. Stacks must be kept at distance from walls to allow inspection, fumigation and avoiding the seed to absorb moisture from moist surface. Restacking once in 3 months or less is important for prolonging seed viability. Use of wooden pallets at the base of stack with cris-cross arrangement of bags is effective for ventilation. Stack cards should be used to show the details of lots stacked and fumigation operations etc.

TYPES OF STORAGE BASED ON STORAGE PERIOD

Commercial seeds: Maximum quantity of seed is stored for one season i.e., 8 or 9 months from harvest till sowing. Orthodox seeds of cereals and pulses are dried upto <14% moisture and oilseed <11% for short term storage of commercial seeds in all-purpose store.

Carry over seeds: It is the unsold seed in the main season. Nearly 20-25% seeds remain unsold in the main season. It has to be stored safely till next sowing time i.e., for 12 months. Properly dried carry over seeds are stored in moisture proof bags at low temperature or in air tight metal bins.

Stock seed: It is handful quantity of the seed of any variety evolved by a plant breeder. Regular multiplication to maintain the variety may deteriorate the variety due to genetic drift. Therefore the stock seeds are stored in polythene bags of >700 gauge thickness at 25% relative humidity at 30°C temperature or at 45% relative humidity at 20°C temperature by using a dehumidifier.

Gene bank: Preservation of germplasm resources of cultivated plants and their related species in considered as seed gene bank. It is essential to check the genetic erosion i.e., gradual loss of genetic variability. NBPGR is the third largest seed gene bank out of 1750 at global level with 0.4 million accession. In general each homogeneous accession minimum 4000 seed and for heterogeneous material 12000 seed is required for storage to preserve the normal genetic variation within a population.

The collected seed is properly dried, cleaned and tested for viability before storage at low temperature. The accession is stored at two differ levels first long term (-196ºC in liquid nitrogen) and second medium term storage. Low moisture and low temperature both are the important factors for long term storage of seed. The seed stored in medium term storage (4-12ºC upto 35% RH) is multiplied at periodical interval i.e., after every 3-4 years and used for regeneration, evaluation and distribution. Relative humidity inside any uncontrolled cold room is usually very high, therefore the seed container should be perfectly tight to prevent intake of water vapour. To store the seed at viable state low ethylene concentration and low oxygen concentration are the most relevant factors apart from seeds dried to lower level.

Storage at low temperature: The use of low temperatures i.e., -18ºC to -20ºC is effective for long term storage of properly dried seed. Temperature between -5°C and +5°C is enough for safe storage of ultra-dry orthodox seeds. The seed is stored in glass jar. In glass jar the presence of oxygen is minimized by filling the containers as much as possible with the seed or by replacing the air with CO_2 because it is heavier and tend to occupy the lower parts of a volume.

Cryopreservation

Storage of seed at ultra-low temperature (–196°C, i.e., the temperature of liquid nitrogen) indefinitely by suspending metabolic activities is known as cryopreservation. Under these conditions all enzymatic processes are practically stopped, therefore any type of biological material can be preserved for a potentially infinite period. Only cosmic rays could place a limit after a few millennia. It is not much popular due to costly infrastructure and running cost and formation of lethal intracellular ice crystals. There are two main types of cryopreservation procedures: equilibrium (conventional slow freezing) and non-equilibrium or ultra-rapid freezing (vitrification).

Slow-freeze procedure: The cytoplasm of a cell contains water, which turns to ice crystals at freezing temperature. When ice is formed inside a cell, the cell gets ragged and dies. Therefore, the fluid in the cell needs to be removed prior to reaching freezing temperatures to avoid cell damage. For an efficient cryopreservation, intracellular formation of ice crystals has to be avoided to prevent damage to cell. Various types of antifreeze (cryoprotectant) fluids can be used to dehydrate the cell before freezing. It is induced by concentrated mixture of osmotic agents and/or substances such as glycerol, proline, abscisic acid, propanediol, dimethyl sulfoxide, ethanol and sugars like sucrose and trehalose etc., after effective dehydration of seed.

The seed is exposed to cryoprotectant in a gradual stepwise fashion to slowly allow equilibration of the cell with the cryoprotectant while releasing water. After removal of moisture from the cells of the seed, the seed with the cryoprotectant are placed in a glass container. The container is placed in a freezer that decreases the temperature gradually. After a few minutes of cooling, Seeding is performed by using a tool (for example, a pair of forceps) which has been pre-chilled in liquid nitrogen to touch the container and cause starter ice crystal formation. The starter crystal initiates controlled ice crystal formation at one spot, away from the cells. Once seeding is complete, the rest of the cooling ramps can proceed. When the container reaches temperatures between -30 and -85 C, the container holding the cells can be plunged directly into liquid nitrogen to complete the cooling to -196 C.

Vitrification Physical process for very fast decrease of temperature to avoid intracellular ice crystallization in the cells and tissues during ultra-freezing by the transition of the aqueous solution of the cytosol into an amorphous glassy state is known as vitrification. It results in protection of tissues from damage therefore they remain viable at –196°C during long-term storage. Vitrification is so ultra-rapid that there is insufficient time for ice formation and the cellular fluid is converted directly into a vitreous or glass phase, without damage to the cell. Programmable rate freezers are not required because the cell is directly plunged into extremely cold liquid nitrogen, achieving an instantaneous glassy state. Vitrification can also be achieved by partial dehydration of sowing material on silica for direct dipping in the liquid nitrogen, to store for an indefinite period without any genetic alteration. One step freezing of seed by directly immersing in liquid nitrogen without slow cooling is an effective method for recalcitrant seed. Once cells are frozen or vitrified, they can be stored indefinitely by immersing them in liquid nitrogen, an extremely cold fluid with a temperature of -196 C (-321 F). To restore metabolic activity in the cell upon thawing, toxic cryoprotectants must be removed from the cell and the normal water balance gradually restored as the cell is returned to a normal functioning temperature.

Under low temperature properly dried seed is stored in vial made of alkaline glass. It softens fast without any crack on heating. A label is placed inside the vial with accession number. Dehydrated seed is placed in the vial. On seed a piece of hydrophobic and dehydrated cotton is placed. On the cotton approximately 3 ml of dehydrated silica gel is added in each vial. The tube is closed with the help of rubber cap and placed for 15-20 days. In the event, the color of silica gel in the vial with the seed not properly dried will change. The seed stored in glass vial should be properly dried. At final stage nearly 5cm top portion of the vial is kept unfilled. The vial is sealed by heating. The sealed vials are immersed in water for 24 hours to check the seal. A set of 25 vials are placed in one glass jar for long term storage.

Storage of whole recalcitrant seed is difficult. Therefore its excised embryo is encapsulated in calcium alginate followed by a desiccation in concentrated sucrose solution before final immersion in liquid nitrogen.

Storage of seed under vacuum: A hand operated vacuum packing system has been developed to facilitate the storage of dry seeds. The system provides a means to seal containers with reduced oxygen content. A sealed container prevents rehydration of seeds, thus extends their viability. The reduced oxygen content suffocates insects that are stored with the seeds, and also extends the viability of the seeds by slowing down the seed respiration rate. The seed under vacuum packaging is stored at low temperature.

TYPES OF STORAGE

Storage at farmers' field: Farmers erect a bin like storage structure with the help of rope prepared by paddy straw. The base of the bin is made by thick layer of rice straw. Properly dried seeds of rice are stored in this bin. The filled rice bin is covered with rice straw and tight in the shape of a cone.

Storage of maize: To store seed of maize the dried un-pealed cob with the tip downward to prevent moisture entry is stored by stacking/ hanging on a rope or stick. This technique is performed throughout the world.

Underground: Underground storage of grain and seed is a very popular method in central part of India for rice, wheat, sorghum and pearl millet. The underground structure is prepared at a place with no sign of water retention. A shallow well is digged and the periphery is covered with dried straw with a thick bed of straw at the base. Properly dried seeds are filled in the underground structure. After filling of seed, the structure is again covered by a thick layer of straw and finally with mud and cow dunk and straw. In the underground storage the seed is protected from environmental fluctuations; and infestation of insects and molds due to low oxygen and high carbon dioxide contents.

Mud jar: A large size jar is prepared with mixture of clay + straw + cow dunk cake with an outlet at the bottom and hopper at the top. The properly cleaned and dried grain/seed is filled in the jar and the openings are closed properly. The jar is erected in the store room or under roof. The grain/seed is drawn from the outlet as per requirement periodically. The top most seed reaches at the bottom prior to sowing and used as seed.

Bagged: Seed is stored in gunny bags and stored as stack in the room by the famers. The seed is vulnerable to environmental fluctuation and infestation if insect pest.

Farm Bins: Farm storage space is needed for three reasons; to hold the crop immediately after harvest, until it can be moved to better storage space or to market. Grain stored in bins maintains quality better in smaller than in larger lots. Farm storage tends to maintain the original condition of grain better than elevator storage, provided grain is not exposed to any moisture increase or to rodents, birds or insects. The farmers have developed the oriental bins by using burnt clay which is in use by the farm families from time immemorial.

■■■■

Chapter 43

INSECT INFESTATION AND ITS CONTROL

1. OBJECTIVE

Determination of insect infestation in the seed sample.

Submitted sample of Foundation and Certified categories are observed visually for insect infestation. Seed samples with insect infestation more than the seed certification standards (Table 1) are rejected.

Table 1. Maximum seed certification standard for insect infestation

Crop	Insect infestation (%)	
	Foundation	**Certified**
Maize and legumes	1.0	1.0
Other than maize and legumes	0.5	0.5

2. IDENTIFICATION, DETECTION AND DISINFESTATION OF INSECT SPECIES ASSOCIATED WITH SEEDS DURING STORAGE

Large number of insect species is associated with seeds during storage. Important store grain pests represent the class *Insecta* (insects) and *Acarira* (mites) are the important store grain pests. Insect pests which damage the seed in tore represent the order *Coleopetra* (beetles / weevils), *Lepidoptera* (moths) and *Hymenoptera.* The important grain mite (saprophytic) belongs to the family *Acaridae.*

3. IMPORTANT STORE GRAIN PESTS

Common name	Scientific name	Order	Family
Primary Pests			
Rice weevil	*Sitophilus oryzae*	Coleoptera	Curculionidae
Maize weevil	*S. zeamais*	Coleoptera	Curculionidae
Granary weevil	*S. granarius*	Coleoptera	Curculionidae
Lesser grain borer	*Rhizopertha dominica*	Coleoptera	Bostrychidae
Khapra beetle	*Trogoderma granarium*	Coleoptera	Dermestidae
Pulse beetle	*Callosobruchus chinensis*	Coleoptera	Bruchidae

	C. maculatus		
Grain moth	*S. cerealella*	Lepidoptera	Gelichidae
Rice moth	*Corcyra cephalonica*	Lepidoptera	Pyralidae
Indian meal moth	*Plodia interpunctella*	Lepidoptera	Pyralidae
Almond / Fig moth	*Cadra (Ephestia) cautella*	Lepidoptera	Phycitidae
Secondary Pests:			
Red rust flour beetle	*Tribolium castaneum*	Coleoptera	Tenebrionidae
Confused flour beetle	*T. confusum*	Coleoptera	Tenebrionidae
Long headed flour beetle	*Latheticus oryzae*	Coleoptera	Tenebrionidae
Saw toothed grain beetle	*Oryzaephilus surinamensis*	Coleoptera	Silvanidae
Rust red grain beetle	*Cryptolestes ferrugineus*	Coleoptera	Cucujidae
Flat grain beetle	*C. pusillus*	Coleoptera	Cucujidae

Angoumois grain moth (*Sitotroga cerealella* Hindi *Patanga*): Damaging stage: larvae; marks of identification : adult small, buff yellowish brown to dull grey in colour, wings narrow and fringed. Larva of whitish colour with yellow head, never free and separated from grain. The larvae bore into the seed of cereals, pass through the pupa stage and emerge through a small round hole cut in the outer layer of the kernel. The moth breeds on the surface of the seed. Damaged endosperm results in reduced seedling vigour. Optimal climatic requirement is 30ºC temperature with >70% relative humidity and 14-16% seed moisture content.

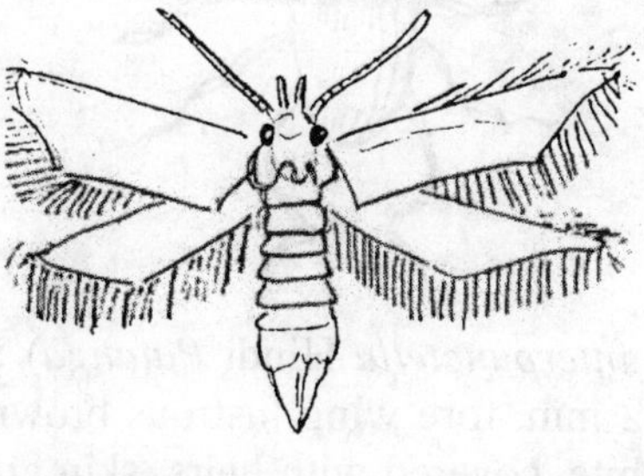

Confused flour beetle (*Tribolium confusum*): Damaging stage: larvae and adult; marks of identification: similar to red flour beetle but larger in size, antennae gradually enlarged towards the tip. Eyes are smaller than red flour beetle. It is one of the most common storage insects in processing plant elevators and store.

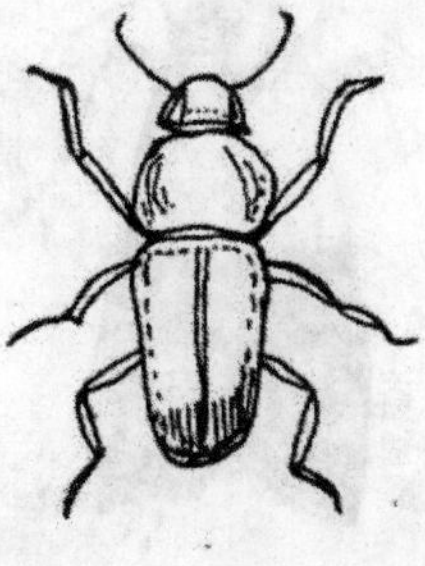

Flat grain beetle (*Laemophilocus minutus*) **:** Damaging stage:- larvae and adult; marks of identification: length 4 mm, flattened, slightly oblong, reddish brown, elongated antennae of about 2/3 as long as the body. Larvae cigar shaped, yellowish white, head reddish brown, spine like appendages on the oral segment.

Granary weevil (*Sitophilus granarius*) **:** Damaging stage: larvae and adult; marks of identification: larger than rice weevil, uniformly shiny black/reddish brown elytra larger oval pits on the thorax (*pronotum*). It cannot fly due to absence of functioning wing.

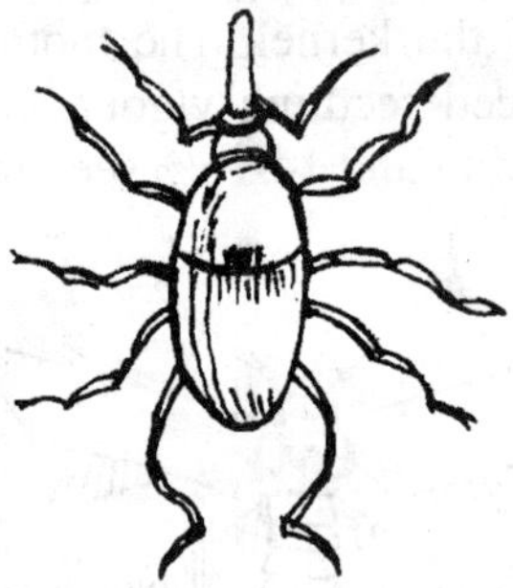

Indian meal moth (*Plodia interpunctella* Hindi *Patanga*) **:** Damaging stage: larvae; marks of identification: wing expends 2 mm, fore wing lustrous brown, yellowish/whitish band across the basal half. Larvae-greyish white, covered with hairs, skin granular. The larvae infest seed of cereals and pulses and may completely web over the surface of the seed. Optimal climatic requirements are 29°C temperature with > 70% RH and 14-16% moisture content of seed.

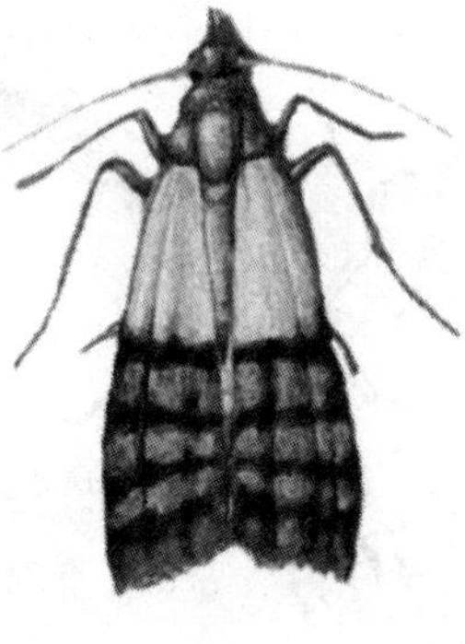

Khapra beetle (*Trogoderma granarium*; Hindi *Pai Khapra)*: Damaging stage: Larva; marks of identification: adult 2-3 mm long, brown in colour, stout in built with grey and light brown marking, oval in shape, body covered with long reddish brown hair and forming a thick tail at the posterior end. It is a primary feeder of all the crops. Larvae feed on germ portion of the seed and reduce the seed into frass (powder form mixed with excreta). Optimal climatic requirement is 34-37ºC temperature with > 50% relative humidity and 11.5% seed moisture content.

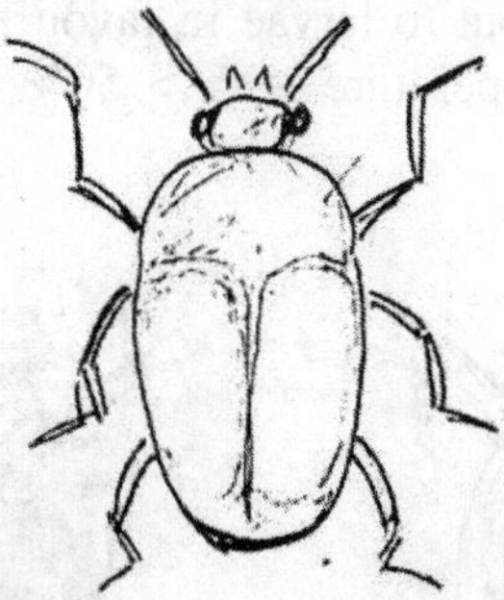

Lesser grain borer (*Rhyzopertha dominica;* Hindi *Ghun*)**:** Damaging stage: larvae and adult; marks of identification: adult 0.08-3 mm long, slender cylindrical form, polished dark brown/black in colour, rough surfaced head scarcely visible from above, larvae dirty white, light brown head, body covered with tiny hairs. It can survive even under drier conditions in comparison to other stored seed insects. It is primary feeder of all the crops. Adults and larvae feed on seed and reduce them to mere shells with many irregular holes. Optimal climatic requirement is 33-36ºC temperature with > 60% relative humidity and 11-145%.seed moisture content.

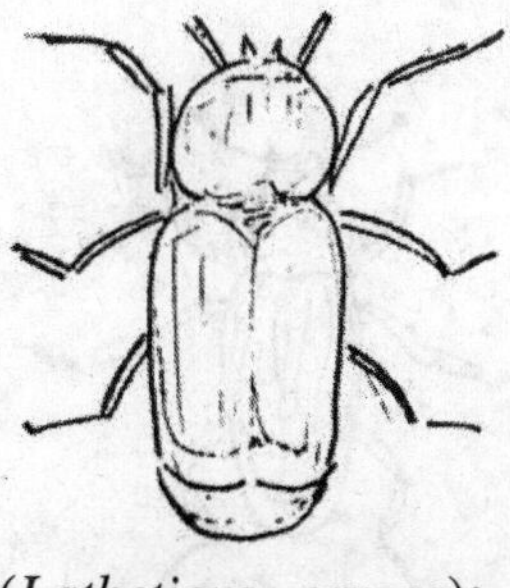

Long headed flour beetle (*Latheticus oryzae*)**:** Damaging stage: larvae; marks of identification: adult 3 mm, slender, flattened and pale brown; larvae yellowish white, dark coloured head and leg.

Pulse beetle (*Callosobruchus chinensis.* Hindi *Dhora*) **:** Damaging stage: adult and grub attack the seed especially pulses during July-August by making a small hole; marks of identification: Length 3-6 mm, appearance heart shape, two irony spots on the middle portion of the dorsal side of the body. Body tapered towards the head. Colour chocolate to dark brown. Grub white, cylindrical, fleshy and wrinkled, usually found inside the grain. Primary feeder; larvae feed and pupate inside the seed. Sowing seed quality of the infested seed is adversely affected. Infested stored seed can be recognized by the white eggs on the seed surface and the round exit holes on seed coat. The eggs hatch out to larvae in favourable conditions in the store. Optimal climatic requirement is 30-35°C temperature with > 50% relative humidity and 12-14% seed moisture content.

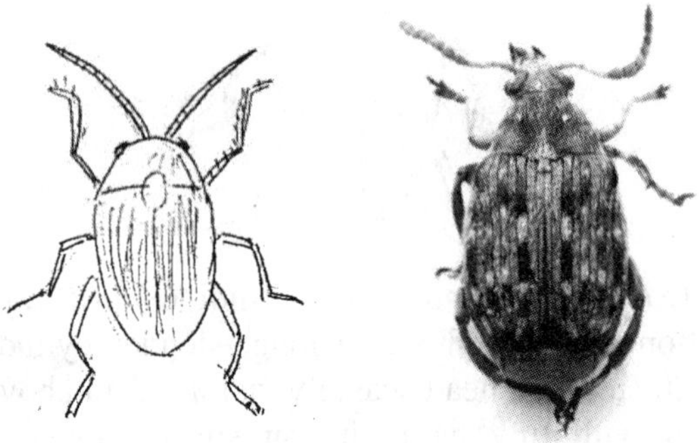

Red flour beetle (*Tribolium castaneum* Hindi *suri*) **:** Damaging stage: larvae and adult; marks of identification: length 3 mm, flattened reddish brown, smooth functional wing. Larva is small, slender to cylindrical in shape. Secondary feeder of all the crops; Adults & larvae feed on already damaged seeds. Optimal climatic requirements are 30-35°C temperature with > 70% RH and 11-16% moisture content of seed.

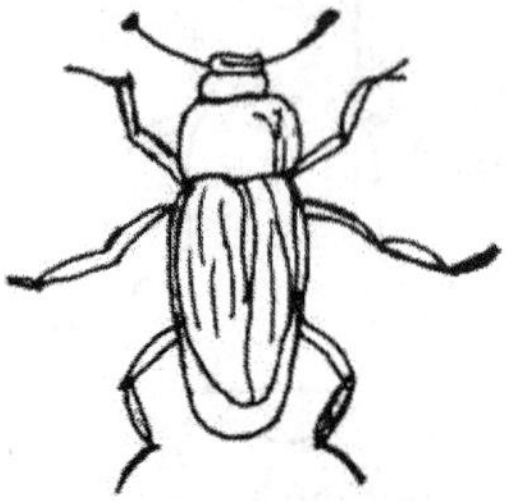

Rice moth (*Corcyra cephalonica Staint*)**:** Damaging stage: larvae; marks of identification: a big adult of greyish brown colour, dirty creamy white larvae with prominent broad yellow head.

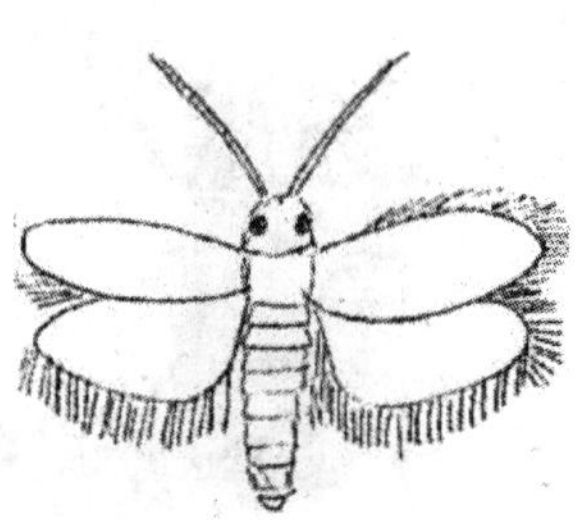

Rice weevil (*Sitophilus oryzae; Hindi Sundwali sursi*): It is one of the most common and destructive stored seed insect that may infest seed of cereal crops in the field before harvest. Larva and adult are the damaging stage. The larva burrows into the heart of the kernel where it feeds and passes through the pupa stage. It emerges as an adult, cutting a small hole through the kernel (weevil exit hole); marks of identification: small snout beetle (3-4 mm) of reddish brown to dark brown colour with sharp and long slender snout and wings with 4 light reddish or yellowish spots. Grubs are small, white, leg less with yellowish brown head and are found inside the kernel. It is a primary feeder. Larvae feed and pupate inside the seed. Seedling vigour of the infested seed is adversely affected. The weevil completes its life cycle inside individual seed and may develop into substantial population before its presence is even noticed. Optimal climatic requirement is 26-30ºC temperature with > 70% relative humidity and 14% seed moisture content.

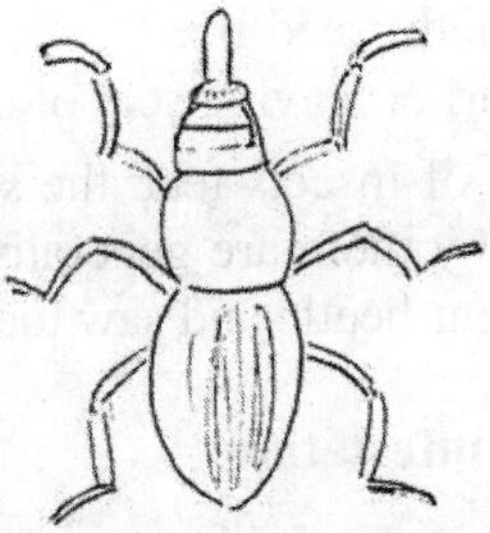

Saw toothed grain beetle (*Oryzaephilus surinamensis*): Damaging stage: larvae and adult; marks of identification: length 2 1/3 mm to 3 mm, small, narrow, flat and reddish-brown to black beetle, six saw tooth like projections on each side of the thorax. Larva is slender in shape and whitish in colour and visible under microscope.

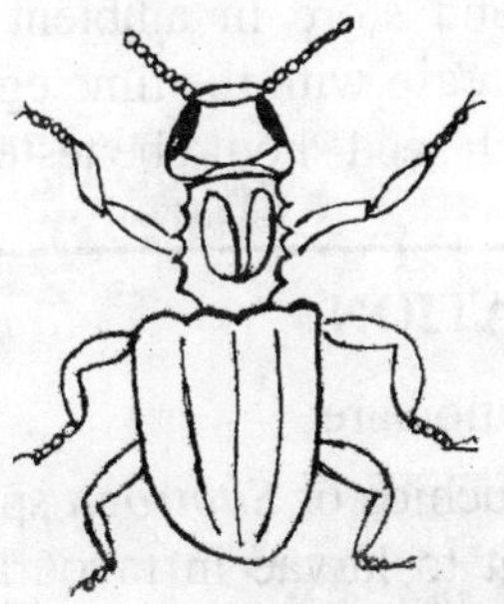

4. CLASSIFICATION OF STORE GRAIN PEST

On the basis of feeding habits:

These insects, depending upon their ability to damage seeds can be divided as Primary and Secondary seed feeders.

A. Primary feeder The primary feeders break the seed coat for feeding. They obtain water from seed and release it into the environment coupled with heat. That forms the hot spot of insect infestation and seed deterioration in the store. *Trogodema granarium, Sitophilus oryzae,*

Callosobruchus spp., Grain moth, *Sitotroga cerealella* and *Rhizopertha Dominica* are the primary feeder.

B. Secondary feeders: These are the insects and mites living on the seeds with cracked seed coat, broken either due to mechanical damage during harvesting, threshing and processing or already damaged by other primary feeders. Most common secondary pests are *Crytolests, Ephestia, plodia, Oryzaephilus, Rhizopertha, Sitophilus, Tribolium, Tenebrionidae, Trogoderma* etc.

On the basis of damage on seed

A. Internal feeder: They lay their eggs in immature seeds at milk stage on plant. They spend a part or entire larval and pupal stages within the seed. The infected seed does not exhibit any external symptoms to show the presence of pests e.g. granary weevil, rice weevil, lesser and larger grain borer, angumois grain moth, seed beetles etc. Some of these insects have short life cycle so the adults are emerged out earlier leaving scar of damage on the seed.

B. External feeder: This group of insects feed the seed from outside and lives with seeds. The insects and the damages caused by them are generally visible eg., cigarette beetle, confused flour beetle, Indian meal moth, red flour beetle and saw toothed beetle.

On the basis of multiplication after infestation

Group I. Insects infests the seed in the field and remain dormant in the store and also do not multiply in store eg., pink boll worm in cotton, midges in millets etc. The number of seeds damaged in field remains constant. Pests are transmitted through the infested seeds used for sowing purpose from one field to other.

Group II. Insects infest the seed in field or store or at both the place. The life cycle of the insect pest continues in both field and store in ambient conditions. Number of insects and intensity of infestation continue to increase with the time eg., weevils, beetles, moths etc. both in field and store. Thus, infestation of such seed should be tested before storage.

5. SOURCES OF INSECT INFESTATION

The main sources of insect infestation are:

Field infestation: Insects like Bruchids or *Sitotroga* sp. lay eggs on immature seed on plant present in field. These eggs hatch out to larvae in favourable conditions when seed reach the stores. This is termed as *hidden infestation.*

Cracks and crevices in storage: In the store not properly cleaned some grains remain in cracks and crevices. Insect present in the store keep on feeding and multiply on these seeds even in empty store. The fresh seed placed in this store gets infested. Infestation by insects from old stocks to fresh stocks is called *cross infestation.*

Containers and bags: Eggs and larvae of insects remain hidden in the corners and mesh of the old used bags. Infestation reappears in the fresh seed when the seeds are stored in such bags without proper cleaning.

Transport: Seeds are harvested, threshed in the fields and then transported to the godowns by trucks, trolleys, bullock carts etc. The Insects which may be present in the joints or corners of the carriers migrate to the seed and cause cross infestation.

6. DETECTION OF INSECT PESTS

Visual inspection: Seeds are spread in an enamel tray and inspected visually with naked eye or under illuminated magnifier depending upon the size of seed. The material is closely scrutinized for the presence of any living or dead insect/ mite and their stages, excreta and damaged seeds. The insect can be detected on the basis of the symptoms of damage caused (Table 1) and morphology of the insect.

Table 1. Symptoms of damage caused by some store grain pests

Symptoms	Crop	Pest
Brown/ damaged seed	Maize, rice, sorghum, wheat, oat, barley	Red flour, saw toothed and flat grain beetle
Circular holes	Pulses, wheat, rice, maize, sorghum and grasses	Pulse beetle, rice weevil, Angoumois moth
Irregular holes	Spices	Lesser grain borer, cigarette beetle
Seeds reduced to husk	Wheat	Khapra beetle
Webbing	Maize, rice, wheat	Indian meal moth
Musty odour and dust	All seeds	Mites

X-ray radiography: Seeds are spread in a single layer on a transparent plate and are exposed to X-ray @ 22 KV, 3 MA for 10 seconds at a distance of 30 cm. On the radiograph, healthy and infested seeds are clearly distinguishable.

Transparency test: Seeds are treated with lactophenol on a boiling water bath for 10-20 minutes (differs from crop to crop). With the reactions of lactophenol seed coat becomes transparent. Each and every seed is observed with the help of magnifier to confirm the presence of insect within the seed. This method is useful for the crops with small seed size. Lacto-phenol should be used with proper care.

7. FACTORS AFFECTING SEED HEALTH IN STORE

Abiotic factors

Temperature: It is one of the important factors that influence metabolic activity of all biological material. Most of the microbes, mites and insects multiply faster and infest seed at higher rate at more than 20°C temperature.

Humidity: Like temperature, humidity is also an important ecological factor that influences longevity of seed and pest activity. Most of the seed under tropical environment deteriorates quickly due to high humidity coupled with temperature.

Seed Moisture: Moisture content in seed is the most detrimental factor that directly affects seed deterioration and pest multiplication. Moisture content below 10% in seed is considered good for storability. For keeping seeds in moisture-proof packets, the moisture content should be further reduced to 4-5%.

8. MANAGEMENT OF STORE GRAIN PEST

Preventive Measures Hygiene or Sanitation; Disinfestation of stores / receptacles and Legal methods

Curative Measures: The infestation of stored grain insect pests can be controlled by the following methods:

(i) Non chemical control methods

(a) Ecological control measures:

Temperature,

Moisture content of grain and

Availability of oxygen

(b) Mechanical control measures:

Screening of grain

(c) Physical control measures:

Heat treatment,

Controlled atmosphere and

Mixing of inert dust

Use of edible oils

(d) Cultural control measures:

Splitting of pulses

(e) Engineering control measures:

Hermetic or air tight storage and

Drying of grain

(ii) Chemical control methods:

(a) Prophylactic treatment

Dosage (1: 100 @ 3L / 100 $m^{2)}$

Frequency of treatment – 15 days interval

Malathion 50EC, 0.05%

Pirimiphos methyl 50EC, 0.05%

Pyrethrum (2% Pyrethrin), 0.02%

(b) Curative treatment:

Knock down chemicals,

Grain protectants and

Fumigants *e.g.*

EDB @ 3 ml / quintal or 22g / cu.m.

EDCT@ 55 ml / quintal or 30-40Kg /100 cu.mt.

Al_3P_2 @ 2-3 tablets of 3 g /metric tonne

EDB + CCl_4

9. DISINFESTATION

Seed treatment: Mixing of Melathian 5% at the rate of 250 g per quintal of seed protects the seed from store grain pests. Treatment of seed with fungicide (Wheat – Vitavax + TMTD 1:1; Maize – TMTD; Sorghum – TMTD + Carbendazim 1:1; Bajra – TMTD) enhances the storability of the crops when packed in poly-lined gunny bags at low moisture content (8%). Seed treatment of wheat, chickpea, pigeonpea, sorghum and cowpea with Thiram @ 0.2 % is effective in maintaining seed germination above standard upto revalidation period.

Deltamethrin 2.5 WP @ 40 ml/ kg seed is effective to control pest infestation in the store. Fabric treatments both by surface and soaking methods with Deltamethrin 2.5 WP @ 125 ppm and Bifenthrin 10 WP @ 125 ppm are effective in protecting the seed from insect infestation. NSK powder @ 5.0 g/ kg seed, neem powder @ 5.0 ml/ kg seed and Deltamethrin @ 40 mg/ kg seed treatments can effectively protect seeds of cereals and pulses from insect infestation.

Seed Fumigation

Exposure of seeds to gaseous form of harmful chemicals (fumigants) to control deterioration of seed from seed borne fungi and insects during storage is known as seed fumigation. Fumigant is a chemical substance, at specified temperature and pressure can exist in gaseous from in sufficient concentration to be lethal to the pest organisms. Apart from other characteristics a fumigant should not affect the viability of seeds, should be non-persistent, non-corrosive and highly diffusible. It is the best method for controlling pest in store. The fumigation should be carried out under expert supervision and proper caution.

Fumigants: ethylene oxide, ethylene dichloride + carbon tetrachloride, chloropicrin, magnesium hydrogen phosphide.

Procedure

The seed brought to equilibrium moisture content at 60 per cent relative humidity is enclosed in air tight container and fumigate at temperature of 27±1.5°C for required duration. After fumigation seed is placed in cloth bags and aerated for 21 weeks under room condition. The fumigation may be repeated after 3 and 6 months as per requirement.

Important points to consider for seed fumigation

Seed moisture level : at or below normal moisture content prescribed as Seed Certification Standard

Type of infestation

Recommended fumigant: use only recommended fumigant based on crop and pest

Doses of fumigant : strict adherence to prescribed dosage

Exposure time : strict adherence to prescribed period of treatment

Temperature : avoid excessive temperature during fumigation

Aeration: immediately after the required period of exposure thorough aeration of the seeds is essential to maintain the vigour and germinability

Number of fumigation: it should be minimum

Fumigant	atmospheric pressure	Duration	Temperature	Seed moisture	Rate kg/ 100 metre^3
Hydrogen cyanide (HCN)	40g/m³	24h	10 to 19°C	< 14 percent	
	32g/m³	24 h	>20°C		
	40g/m³	3 h	>20°C		
Do not fumigate at temperatures below 10°C and seed with moisture content more than 14%.					
Ethylene dichloride/ carbon tetrachloride (EDCT)	480g/m³	24 h	20°C		3:1 per 1000 cubit
*Methyl bromide	24g/m³	24 h	10 to 19°C	Dry seed	10.5
	16g/m³	24 h	>20°C		
	40g/m³	3 h	>20°C		
Repetition of fumigation should be avoided and fumigation above 25°C not advisable.					
Carbon disulphide	160g/m³	24 h	>20°C		
Highly flammable					
Phosphine formulation as *aluminium or magnesium phosphide)		7 days	15 ±2°C		30 tablets
		6 days	16 to 20°C		
		5 days	21 to 25°C		
		4 days	above 26°C		
Recommended for wide range of crops with minimum ill effect on germination					

*Banned by Government of India

The most commonly used seed fumigates are tablets/ powder of aluminum phosphide. However it is banned by the Government of India. Fumigation with aluminum phosphide protects the seed without affecting its viability. It produces phosphine gas when come in contact with water. On contact with normal air moisture, it slowly begins to liberate phosphine gas. The liberation rate of posphine gas depends on moisture and air temperature levels.

In addition to this significant hazard, the fumigant is a very small molecule that is slightly heavier than air. Because of these properties, the gas is hard to contain unless the seed bin is tightly sealed. Phosphine can move through plastic, concrete, wood, and any crack or seam in a metal bin.

Fumigation with phosphine at higher seed moisture level may cause complete loss of viability. Mutagenic effects and chromosomal aberrations in wheat and onion seeds due to phosphine have been reported. In rice, high phosphine dose resulted in increased dihydrogenase activity, seed leachate electrolytes along with decreased respiratory enzymes. In most of the crops, the germination and vigour is affected adversely with excessive fumigation doses and prolonged exposure period with high seed moisture content. Therefore, the movement of moisture in the store should be avoided, as seed quality is affected adversely if phosphine fumigation is done under more than 12 per cent seed moisture content.

To use fumigants, methods of application, dosages, temperature, moisture and exposure periods should be strictly followed along with all the recommended precautions. Fumigants are not fully effective at temperatures below 15°C. Fumigant should be placed at mid height of the storage structure as the gas is heavier than air and travel downwards upto 8 feet. It should never be placed at the bottom. Before fumigation, it should be ensured that the storage structure is air tight. Methyl bromide should not be used for fumigation of seed. E.D.B. ampoules @ 3 ml/q of seed in air tight storage structure may be applied to control insect infestation. However, fumigation with EDB reduces the germination of many cereals and legume crops.

Seed treatment with inert material: dry mixing of activated clay, ash, red earth or sand with seed at 1:100 ratio, minimize the infestation of Bruchids in store due to abrasive action.

Pheromones: Pheromone loaded traps are placed in the store to catch the insects like *Tribolium* and *Sitophilus*. It is effective in detecting hidden infestation. Synthetic pheromones are available in the market for red and confused flour beetles, lesser gain borer, khapra and few other insects species. Pheromone traps are of two types

1. Pheromones trap with sticky substance to catch the insects.
2. Conical shape traps with perforated top for entrance of insect.

Chemical control: Spray of following chemicals on floor, wall and ceiling to kill any insect that has remain in the store or insects that may damage the stored seed.

0.1% Malathion 50 EC (1 part of insecticide + 500 parts of water) or

0.1% Fenvelrate 20 EC (1 part of insecticide + 200 parts of water) or

0.1% Cypermathrin 25 EC (1 part of insecticide + 2500 parts of water)

Disinfestation of store with the spray of 0.5% Malathion 50 EC 1:100). Spray the stores with DDVP @ 0.25% (1:300)/ 100m^2 area 3 liters

Eggs and larvae of insects remain hidden and feed on the seed fragments in the off season. Whenever the seeds are filled up in such bags, infestation reappears. Therefore, seed should be stored in new gunny bags. In the event old bags have to be used then it should be dipped in boiled water for 15 minutes and then treated with10% malathion solution before storage of seed.

The insecticides and their concentrations recommended for prophylactic treatment of stored seed are Malathion 50 EC at 0.5% and Pxrimiphos methyl 50% EC at 0.5%. These insecticide treatments are to prevent the insect infestation and cross infestation.

Biological control: Store grain pests are attacked by many parasites and predators. Use of these enemies offers new dimension in the management of store grain pest. A parasitic wasp, *Bracon hebetor* attacks late stage larvae of *Cautella* and *Plodia interpunctella* and suppresses their populations.

Host resistance to insect-pest: Variability in many crops exist for degree of infestation of store grain pest. Genetic ability of the variety to resist infestation of store grain pest has yet to be commercially exploited.

Temperature manipulation: Application of high or low temperature offers a non-chemical means of disinfestations. Sublethal temperature can affect reproduction, growth, development, feeding and movement of insect. However, temperature adverse to survival of insect population in store must not affect the seed viability.

Controlled or modified atmospheres: Insect infestation may be control by altering the concentration of gases *viz.,* oxygen, carbon dioxide and nitrogen in the store. Deficiency of oxygen effectively reduce the population of insect pest in the store e.g., store with only 2% oxygen create suffocation for *Sitophilus* spp. without adversely affecting the sowing seed quality of wheat, rice and barley.

Physical barriers: The packaging materials *viz.,* ethylene tetrafluorethylene, polyester, polyvinylchloride, polypropylene and polycarbonate are resistant to insects. Insecticides are also inserted into packaging material to make them insect resistant. Inert matter *viz.,* sand, ash, crop seed of small size *viz.*, minor millets, with repellent nature *viz.*, mustard are traditionally mixed with seeds as physical barrier for insect infestation.

Threshold level: One pair of pulse beetle (bruchids) can cause injury to seed of pulses and lesser grain borer (*Rhizopertha dominica*) to cereals above the MSCS therefore, the seed should be kept free from these insects.

■■■■

Chapter - 44

RIGHTS IN PROTECTION OF PLANT VARIETIES AND FARMERS' RIGHTS ACT

INTRODUCTION

Under the TRIPS agreement it is obligatory on part of a Member to provide protection to new plant variety either through patent or an effective *sui generis* system or a combination of these two systems. In many countries such varieties can be protected through Breeders' Rights, patents and UPOV Convention. UPOV is an abbreviation of *Union pour la Protection des Obtentious Vegetals* (Union for protection of new varieties of plant). It is an international convention which provides a common basis for the examination of plant varieties in different member States of UPOV for determining whether a plant variety merits protection under UPOV or not. India was therefore under an obligation to introduce a system for protecting new plant variety. India opted for *sui generis* system for protection of varieties by three important notifications viz., act, rules and legislation. All the three are enclosed in the book.

In place of patent protection is provided to the varieties of crops under PPV and FRA in India. A patent in an exclusive right granted by a country to the owner of an invention to make, use, manufacture and market the invention, provided the invention satisfies certain conditions stipulated in the law. Exclusivity of right implies that no one else can make, use, manufacture or market the invention without the consent of the patent holder. This right is available only for a limited period of time. However, the use or exploitation of a patent may be affected by other laws of the country which has awarded the patent.

The Protection of Plant Varieties and Farmers' Rights Act, 2001 received the assent of the President on the 30th October, 2001. The Act has 97 sections distributed over 11 chapters. The act contains provisions for protecting the larger public interest with safeguards to farmers by giving farmers' rights, researchers by providing researchers' rights alongwith an effective system of protection of plant breeders' rights. The farmer's rights include his traditional rights to save, use, share or sell his farm produce of a variety protected under this Act provided the sale is not for the purpose of reproduction under a commercial marketing arrangement. The Protection of Plant Varieties and Farmers' Rights Authority has been set up at New Delhi to administer the Act.

The Protection of Plant Varieties and Farmers' Rights Rules, was notified by the Department of Agriculture and Cooperation, Ministry of Agriculture, Government of India New Delhi on 12th September 2003. It contains 76 clauses distributed over IX chapters with four schedules, 12 plant variety registry forms and 33 plant variety forms.

Protection of Plant Varieties and Farmers' Rights Regulations was notified by the Department of Agriculture and Cooperation, Ministry of Agriculture, Government of India, New Delhi on 7th December, 2006. It contains 13 sections distributed over four chapters.

Objectives of the Act

The PPVFR act was implemented to stimulate investments for research and development both in the public and the private sectors for the developments of new plant varieties by ensuring appropriate returns on such investments. It will also facilitate the growth of the seed industry in the country through domestic and foreign investment to ensure the availability of high quality seeds and planting material to Indian farmers. Apart from this the act will also recognize the role of farmers as cultivators and conservers and the contribution of traditional, rural and tribal communities to the country's agro-biodiversity by rewarding them for their contribution through benefit sharing and protecting the traditional right of the farmers.

The act is implemented by PPV&FR Authorty for protection of new plant varieties (if required) and safeguard rights to breeder, farmer and researcher (Fig 1).

Protection: Protection means no body other than the concerned breeder or institute may commercialize the registered variety or produce, sell, market, distribute, import or export the seed for commerce of the registered variety without prior authorization form the breeder/institute developed the variety upto 15 years for crops and 18 years for tree and vines from the date of registration.

Protection of new plant varieties

On request of the breeder the act provides protection to new varieties that may be extant varieties, including Farmers' varieties, the varieties of common knowledge and essentially derived varieties. The each class is defined under the chapter variety.

Varieties with following bottlenecks can not be protected under the

(i) not capable of identifying such variety; or

(ii) consists solely of figures; or

(iii) is liable to mislead or to cause confusion concerning the characteristics, value, identify of such variety, or the identity of breeder of such variety;

(iv) is likely to deceive the public or cause confusion in the public regarding the identity of such variety;

(v) is comprised of nay matter likely to hurt the religious sentiments respectively of any class or section of the citizens of India;

(vi) is prohibited for use as a name or emblem for any of the purposes;

(vii) is comprised of solely or partly of geographical name.

The Act does not provide protection to plant variety that carries **genetic use restriction technology** (GURT) and **'terminator gene'**.

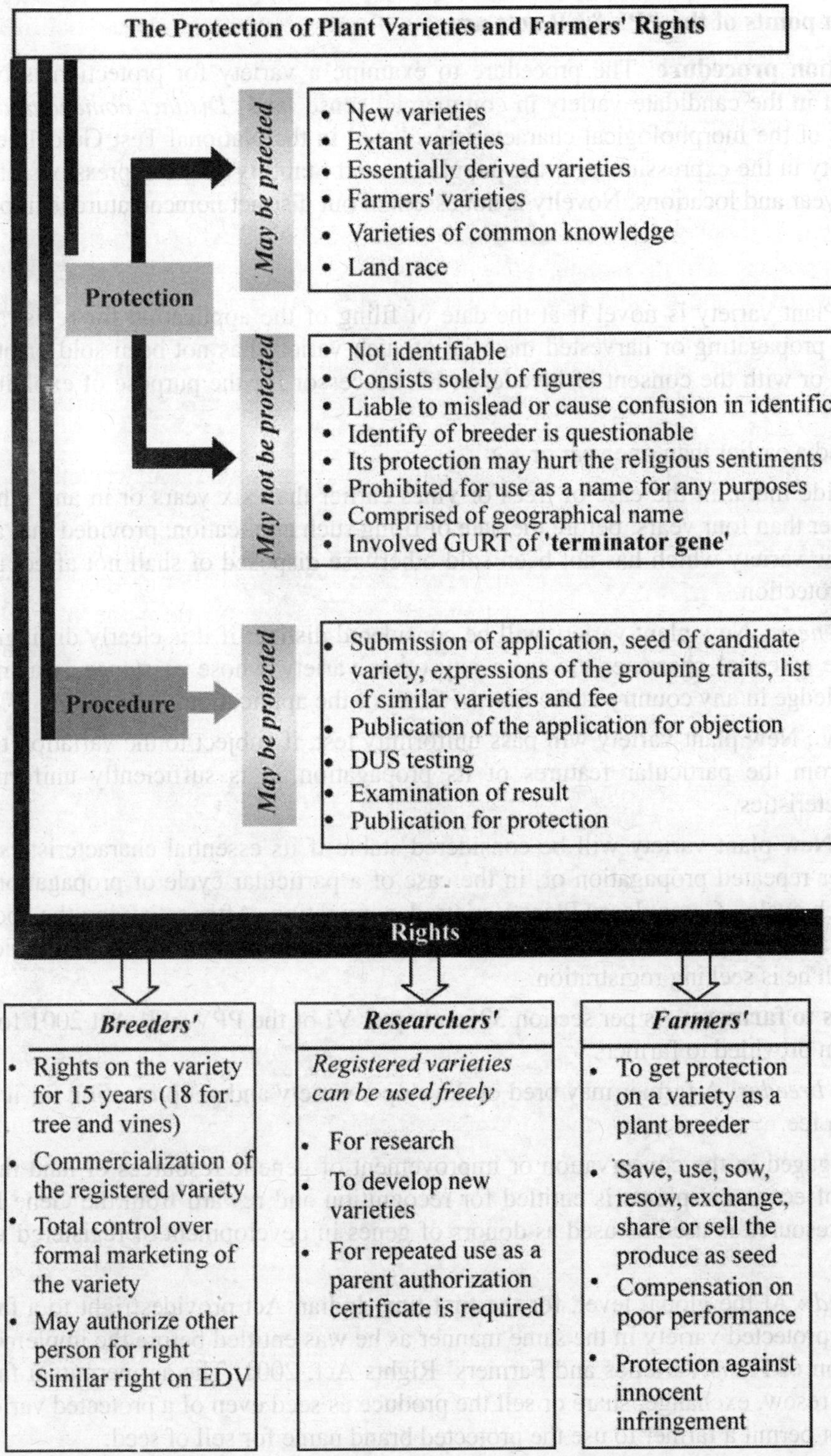
The Protection of Plant Varieties and Farmers' Rights
Protection
May be pretected
• New varieties
• Extant varieties
• Essentially derived varieties
• Farmers' varieties
• Varieties of common knowledge
• Land race
May not be protected
• Not identifiable
• Consists solely of figures
• Liable to mislead or cause confusion in identification
• Identify of breeder is questionable
• Its protection may hurt the religious sentiments
• Prohibited for use as a name for any purposes
• Comprised of geographical name
• Involved GURT of 'terminator gene'
Procedure
May be protected
• Submission of application, seed of candidate variety, expressions of the grouping traits, list of similar varieties and fee
• Publication of the application for objection
• DUS testing
• Examination of result
• Publication for protection
Rights
Breeders'
• Rights on the variety for 15 years (18 for tree and vines)
• Commercialization of the registered variety
• Total control over formal marketing of the variety
• May authorize other person for right
Similar right on EDV
Researchers'
Registered varieties can be used freely
• For research
• To develop new varieties
• For repeated use as a parent authorization certificate is required
Farmers'
• To get protection on a variety as a plant breeder
• Save, use, sow, resow, exchange, share or sell the produce as seed
• Compensation on poor performance
• Protection against innocent infringement

The important points of the PPV&FR acts are

Examination procedure: The procedure to examine a variety for protection is based on *Novelty* present in the candidate variety in commercial sense and *Distinct nomenclature* based on expressions of the morphological characteristics listed in the National Test Guidelines of the crop, Uniformity in the expression over the population and Stability in the expressions of distinct traits over the year and locations. Novelty is not essential but distinct nomenclature is important.

NDUS Test

Novelty: Plant variety is novel if at the date of filing of the application for registration for protection, the propagating or harvested material of such variety has not been sold or otherwise disposed of by or with the consent of breeder or his successor for the purpose of exploitation of such variety-

(i) in India earlier than one year or

(ii) outside India, in the case of trees or vines earlier than six years or in any other case, earlier than four years, before the date of filing such application: provided that a trial of a new variety which has not been sold otherwise disposed of shall not affect the right to protection.

Distinctiveness: New plant variety will be considered distinct if it is clearly distinguishable by at least one essential characteristic from any other variety whose existence is a matter of common knowledge in any country at the time of filing of the application.

Uniformity : New plant variety will pass uniformity test, if subject to the variation that may be expected from the particular features of its propagation, it is sufficiently uniform in its essential characteristics.

Stability: New plant variety will be considered stable if its essential characteristics remain unchanged after repeated propagation or, in the case of a particular cycle of propagation, at the end of each such cycle. Compulsory Plant Variety denomination: After satisfying the above four essential criteria every applicant shall assign a single and distinct denomination to a variety with respect to which he is seeking registration.

The rights to farmers: As per section 39 in chapter VI of the PPV&FR Act 2001 following rights have been provided to farmers

As a plant breeder: A farmer may bred or develop a variety and get protection for it without any registration fee.

Farmer engaged in the conservation or improvement of genetic resources of land races and wild relatives of economic plants is entitled for recognition and reward from the Gene Fund, if the conserved resource is used as used as donors of genes in development of registered varieties under the Act.

Use of seed : At the global level, for the first time Indian Act provides right to a farmer to use the seed of protected variety in the same manner as he was entitled before the implementation of the Protection of Plant Varieties and Farmers' Rights Act, 2001. The act permits a farmer to save, use, sow, resow, exchange, share or sell the produce as seed even of a protected variety. But the Act does not permit a farmer to use the protected brand name for sell of seed.

Compensation: The expected performance of the protected variety under given conditions has to be disclosed to the farmers. In the event of failure in the performance of the protected variety under given conditions, the farmer(s) may claim compensation as per the prescribed manner.

Protection against innocent infringement: Section 43 specifies that the farmer cannot be prosecuted for infringement of rights specified in the Act if he can prove in court that he was unaware of the existence of such a right.

The rights to plant breeders

An exclusive right on the breeder or his successor, his agent:

To commercialize the registered variety either in his/her own person or through anyone he designates.

To produce, sell, market, distribute, import or export the variety. In short, Breeder has full control over formal marketing.

The breeder of the protected variety has rights for 15 years (18 years for tree and vines) from the date of registration.

A breeder may authorize any person to produce, sell, market or otherwise deal with the variety registered under this Act.

The breeder of essentially derived variety has the same right as of new variety only after authorization by the breeder of the initial variety.

Rights of researchers

Under section 30 that falls in chapter IV the right of research is mentioned. it provides following rights to a researcher

Researchers can freely use any registered varieties for research even to develop new varieties.

Authorization form the original breeder of the protected variety is required for repeated use of the protected variety as parental line to develop new variety.

Benefit sharing

Any citizen / group of citizens of India / governmental organizations / NGO / firm formed or established in India can claim PPV&FR Authority to assess and determine claim Compensation may be in any form for a commercialized protected variety.

National Gene Fund: To credit *inter alia* benefit shared under the Act at local community level to use the fund for facilitating sustainable use of genetic resourcesl

Compulsory license

After three years from the date of registration of the protected variety, any person can apply to PPV&FR Authority for undertaking production, distribution, and sale of the seed or other propagating material in the event of non availability or short supply of the seeds or other propagating material or non availability at a reasonable price by the Breeder / institution /

licensee, then The PPV&FR Authority in consultation with the Central Government and after hearing both the parties may pass an order for the registered proprietor to grant the license to the applicant for undertaking production, distribution, and sale of the seed as per the rule for determined period of time. This is known as compulsory licensing.

The compulsory seed production license may be provided when Breeder / institution / licensee, of the protected variety

The planting material is not supplied or short supplied.

Charges prohibitively high prices for planting material.

Plant Variety Tribunal may also provide compulsory license to expedite disposal of related legal disputes

■■■■

Chapter 45

NDUS TEST

OBJECTIVE

Testing of variety for providing protection under Protection of Plant Variety and Farmers' Right Act

The Act provides protection to any variety based on its Novelty and Distinctiveness. These two things are tested by Novelty, Distinctiveness, Stability and Uniformity Test of the variety. The variety submitted for protection is considered as Candidate variety. It may be a new variety, Essentially derived variety, Farmers' variety or land race. A new variety is registered under the act after NDUS test.

Novelty: Plant variety is novel if at the date of filing of the application for registration for protection the propagating or harvested material of such variety has not been sold or otherwise dispose-off by or with the consent of breeder or his successor for the purpose of exploitation of such variety-

(i) In India earlier than one year or

(ii) Outside India, in the case of trees or vines earlier than six years or in any other case, earlier than four years, before the date of filing such application: provided that a trial of a new variety which has not been sold otherwise dispose-off shall not affect the right to protection.

Distinctiveness: New plant variety will be considered distinct if it is clearly distinguishable by at least one essential characteristic from any other variety whose existence is a matter of common knowledge in any country at the time of filing of the application.

Uniformity: New plant variety will pass uniformity test, if subjected to the variation that may be expected from the particular features of its propagation, it is sufficiently uniform in its essential characteristics.

Stability : New plant variety will be considered stable if its essential characteristics remain unchanged after repeated propagation or, in the case of a particular cycle of propagation, at the end of each such cycle. Compulsory Plant Variety denomination: After satisfying the above four essential criteria every applicant shall assign a single and distinct denomination to a variety for which the registration is requested.

Initially 35 crops are covered such as rice, wheat, maize, sorghum, pearl millet, chickpea, mungbean, urdbean, field pea, rajmah, lentil, pigeon pea, rapeseed mustard, groundnut, soybean, sunflower, safflower, castor, sesame, linseed, cotton, jute, sugarcane, lucerne, berseem, tomato, brinjal, okra, cauliflower, cabbage, potato, onion, garlic, rose and chrysanthemum. The DUS test guidelines are available for all the 35 crops.

PROCEDURE

Submission of application

Applicant has to submit the application form to PPV&FR Authority for protraction of the variety. Expression of the grouping traits (Fig. 1) of the candidate variety as prescribed for the crop has to be submitted by the concerned breeder/institute (point 5) alongwith the name of reference variety (example variety) with the same expression and note. Reference variety for expression of each characteristic is mentioned in the *Guidelines for the conduct of test for distinctiveness, uniformity and stability* of the respective crop published by PPV& FR Authority. The name and denomination of the variety(ies) similar to the candidate variety and expression of the characteristics in which the similar variety is different then the candidate variety has to be submitted by the concerned breeder as prescribed in the point 6 of the proforma.

Submission of seed

Applicant submits the prescribed quantity of healthy and vigorous seed that fulfil the Indian Seed Certification Standard, of the variety (candidate variety) for DUS testing for protection. In general the seed submitted for protection of the variety is untreated. If the treatment is unavoidable then details of the treatment have to be provided.

Charges for registering a plant variety

Applicant has to submit the required fee to the Registrar for seeking protection of the variety.

Address of the authority

Protection of Plant Varieties and Farmers' Rights Authority, Ministry of Agriculture, Government of India, NASC Complex, DPS Marg, Opposite Village Todapur, New Delhi 110 012

Grouping of varieties

Characteristics in which the documented states of expression even over the location can be used to select, either individually or in combination with other such characteristics, varieties of common knowledge that can be excluded from the trial used for examination of distinctiveness. The traits in which the expression over the location is as per documentation and can group the similar varieties effectively to organize the trial is considered as grouping trait. Grouping traits are mainly qualitative or pseudo-qualitative in nature. The candidate variety is grouped to facilitate the assessment of distinctness based on the listed traits. Level of expression of these characteristics is sufficient to establish distinctiveness without any influence of environment. Within variety variation should be minimum for grouping traits. For each and every crop grouping traits are listed. The candidate variety is compared with the most similar varieties based on the grouping traits.

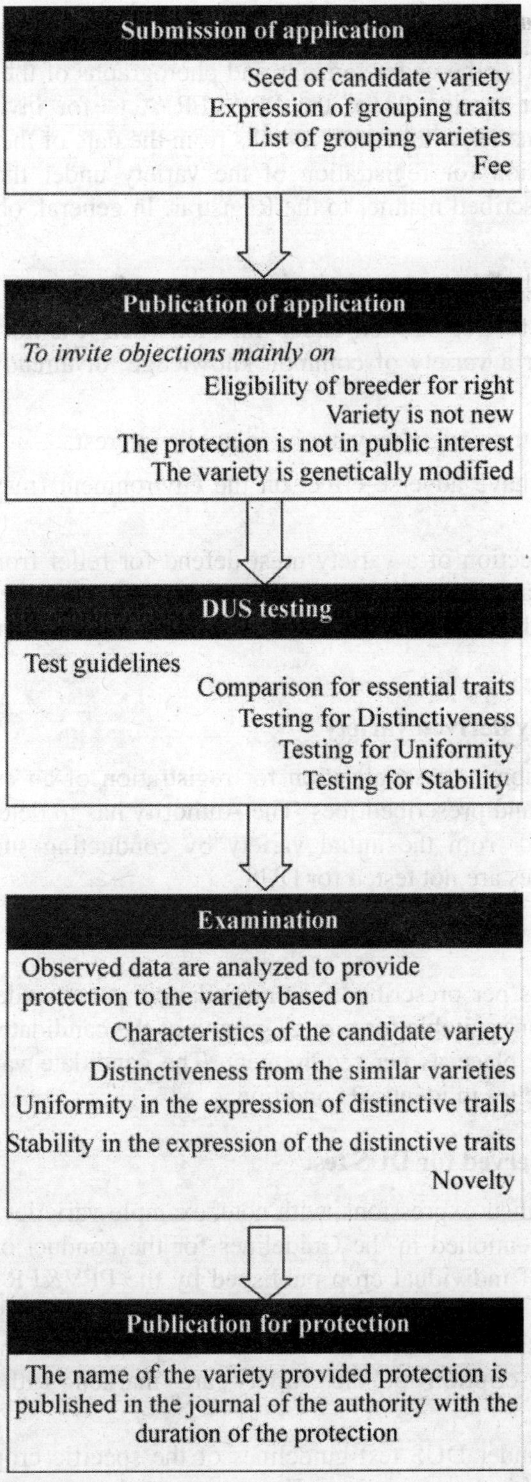

Fig. 1. Steps involved in providing protection to a variety

Publication of application

The Registrar advertises the characteristics and photographs of the candidate varieties in the prescribed manner as per section 21 of the PPV&FR Act for inviting objections from the interested persons in the matter within three months from the date of the advertisement. Interested person may lodge objection for registration of the variety under the act on payment of the prescribed fees in the prescribed manner to the Registrar. In general, objections are concerned in case

- Applicant is not eligible for breeder's right;
- Variety is not fit for registration under this Act when it is not new; or a repetition of a known variety; or a variety of common knowledge; or already registered; or protected variety etc.
- The protection of the variety may not be in public interest;
- The variety may have adverse effect on the environment (mainly genetically modified varieties).

Applicant seeks protection of a variety must defend for relief from objection by providing satisfactory justification case within two months after receiving the objection. The application for protection may be rejected in the event no satisfactory reply from the applicant is received within the prescribed time.

Registration of essentially derived variety

An applicant may submit the application for registration of an essentially derived variety with required documents and prescribed fees. The Authority has to determine that the essentially derived variety is derived from the initial variety by conducting such prescribed procedure. Essentially derived varieties are not tested for DUS.

Conduct of tests

The crop is grown as per prescribed recommendation at two independent similar growing seasons and in two locations suitable for normal growth of the candidate variety. The variety may be tested at an additional place as per requirement. The candidate variety is grown alongwith reference and similar varieties in identical condition.

Characteristics to be observed for DUS test

List of the traits and their expressions, with note, example varieties, stage of observation and type of assessment are mentioned in the Guidelines for the conduct of test for distinctiveness, uniformity and stability of individual crop published by the PPV&FR authority. For each crop number of characteristics has been listed for testing the distinctiveness. Expression of some of these characteristics that should be used in every growing period for the examinations of all the varieties included in description of the variety are marked with asterisked as essential characteristics.

All the traits listed under DUS test guidelines of the specific crop have to be observed at proper stage as mentioned in the guidelines. Expression of the characteristics is never observed on plants in border rows. The traits are divided into two groups i.e., observation based on measurement (M) or by visual assessment (V) both types are further divided into two groups observation on single plant/part of plant (S) and observation on group (G) of plants.

Qualitative traits viz., growth habit, colour, shape, hairiness etc are mostly observed visually whereas, quantitative traits like height, width, weight, etc are measured. Phenological traits are measured on time scale viz., days to flowering, maturity etc. Few quantitative traits are measured on group of plants as qualitative trait viz., hairiness, growth habit etc., similarly few traits like leaf shape, leaf size recorded by observation on group of pants are considered as peudo-qualitative characteristics.

Test of Distinctiveness

The candidate variety must be distinct from all the varieties whose existence is a matter of common knowledge. For the assessment of distinctness, expression of the traits is recorded on prescribed number of plants of the candidate variety. The expression of the trait is matched with the reference varieties for proper denomination and compared with the designated similar varieties but not with all the varieties of common knowledge to establish the distinctiveness.

A variety is considered as distinct when the difference in the expression of the characteristic is *consistent* and *clear*. Consistent difference means the difference in the expression of the trait under consideration has distinct expression at two independent occasions it may be over two locations or two seasons from the varieties of common knowledge. Clear difference indicates difference in the expression for qualitative traits with main emphasis on essential traits.

Test of Uniformity

To consider the candidate variety uniform, the variation in expression of the essential traits should be minimum to permit accurate description of the variety and assessment of distinctness and stability. Testing of uniformity involves certain tolerance however its limits may vary according to the reproductive behaviour of the crop. Tolerance limit is low for self pollinated, vegetatively propagated, inbred lines, parents of hybrids, but high for cross pollinated crops and open pollinated, composite and synthetic and hybrid varieties.

Assessment based on visually observed traits: All the plants of self pollinated, vegetatively propagated, inbred lines, parents of hybrids are expected to be uniform therefore the uniformity may be assessed by the number of offtypes present. In DUS testing a plant is considered as offtype when it is clearly distinguishable from the plant of the same variety based on the expression of the listed designated characteristics. Offtpyes in self pollinated crops are observed mainly based on visually observed designated qualitative traits. The number of off type should not exceed the tolerance limit in a fixed population size to consider as uniform and stable.

Assessment based on measurable traits: Varieties of cross-pollinated crops exhibit wide variation mainly for distinguishable quantitative (required measurement) traits, therefore it becomes essential to count the offtypes. The candidate variety of the crop of cross pollinated nature should not be significantly less uniform than the comparable varieties. For cross pollinated crops the number of offtypes recorded based on the distinguishable qualitative traits assessed visually in a fixed population crops is considered to judge the uniformity. The uniformity is assessed by computing the standard deviation (sd) from the values of the prescribed number of the plants per replication.

Test of Stability

A variety is considered as stable, based on unchanged expression over the year and location for the essential traits enlisted in the national test Guidelines of the crop under test. In general homogeneous (uniform) population of the candidate variety is also considered as stable. Protection can not be provided to the candidate variety in the event of instability in the expression of the essential characteristics. For hybrid variety the stability of the expression is tested on the F_1 plant i.e., plant formed by the certified seed provided to the farmers for commercial cultivation.

■■■■

Chapter 46

VCU TEST

OBJECTIVE

To test Value for Cultivation and Use (VCU) of a plant variety

NDUS test determines uniformity and stability of the expression of distinguishing traits for protection of a variety. Economically important traits based on which the variety is selected for cultivation by farmers is not tested under DUS. Therefore, in European and some other countries apart from DUS test the variety is also tested for its Value for Cultivation and Use for listing in National Catalogue of Plant Varieties, which is a prerequisite to place a variety in the commercial market. It shows that a new variety has a value for cultivation and use over already registered varieties.

The National Catalogue is compiled based on the satisfactory Value for Cultivation and Use of the proposed variety (Table 1). VCU certificate requires results for two successive years of yield performance (productivity), expression of biological characteristics, quality of the obtained produce, chemical and technological characteristics, resistance to pests and diseases and any other claimed commercially important characteristics.

Table 1. Factors considered to test value for cultivation and use of field crops

Value for cultivation	Value for use
• Yielding ability	• Nutritional quality: pulses, oilseeds, cereals
• Duration for maturity	• Nutritional quantity: pulses, oilseeds, cereals
• Plant geometry	• Cooking quality: rice, wheat
• Response to management	• Ability to fix nitrogen: legumes
• Tolerance against abiotic stresses	• Fitness for processed food: maize, potato etc.
• Resistance against biotic stresses	• Fibre quality: cotton, Jute etc.
• Any other characteristics	• Chemical characteristics
	• Technical characteristics
	• Any other characteristic

In India, VCU test is not applicable at present and the variety is recommended for cultivation based on well recognized multilocational performance/testing over the years in All India Coordinated Crop Improvement Project of different crops and adaptive trials in the state or zone.

VCU test of potato in European Seed Association (ESA)

A variety of potato has to be included in the Common Catalogue to place it on the market in Europe. The testing for Value for Cultivation or Use (VCU) includes following key principles as minimum standards

Class

- Testing grouped the potato varieties into two classes based on its use

  Starch types

  Consumption types

Days to maturity

- Each variety is ranked in maturity groups in scale of 1-9

  Very early – 1

  Very late – 9

Yield

- Within each class the relative yield of each candidate variety is compared with well known and consistent local standard varieties over the years.

Disease

- VCU testing of potato varieties include assessment of resistance to the following diseases based on harmonized protocols:

  Late blight (foliage blight)

  PVY-n (including PTNRD)

  PCN (*G. rostochiensis, G. pallida*)

  Wart disease (race 1)

Additional diseases relevant to the country

Quality: VCU test includes following quality traits

Sugar content – based on fry colour using a photographic scale

Dry matter content – based on under water weight

Cooking type – based on EAPR standards

Additional characteristics

Reliable and cost effective VCU testing under best optimum management for two successive years by adopting scientifically recognised statistical design maximises precision and minimises bias. This is the responsibility of official authorities. A seed sample of the candidate variety for VCU test is used for authentication by a DUS Test in both the years. Result of Plant Varieties Value for Cultivation and Use Testing is notified.

No VCU test of GM candidates is undertaken until it has been released by the competent authority for given environment. Application of such GM candidate variety is submitted to GM Regulatory Authorities. The authority is responsible to conduct trial on GM varieties in safe mode.

■■■■

Chapter 48

PROCEDURE AND POLICY FOR IMPORT AND EXPORT OF SEED

OBJECTIVE

The Export and Import Policy (EXIM) Committee is constituted in the Division of Seed to deal with application for exports/imports of seeds and planting materials in accordance with the New Policy on Seed Development and EXIM regulation. The export and import of seeds/planting materials are governed by the provisions of EXIM on 1st April 1997- 31st March 2002 of Ministry of Commerce, Government of India. The functions of the EXIM Committee are to analyze applications and furnish recommendations to Plant Protection Advisor (PPA) / Directorate General of Foreign Trade (DGFT) for issue of license for import/export of seeds and planting material.

IMPORT OF SEED

The objective of the import is

New Policy on Seed Development-1988

The Government of India introduced a New Policy on Seed Development in October 1988 with the objective to provide the best safe and suitable planting material available anywhere in the world to Indian farmers, to increase productivity, farm income and export earnings, while ensuring that there is no deleterious effect on environment, health and bio-safety. The policy liberalized the parameters governing the import of seed and planting material.

The Policy reiterate that all imports of seed and planting material would be regulated under the Plant Quarantine Order 2003 and amendments made their in under, Import licenses would be granted by DGFT/PPA on the recommendations of DAC.

While importing seed and planting material, care is taken to ensure no compromise on the requirements under prevailing plant quarantine procedures to prevent entry of exotic pests, diseases and weeds detrimental to Indian agriculture.

Import of seed and planting materials, etc. is allowed freely subject to EXIM Policy guidelines and the requirements of the Plants, Fruits and Seeds (Regulation of import into India) Order, 1989 as amended from time to time. Import of parental lines of newly developed varieties is also encouraged.

PROCEDURE

Submission of application

Importer has to submit 20 copies of application for import in the prescribed formats. The EXIM Committee scrutinizes the proposal every month subject to purpose of proposals for import/export of seeds and planting material.

Testing under ICAR

A small prescribed quantity of seed of variety (oilseeds, pulses, fodder and cereals) under consideration is submitted by the importer for multilocation testing under All India Coordinated Trial of ICAR. The new variety is tested for one crop season in ICAR's multilocation trials with special reference to their resistance to seed/soil borne diseases. The results of the trials and evaluation are communicated to the Department of Agriculture and Cooperation (DAC) within 3 months of the end of the crop season. Within 30 days after the receipt of report DAC rejects or approves the application and inform to DGFT for grant of commercial import license. If the performance of the variety is found to be satisfactory in the multi-location trials, the importer can apply for bulk import of the seed of the new variety.

Sample to NBPGR

All importers are required to submit the specified quantity of the imported seeds to the gene bank of NBPGR at cost price for testing/accession.

Bulk import of seeds

New variety that has performed satisfactorily in ICAR trials is allowed for import only for a period of 2 years. Within the period of 2 years, the importer must make arrangements for commercial multiplication of the seed in India through transfer of parental lines or Breeder seed and the seed production technology.

Quarantine testing of bulk import

To ensure no entry of exotic pests, diseases and weeds into the country the imported seed and planting material are tested under prevailing plant quarantine procedures. After the arrival of consignment at the port of entry, quarantine checks are undertaken that include visual inspection, laboratory inspection, fumigation and grow-out test. For the purpose of these checks, samples are drawn and the tests are conducted concurrently. The import of seed is tested under quarantine checks and within three weeks, it is cleared or rejected by PPA. During quarantine, the imported consignment is kept in a bonded warehouse at the cost of the importer. The rejected consignment is destroyed. Every effort is made to prevent the entry of exotic pests, diseases and weeds in the country those are harmful to the agriculture.

Permission

Imports of all the category of seeds require a permit granted by the Plant Protection Advisor to the Government of India, issued within the minimum possible time frame. Import of seed, planting material and living plant by ICAR, etc. is allowed without a license in accordance with conditions specified by the Ministry of Agriculture.

Quality

Seeds and planting materials imported for sale into the country should meet Indian seed standards of seed health, germination, genetic and physical purity.

Field crops

Import of seed of wheat, rye, barley, oat, maize, rice, millet, sorghum, pearl millet, finger millet, other cereals, soybean, groundnut, linseed, cotton, castor, sesame, mustard, safflower, clover, palmnut, etc. is allowed without license subject to the New Policy on Seed Development, 1988 and in accordance with import permit granted under PQ Order, 2003 and amendments made their under. In the case of wheat and rice, ICAR has effective research collaborations with International Research Institutes. Therefore import of seed of particular variety beyond two years for these crops is not permitted.

Horticultural Crops

Import of seeds/tubers/bulbs/cuttings/saplings of vegetables, flowers and fruits is allowed without a license in accordance with import permit granted under Plant Quarantine Order, 2003 (PQ Order).

Import of seeds/tubers of potato, garlic, fennel, coriander, cumin, etc. is allowed in accordance with import permit granted under PQ Order, 2003 and amendments made there under.

Import of seed/planting material of fruits are permitted selectively by the Department of Agriculture and Co-operation (DAC) on a case to case basis on the recommendation of the State Director of Horticulture/Agriculture subject to post-entry quarantine and such regulations as may be laid down by the Plant Protection Advisor. Free import of vegetable/flowers seeds is permitted to specified categories of imports. No post-entry quarantine is insisted upon for the import of seed of vegetable/flowers/ ornamental plants. Import is permitted after visual inspection, fumigation, laboratory tests and grow-out test. Import of tubers and bulbs of flowers are subjected to post-entry quarantine.

EXPORT OF SEED

Under the EXIM Policy 1997-2002 seeds are classified into the following categories:

(a) *Seed in the restricted list of exports:* Breeder and Foundation category of all the crop varieties and following crop species are restricted for export

Forest	Field crops	Fodder	Horticultural
All forest tree species	Rice all wild varieties	All fodder	Onion
Nuxvomica	Wheat all wild varieties	Egyptian clover	Saffron

Rubber	Oilseeds all Certified seed	Lucerne	Pepper
Neem	Pulses all Certified seed	*Medicago sativa*	Wild varieties of ornamental crops
Sandalwood	Soybean	All green manure crops except *Dhaincha*	
	Linseeds		
	Mesta seeds	Persian clover	
	Cluster bean	Red sanders	
	Cotton, except of other countries	Russa grass	
	Niger channelized through TRIFED, NAFED	Tufts	
	Groundnut export is subjected to contract with APEDA		

APEDA : Agricultural and Processed Food Products

NAFED : National Agricultural Co-operative Marketing Federation of India Ltd

TRIFED : Tribal Co-operative Marketing Development Federation of India Ltd

(b) Seeds whose exports are allowed without license.

Procedure

In order to encourage the export of seed, Government of India, Ministry of Agriculture, DAC issued circular dated 17.11.97 on policy regarding export of seeds produced against export order and fixing of quantitative export ceiling for seed.

i. The entire quantity of seed of any crop/variety produced by any Public or Private Sector Seed Company or other exporter is allowed to be exported against export order.

ii. The exporter has to inform the Division of Seed, Ministry of Agriculture, Government of India about export order, production programme and other details at least two months before harvesting. The information with recommendation is forwarded to the DGFT (Directorate General of Foreign Trade) and EXIM Committee of DAC for export of seed.

iii. Permission for export against the export order is granted for all the privately developed notified/non-notified varieties of all the crops.

iv. The export of seeds of crops under restriction is allowed on case-to case basis under license issued by DGFT on the basis of the recommendations of DAC.

v. The permission for export for all the privately developed notified/non-notified categories is also granted for the crops under restricted list and even when advance notice is not given.

vi. The export is not permitted when the seed is essential for meeting the requirement in an emergency caused by a natural calamity in the country. In such a case a view is taken by the EXIM Committee for the quantity of seed which may be allowed for export.

vii. In almost all the cases permission for export is granted with liberal approach for prescribed quantity. In the event of quantify higher than the prescribed limit, the case is referred by DGFT to Department of Agriculture & Cooperation.

Diverse agro-climatic conditions allow production of high quantity of quality seed of almost all the crops in India. Government of India evolves a long term policy to enhance the share of India in global seed export from less than 1% to 10% by the year 2020 by encouraging custom based production of seeds for export. To promote export of seed following initiations have been taken

- Establishment of Seeds Export Promotion Zones.
- Prepetition of data bank on the International Market.
- Identification of export potential of Indian varieties in different parts of the world.
- Preparation of data base on availability of seeds of different crops to assess impact of exports on domestic availability of seeds.
- Initiation of promotional programme to enhance the quality of seeds as per requirement of International Market.
- Promotion of ISTA accreditation of seed testing laboratory for certification and testing as per international norms and requirements.
- To provide a boost to seed export, India has decided to participate in the organization for Economic Cooperation and Development (OECD) Seed Schemes for fodder, legumes, oilseeds, fibre species, cereals and vegetables.

■■■■

Chapter 48

ESTABLISHMENT OF SEED TESTING LABORATORY-TENTATIVE PLAN

OBJECTIVE

Equipment, instrument, glass and plastic wares, chemicals etc., required for establishment of seed testing laboratory.

Literature

Blue book on Indian Seed Certification Standard Ministry of Agriculture, GOI	
Rule book on seed testing	ISTA
Hand book of seed testing	Ministry of Agriculture, Government of India
Seed Analyst manual	Ministry of Agriculture, Government of India
Seed inspector manual	Ministry of Agriculture, Government of India

Descriptors of crop varieties

Notifications of crop varieties

Herbarium

- Objectionable weed
- Objectionable diseases
- Objectionable insect pests
- Objectionable parasite
- Authentic seed of crop varieties

Test	Equipment	Others	Chemicals
Sampling	Trier of various sizes	Containers	
	Top pan weighing balance	Plastic boxes	
	Seed divider	Cloth bags	
	Boerner type	Paper bags	
	Gamete type	Sealing equipment	
	Raffle type	Sticker	
		Tray with cups	
		Spoon	

Physical purity	Diaphanoscope Seed divider Analytical weighing balance Seed blower Air screen cleaner Hand sieves Binocular microscope	Petri dish Forceps Wax paper bags Magnifying lens Reflected light source	
Germination	Germinator Autoclave UV lamp Seed counting board/machine Vacuum counter	Paper towel Sand Filter paper Wax paper Rubber band Plastic tray Bucket	Water
Dormancy	Seed scarifier Low temperature machine Oven Analytical weighing balance	Measuring cylinder Beakers Volumetric flask Piercing device Filing device Clipping device Razor blade	Potassium nitrate Gibberellic acid $Na_2HPO_4.2H_2O$ $NaH_2PO_4.2H_2O$ H_2SO_4 HNO_3
Test	Equipment	Others	Chemicals
Viability	Phase contrast microscope Incubator pH meter Water distillation unit Refrigerator	Knife Needle Petriplate Beaker Magnifying lens Bottles	Tetrazolium salt Water KH_2PO_4 $Na_2HPO_4.2H_2O$ Lactophenol $AlK(SO_4)_2.12\ H_2O$
Moisture	Oven Analytical Weighing balance Universal moisture meter Grinding mill Sieves	Non corrosive container Desiccators	
Seed health			
Pathogen	Diaphanoscoe Stereo binocular microscope with cool lamps Incubator Black light fluorescence lamp	Petriplates (90mm) Beakers Filter paper (Whatman No.1) Sand	Water Sodium hypochlorite Agar agar Malt extract Potato dextrose sugar

	Water distillation unit Water de-ionizing unit Autoclave Seed blower NUV tubes	Plastic trays	2,4dichlorophenoxyacetic acid 2,4-D NaOH
Insect pest	Diaphanoscoe Sieves	Wax paper bags Magnifying lens	
Parasite	Diaphanoscoe Seed blower Stereoscopic binocular microscope	Wax paper bags Magnifier lens	
Genetic purity			
Seed level	Diaphanoscoe UV lamp pH meter	Tray Petriplate Beaker Filter paper Measuring cylinder Glass bottles Herbarium of the seed varieties under test	Phenol NaCl HCl Lugol's solution Methyl alcohol Sodium hydroxide KOH Bleach Guaicol Hydrogen peroxide Copper sulphate NH_4OH_2
Plant level	Net house Fenced field	Diagnostic characteristics of the variety	

Other

Store	Dehumidifier Iron bin Hygrometer Two stores one for short term and second to store seed for three years

Schematic Plan of Seed Testing Laboratory

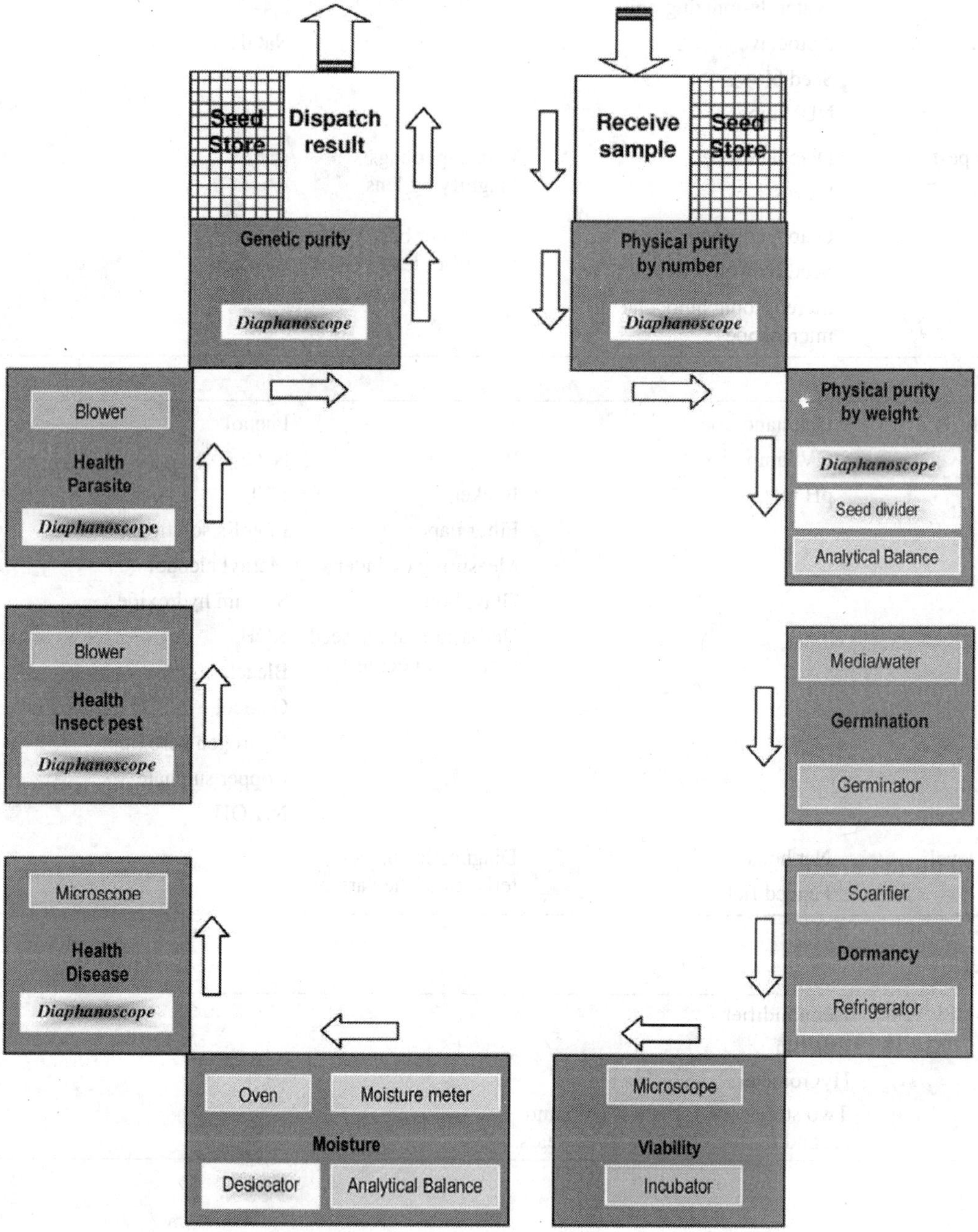

GLOSSARY

Abaxial : On the side of an organ away from the axis (Dorsal).

Aberrant : Departing from the normal.

Abnormal seedling : A seedling which does not possess all normal structures required for growth, nor shows the capacity to form normal plant.

Abortion : Loss of reproductive structure during development.

Abortive : Defective or barren.

Abscission : The shedding of leaves, flowers, fruits or other part of a plant after formation of an abscission zone.

Abscission zone : Zone constitutes of a layer of thin-walled cells at the base of leaf, flower, fruit, seed or other part of plant which separates the part from the plant.

Absolute humidity : The amount of water vapour present in a given volume of atmosphere, usually indicated in g water per kg dry air, or g water per m^3 air.

Absorption : Intake of water molecules from the atmosphere by cells or tissue.

Accelerated ageing (AA) : Exposing seed to high temperature and humidity to enhance the process of natural ageing to predict vigour and storage life.

Accessions : Individual plant entries in the germplasm collection.

Acclimatization : The adaptation of an individual to a changed environment.

Accrescent : Continuing to grow.

Acervulus(pl = acervuli) : A compact sub-epidermal fungal fruiting body consisting of mat of hyphae, which gives rise to short conidiophores, conidia and sometimes setae on the upper surface.

Aceto-carmine : A stain consisting of 55 solution of carmine in 45% acetic acid used in the preparation of chromosome squashes

Achene : A small, dry, one-chambered, one-seeded fruit with a thin, dry wall that does not split open at maturity.

Achlamydeous : Flowers without perianth.

Acicular : Needle like.

Acid scarification : Treatment of seed with acid to break the dormancy caused by hard seed coat.

Acre : A land area that can be cultivated by a pair of bullock in a day. 1 acre=4840 sq yard=4000 square meter.

Acron : A dry indehiscent, one seeded fruit with a hard and woody shell.

Acropetal : Towards the apex.

Actinomorphic : A flower having radial symmetry, regular.

Acuminate : Narrow, tapering ending in a pointed sharp apex.

Acute : Tapering more broadly than acuminate.

Adaptation : The process by which individuals, populations or species undergo modification to function more perfectly for survival under given environmental conditions.

Adaxial : Towards the axis (Ventral).

Adelphous : Combined in groups.

Admixture : Something added to seed other than the kind or variety specified.

ADP : Adenosine diphosphate

Adsorption : The accumulation and adhesion of a thin layer of water or gases on the surface of another substance.

Adventitious : Development of a part of plant from an irregular or unusual position.

Adventitious embryony : Formation of embryo from somatic rather than reproductive tissues.

Aerial mycelium : Mycelium that grows on the surface of substrate.

Aerobic : Biochemical process which implies presence of oxygen and in which oxygen is consumed.

Aestivate : To pass through a hot, dry season in a tropical condition.

Aestivation : The arrangement of the parts of the floral envelopes in bud.

After-ripening : The physiological maturation of seed after harvest. It is necessary for immature seeds to become germinable by terminating the dormancy of the seed

Agamogony : Formation of diploid embryo sac rather than the normal haploid embryo sac due to abnormal meiosis during megasporogenesis.

Agamospermy : Development of seed from saprophytic tissue.

Agar medium : A complex polysaccharide extract of certain seaweeds dissolves in hot water to form a stiff jelly on cooling. Seeds/pollens are tested for germination on 1% agar medium.

Aggregate : Sum total of a collection. Crowded together.

Aggregate fruit : Many-seeded fruit derived from an *apocarpous* ovary. With the ripening of the fruit a single unit is formed.

Agro-ecosystem : A comprehensive approach to factors associated with agriculture, ecology, and their interactions.

Air screen cleaner : Equipment that cleans seed with the pressure of air.

Airlift elevator : A machine that lifts seed by vacuum or pneumatic conveyor.

Albino : Mutant form of plants with degrees of chlorophyll deficiency from complete to banding.

Albumen : Layer of nutritional tissue between embryo including endosperm and seed-coat.

Albuminous seed : Containing albumen; A seed with endosperm as store of food.

Aleurone : A layer of diploid cells located on the periphery of the endosperm rich in nutrients.

A-line : The male sterile parental line used as female for production of hybrid seed.

Allogamy : Cross pollination. Pollen grains from flowers of one plant pollinate the flowers of other plants.

Allozymes : Allelic forms of an enzyme that can be distinguished by electrophoresis.

Alteration of generation : Reproductive cycle in which a haploid phase alternates with a diploid phase.

Alternate : Borne on the axis at different levels at regular intervals in in a zigzag manner, neither opposite nor whorled

Ambient conditions : Normal surrounding temperature, humidity and light without the use of artificial means.

Amino acid : Organic acids containing one or more amino ($-NH_2$) groups, at least one carboxyl (COOH) groups and sometimes sulfur. Many amino acids are linked together by peptide bands to form a protein molecule. In all 21 amino acids are found universally.

Amphimixis : Sexual reproduction.

Amphitropous ovule : Slightly curved ovule with its micropyle near the funicular attachment.

Amplexicaul : The arrangement in which the leaf encloses the stem at its base; the petiolar base surrounds the stem half way.

Amylase : The enzyme responsible for catalyzing the breakdown of starch into sugars.

Anaerobic : Biochemical process which does not require presence or involvement of oxygen.

Analysis of variance : A statistical tool that allows the partitioning of the total variation observed in an experiment among several statistically independent possible causes of the variation.

Analytical purity : The proportion of intact seeds of the cultivar in a seed lot.

Anatropous : A type of ovule arrangement where the ovule is completely inverted in such a way that the nucellus remains straight and the micropyle is directed towards the point of origin of the stalk or funicle..

Androdioecy : A sexual dimorphism with plants having either bisexual flower or male flowers only.

Androecious : Plant with male flower only.

Androecium : Collective term for the male part of a flower.

Andromonoecious : Presence of hermaphrodite flowers and flowers with male part only on the same plant.

Andtoecium : Collective term for anther stamens and staminodes.

Anemophily : Pollination by wind.

Angiosperms : [Angio=cover, sperm=seed]. Flowering plants; sub-division of spermatophytae, having their seeds enclosed inside the ovary. After fertilization the ovary becomes a fruit that enclose the seed formed by ovule.

Anisoploid : Mixture of diploid, triploid and tetraploid plants obtained from seeds harvested from a mixture of diploid and tetraploid plants.

Annual : The type of plant, which completes its lifecycle from germination of seed to production of seed within one, growing season in a year and dies.

Annuoline : The fluorescent protein pigment, which is exuded by the roots of ryegrass seedlings.

Anoxia : Dying from lack of oxygen.

Anterior : The lower one (sepal or petal).

Anther : Pollen-bearing part of the stamen.

Anther culture : The culturing of anthers *in vitro* for generating haploid plantlets.

Anthesis : The time or stage for opening of bud and to release pollen.

Anthocyanin : A red-purple soluble colouring matter present in plant tissues.

Anthracnose : A disease of plant with symptoms viz., limited lesions and necrosis caused by acervular fungus.

Antibiosis : Antagonistic association in which one organism has an injurious effect on normal growth and development of another.

Antibiotic : A bactericidal substance.

Anti-body : A protein produced by plasma cell of vertebrate animal in response to foreign substances (antigen) and capable of coupling specifically with its homologus antigen.

Antigen : A foreign substance that stimulates the production of homologus antibodies on introduction to the cell of vertebrate animal.

AOSA : Association of Official Seed Analysts American based seed testing system used in North and South America.

Apetalous : A flower without petals.

Apical : Referring to the apex situated at the end.

Apical dominance : Inhibition of lateral buds by production of high levels of auxins in the apical meristem.

Apical meristem : Cells present at the tip of a shoot that function in cell division and differentiation.

Apical placentation : A type of free-central placentation in fruit where the seeds are attached near the top of the central ovary axis.

Apocarpous : Carpels in a flower free from each other.

Apogamy : Development of embryo from synergids or antipodal cells without fertilization.

Apomixis : Formation of seed without fertilization.

Apospory : Formation of embryo sac from a vegetative cell of the ovule.

Appresorium : A swelling on the tip of a germ-tube for attachment in early stage of host penetration.

Appressed : Lying close and flat against erect type.

Archesporial cell : The cell of the nucellus, which differentiates and gives rise to cells undergo meiosis and produce the megaspore mother cell.

Aril : : A fleshy or papery covering or appendage as an outgrowth on the hilum of a seed

Arilloid : Outgrowth on the seed from other parts than the funiculus.

Arista : An awn.

Aristate : Awned

Articulate : Jointed, having a node.

Artificial aging : Exposing seeds to unfavourable conditions.

Artificial seed : Synthetic seed

Ascending : Becoming erect from a prostrate base.

Ascospore : A spore produced in an ascus.

Ascus (pl = asci) : A saclike reproductive cells of perfect stage of Ascomycetes (a group of fungi) in which ascospores are produced.

Aseptate : A cell without cross walls or septa.

Asexual : Reproduction. : Reproduction process from vegetative part without involving reproductive part and formation of seed.

Aspiration : The process of removal of lighter material from the seed mass

Aspirator : A seed cleaning machine to separate lightweight materials from seed mass with the pressure of air.

Astered embryo type : A type of embryo in which the terminal cell of the proembryo divides by a longitudinal wall and both the basal and terminal cells contribute to embryo development.

Attenulate : Drawn out into a long slender point.

Auricle : The claw-like or ear shaped appendage at the base of the leaf blade and leaf sheath. It clasps the stem in some cereals and grasses.

Auriculate : Presence of auricles.

Autoclave : A chamber for sterilizing equipment or material with high pressure steam.

Autogamy : Self-fertilization.

Auxin : A growth regulator that stimulates cell growth longitudinally i.e., root development and seed germination. Natural auxins are indole derivatives bio-synthesized from tryptophan.

Avirulent : Inability of a pathogen to overcome host resistance.

Awn : A long, stiff and hairy terminal appendage of the glume or lemma in cereals and grasses.

Axil : The angle between a leaf and the stem, which bears it; the angle formed between any two organs.

Axil placentation : A type of ovule attachment in which they are attached along the central axis at the junction of septa.

Axile : Situated round an axis.

Axillary : Borne on the axis in the angle between any two organs.

Axis : The central part of the longitudinal support in which organs or parts are arranged.

Baccate : Berry like

Back cross : A cross of a hybrid with one of its parents

Bag closer : A machine used for sewing filled seed bags

Bagger weigher : A machine that fills and weighs a bag in a single operation

Bagging test : A selfing test in which flowers are covered with a bag before anthesis and opened a few weeks later to find out mode of pollination based on seed formation

Basal application : Application of fertilizer to the soil before or during sowing of seeds

Basal placentation : Central placentation where the seeds are attached at the bottom of the central ovary axis

Base collections : Seed stored for medium or long term in gene bank

Basic number : The number of chromosomes in ancestral diploid ancestors of polyploids

Basic seed : Final product of a breeding method notified as a variety to start seed multiplication chain

Basifixed : Fixed to the filament (stalk) at the base

Batch : A quantity of harvested crop put into bin or container on a repetitive basis for drying or any other operation

Beaked : Provided with a solid narrow tubular beaklike prolongation

Berry : A simple fleshy or pulpy and many seeded indehiscent fruit developed from a single pistil and with no hard layers in the pericarp

Between paper : The method in which seeds are placed between two layers of papers to test the germinability of a seed lot

Biennial : A plant that requires two years to complete its life cycle from seed to seed

Bin sampler : A large sized trier used for drawing samples from bins

Bin : An enclosed structure generally made up of metal to store a seed

Biochemical : Chemical under active process in a living organisms

Biological control : Employment of natural enemies or their products for controlling plant pathogens and maladies caused by them

Biological value : An index of the portion of absorbed nitrogen retained in the body for growth or maintenance or both

Biological yield : The total yield of plant material

Biomass : Renewable organic matter formed through the process of photosynthesis

Biotechnology : The application of recombinant DNA cell tissue culture and other methods used to develop new and improved plants and plant products

Bipinnale : Twice pinnate (compound leaf)

Bisexual : Containing both stamens and carpels

Bisexual : Same as *hermaphrodite*

Bitegnic testa : Testa composed of two integumentary layers

Black layer : An abscission layer formed in the hilum region at physiological maturity of the seed

Blade joint : The flexible union between the leaf blade and the leaf sheath

Blade : The expanded portion (lamina) of a leaf

Blend : Seed mixtures of different grades of a cultivar

Blight : A kind of plant disease which is characterized by extensive spotting discolouration or destruction of leaves flowers stem or entire plant and is often coupled with the name of affected plant e.g., tomato or potato or celery leaf blight

Blinding : Blocking of screen mesh by stucking of seed in the holes

B-line : The fertile counterpart parent of A-line the maintainer line

Bloom : A whitish powdery and glaucous covering of the surface often of a waxy nature

Blotter test : A method of seed health inspection in which seeds are incubated on moistened blotter paper to test the presence of associated microorganisms especially fungi

Bolting : Formation of an elongated stem or seed stalk in biennial plants during the second season of growth

Boot : The development stage at which an inflorescence is enclosed by the sheath of the upper most leaf (grasses)

Border rows : The recommended number of rows of male parental line grown on all the sides of the field growing two different parents for hybrid seed production

Brachytic effect : Reduced length of inter-nodes in a cereal culm

Bract : Modified leaf like structure at the base of a flower or floral axis

Bracteate : With bracts

Bracteole : Small bracts occurring on the pedicel of the flower

Bran : The outer layers of a cereal grain removed in milling consisting of pericarp testa and aleurone layer

Breeder seed : Progeny of nucleus seed multiplied under the supervision of plant breeder to provide genetically pure seed for foundation seed production

Breeder's stock seed : The seed harvested from a nucleus or a breeder's stock field

Breeder's stock : A stock of highly pure seed of a cultivar in the custody of the plant breeder or plant breeding institution which originated the cultivar

Breeders rights : Varietal protection the legal rights of a breeder owner or developer in controlling seed production and marketing of crop cultivars

Breeding : The art and science of improving plants/animals genetically

Brick gravel test : A type of seedling emergence test utilizing uniformly crushed brick gravel through which seedlings emerge to exhibit their vigour

Bristle : A stiff hair like growth at tip of an awn

Bruchid : Group of seed-infesting beetles attacking seed of eg Leguminosae

Bucket elevator : Equipment used to lift seeds to the top of bins or machines for cleaning sorting and sacking

Bud : The rudimentary state of a stem branch or flower

Buffer : A substance or condition that counteracts changes caused by external impact

Bulb : An enlarged fleshy thick underground part of a stem surrounded by a mass of scale leaves and roots develop from the base

Bulbil : A small bulblet produced above the ground

Bulk method : In this method F_2 and the subsequent generations are grown in bulk usually without artificial selection

Bulk : To grow and maintain different plants in a population without applying selection

Bulliformi : Large thin walled epidermal cells of most gramineae

Bunt : A smut fungus that sporulates only in the ovaries also the disease caused by these fungi eg rice bunt wheat (Karnal) bunt

Bunt ball : Sorus of smut fungi that replaces the kernel

Caducous : Falling off very quickly.

Calibration : Correlation between reading of an instrument and a standard e.g. A moisture meter and the standard oven-drying method for determination of moisture content.

Callus : A mass of undifferentiated cells, originating from an anther, microspore, plant tissue explant or other cellular source, when cultured *in vitro*.

Calyx : Collective term for the sepals the outermost green floral leaves that forms outer floral envelope of a flower.

Calyx-tube : The tube formed by the cohesion of the calyx leaves (sepals).

Campanulate : Bell-shaped.

Campylotropous ovule : An ovule slightly curved and the micropylar end is pointed slightly downward so the funiculus and micropyle are close together.

Capitate : Knob like (swollen).

Capitulum : A dense inflorescence of usually sessile flowers; head of flowers

Capsule : A dehiscent fruit with a dry pericarp usually containing many seeds composed of two or more locules that split at maturity to release their seeds.

Carpel : Female reproductive organ of flowering plants. One or more carpels are united to form pistil.

Caruncle : A fragile appendage present on seed as outgrowth of the outer integument near the micropyle e.g. Castor.

Caryopsis : A dry, indehiscent one seeded fruit with tightly fused pericarp and integument e.g. Wheat.

Catalase : An enzyme which catalyzes the degradation of hydrogen peroxide to water and the oxidation by hydrogen peroxide of alcohol to aldehydes during seed germination.

Catalysis : It is the acceleration (increase in rate) of a chemical reaction by means of a substance, called a catalyst, that is itself not consumed by the overall reaction.

Catalyst : A substance that can induce or accelerate a chemical reaction without undergoing any change itself.

Catkin : A spike inflorescence of a single unisexual flower arising from the peduncle.

Cauline : Belonging to the stem.

Cell : Basic unit of structure and function of all living things consisting of a mass of protoplasm and nucleus.

Cellular endosperm : A type of endosperm in which the early development is characterized by cell wall formation accompanying each nuclear division.

Cellulose : A long-chain complex carbohydrate compound (polysaccharide) which forms cell walls and woody parts of plants.

Centrifugal divider (*Gamet type*) : Equipment for dividing a sample by centrifugal force, which mixes and scatters the seeds over the dividing surface.

Cereals : Species of the grass family cultivated for their edible Caryopsis.

Certification agency : An agency established as per seed Act 1966 to certify seeds of any notified variety.

Certification sample : A sample of seed drawn by a certification agency or by an authorised representative of certifying agency.

Certification tag : A tag or label with a specified design, which constitutes a certificate granted by the certification agency.

Certification : The official process of approval of the genetic identify, purity, and quality of seed of any cultivar by a legally constituted independent agency, in accordance with the seed laws and description of the cultivar.

Certified seed : Progeny of foundation seed produced by institute, government farm and registered cultivator under the supervision of certification agency by maintaining specific genetic identify and purity as per certification standard.

Chaff : Short hairs present on the grain at opposite ends of embryo.

Chaffy grass divider : A sampling device used to divide a sample of chaffy grass seed.

Chalaza : The part of the ovule opposite the micropyle from where the integument originate and integuments fuse with funiculus.

Character : The expression of a gene as revealed in the phenotype; an identifiable hereditary property.

Check : A row or plot of standard variety included in nursery or plot tests for comparison.

Chemical Hybridizing Agents (CHA's) Chemicals, which cause pollen abortion by render the treated plants male-sterile, without affecting the ovule fertility.

Chemosterilants : Chemicals used to make flower sterile.

Chenopodial embryo type : A type of embryo in which the terminal cell of the proembryo divides by a transverse wall and both the basal and terminal cells contribute to embryo development.

Chilling injury : Injury to seed due to low temperature.

Chilling : Literally meaning cooling down. Low temperature treatment of seed under cold moist environment break dormancy.

Chimera : An individual composed of cells of two or more expression, mainly colour.

Chlamydospore : A thick-walled, non-sexual (asexual) resting spore developed by thickening of hyphal cell or cells.

Chlorosis : Failure of chlorophyll development because of a nutritional disorder or disease.

Chromatography : A process of separation of materials into bands of different colours based on preferential adsorption of constituents on strip of filter paper or column of silica.

Chromosome aberration : The lack of complete chromosome complement.

Chromosome : A structural unit in nucleus, which carries genes in a linear constant order.

Ciliate : Possessing hairs.

Circinotropous ovule : An ovule in which the funiculus is very long and completely encircles the ovule.

Circumscissle capsule : A capsule, which at maturity splits open at the middle so that the top comes off, likes a lid.

Cirrhus (or cirrus) : Mass of spores laden with mucus forcing out of an ostiole (mouth) in a ribbon like fashion. *(pl = ciri or cirrhi)*

Citric acid cycle : It (also known as tricarboxylic acid cycle, TCA cycle, Krebs cycle) is a series of enzyme-catalysed chemical reaction of central importance in all living cells that use oxygen as part of cellular respiration. In aerobic organisms, the citric acid cycle is part of a metabolic pathway involved in the chemical conversion of carbohydrates, fats and proteins into CO_2 and water to generate a form of usable energy.

Clavate ear : A spike with compact enlarged top giving a club shape appearance.

Claw : The narrow stalk of some petals.

Cleistogamy : Pollination and fertilization in an unopened flower (bud).

Clonal selection : A method of selection based on clones, which does not involve sexual reproduction.

Clone : A group of plants originating by vegetative propagation from a single plant.

Coalesce : Running together or joining of fructification or spots.

Coat imposed dormancy : When seed dormancy is imposed by the structures surrounding the embryo.

Cob : The rachis of a maize inflorescence in which kernels develop.

Coccus : A part of a schizocarpic fruit, each of which is usually derived from a single carpel of the ovary. Compound, composed of two or more similar parts, generally referred to a compound leaf, compound inflorescence and compound ovary.

Code : A particular triplet sequence of bases in DNA specifying an amino acid.

Coefficient of variations : The standard deviation expressed as a percentage of the mean.

Cold test : Testing of seeds for germination in moist soil at 5-10°C up to one week and then at warmer temperature.

Coleoptile : A transitory membrane covering the shoot apex of cereals to protect plumule as it emerges through soil.

Coleorhiza : A transitory membrane covering the emerging radicle of cereals for its protection at early stage.

Collor rots : Rotting of the stem near the soil surface due to infection by a pathogen.

Colony : A group of hyphae or mycelia belonging to one species growing together.

Colour separator : A machine that separates seeds based on their colour differences.

Combine harvester : A machine that cuts the crop and threshes in one operation.

Commercial hybrid : First generation of hybrid.

Commercial seed : Seed that is intended for crop production but has not been produced under a recognized certification scheme.

Compensation point : The level of photosynthetically active radiation or CO_2 required for equaling respiration losses.

Complete (flower) : Angiosperm flower containing both stamens, pistil, sepals and petals.

Composite sample : A sample formed by combining and mixing of all the primary samples taken from the different parts of al ot.

Composite variety : It is developed by inter-mating of selected open pollinated varieties, parent or genotypes for three-four generations.

Composite : A mixture of genotypes from several sources, maintained by open pollination.

Compound microscope : An optical instrument to magnify objects invisible to naked eye. It consists of two lenses of different focal lengths mounted on the same viewer.

Concentric : One circle within another with a common center e.g. In symptoms of *Phytophthora* sp., *Alternaria* sp.

Conical divider : Equipment for dividing seed samples. (=Boerner type)

Conical : More or less cone shaped.

Conidiophore : A specialized hypha which bears conidium / conidia i.e., an asexual spore of a fungus.

Connate : United similar parts *(e. G.* Sepal to sepal).

Connivent : The bending toward~ each other of two or more similar organs.

Constancy of destination : The theory of embryo development, which suggests that each part of the mature embryo inevitably, arises from predetermined cells of the proembryo.

Container : Things in which seeds are packed.

Contaminant : The material mixed with the seed, which affects the genetic and physical quality of seeds.

Contamination : Intermixed with a pathogen, (of spores) not pure.

Control of disease : To check or reduce the presence of or the affect of plant disease.

Conveyers : The equipment used for moving the seeds.

Convolute : Rolled up from one or both margins.

Cool germination test : The effect of low temperature (18°C) on seed germination and seedling growth rate (cotton).

Cordate : Heart shaped usually referred for leaves.

Coriaceous : Leathery.

Corm : A bulky, short, vertical underground stems which stores starch as food.

Corolla : The whorl of petals.

Correlation : A mutual relationship between two things such that an increase or decrease of one is generally associated with an increase or decrease of the other.

Cortex : The outer portion of the stem or root of a plant

Corymb : An indeterminate inflorescence in which the lower pedicels arising from the peduncle are successively longer than the upper ones, giving a flat-topped appearance.

Corymbose: Corymb like arrangement.

Costate : Ribbed.

Cotyledons : In seeds, the embryonic leaves, which in many species have absorbed the entire or the major part of the nucellus and endosperm, hence becoming the principal nutrient storage tissue. Monocotyledons have one, dicotyledons two and conifers often many cotyledons. During germination the cotyledons may remain underground (hypogeal germination) or be pushed above the soil to become the first photosynthesising leaves (epigeal).

Coumarin : A chemical growth inhibitor, which has capability of inhibiting germination.

Crassinucleate : The nucellar condition where the embryo sac originates and develops rather deep under the nucellar epidermis.

Crease : A furrow extending throughout the length of grain (wheat).

Crib : A structure with a roof and open sides for drying maize cobs.

Critical leaf area : A leaf area above which detrimental mutual shading reduces net photosynthesis rate.

Crop ecology : The study of the relationships between crop plants and their environment.

Cross protection : A phenomenon where infection of a plant with one viral strain protects against super infection by another related strain.

Cross-fertilization : Fertilization by pollen from another plant.

Cross-pollination : Pollination by pollen from another plant.

Crown : The persistent base of a tufted perennial herbaceous plant.

Crucifer embryo type : A type of embryo in which the terminal cell of the proembryo divides by a longitudinal wall and the basal cell plays no or minor part in subsequent embryo development.

Cryopreservation : Long term storage of seeds at ultralow temperature, typically between -150°C and -196°C. In normal cryopreservation the sample is pre-treated with a cryoprotective substance, followed by slow, controlled freezing.

Cryoprotection : Application of certain chemical compounds to seed before freezing to protect them from freezing injury.

Culm : Stem of grasses consisting of nodes and internodes.

Cultigen : Any variety of horticultural crops under cultivation with no known wild state.

Cultivar : A variety, strain, race, or hybrid that has originated and persisted under cultivation or specifically developed for cultivation through selection or hybridization (cultivated varieties).

Cuneate : Wedge-shaped.

Curd : A mass of flower primordium meristems as in cauliflower.

Cuspidate : Ending in a sharp rigid point.

Cuticle : The outer film of epidermal cells, impermeable to water composed of waterproof fatty substance (cutin) on the surface of seeds, leaves and stems.

Cutin : A complex fatty or waxy substance found on the surface of certain seeds or leaves often making them impermeable to water.

Cyclic : Parts arranged in whorls.

Cyme : The main axis terminates in a flower; A centrifugal inflorescence on which the secondary or lateral branches continue to grow and extend beyond main axis.

Cytokinins : A group of plant hormones that promote primarily cell division as well as cell enlargement, differentiation, dormancy, retardation of leaf senescence.

Cytology : A study of cell structure and components in relation to their functions.

Cytoplasm : The protoplasm of a cell excluding the nucleus.

Cytoplasmic genetic male sterility : Cytoplasmic male sterility for which a restorer gene is available.

Cytoplasmic male sterility : Sterility due to factors that are carried only through the female and are not diluted or lost in successive generations of reproduction.

Damaged seed : Broken part of a seed smaller than half of its normal size.

Damaged seedling : Seedling with no cotyledon/ constricted seedlings or seedlings with split/ cracks / lesions on the essential structures.

Damping off : A seedling disease caused by seed and soil borne microorganisms in which the seedlings die immediately after seed germination or emergence. Damping off in seedlings often causes rot of the stem near the surface of the soil.

Day length : The number of hours of light in each 24 hours cycle.

Dead heart : Symptom characterized by death of the growing point and young leaves.

Dead seed : Seeds, which at the end of the test period are neither hard nor fresh and have not produced seedlings.

Debearder : A preconditioning machine used to clip oats, separate and polish vegetable seeds, decorticate sugar beet seeds, debear barley etc.

Decayed seedling : Seedling with any of their essential structures so diseased or decayed that normal development is not possible.

Deciduous : Shedding off leaves at a particular growth stage.

Decompound : Branched repeatedly.

Decortication : Removal of pith and bark from fibrous and other tissues.

Decumbent : Having the lower parts prostrate.

Decurrent : Prolonged downwards from the base.

Defoliants : Chemicals or method of treatment to get the leaves falls from plants and the fruits remain attached.

Deformed seedlings : Seedling with weak or unbalanced development of essential structures.

Degeneration : The progressive decrease in vigour of successive generations of plants usually caused by unfavourable growing conditions or diseases.

Dehiscence : The splitting open of an anther or a dry fruit to discharge its content. In fruit terminology dehiscent fruits are those that split open at maturity upon drying, usually while still attached with plant.

Dehydration : Loss of water from tissue during drying of seed.

Dehydrogenase : Group of enzymes catalyzing reactions involving transfer of hydrogen from a substrate to a hydrogen acceptor. The reaction is typical of biochemical processes in living plants, and the activity of dehydrogenase is taken as an indication of seed viability through TTZ test.

Delinting : Removal of fuzz from cotton seeds.

Denaturation : Alteration in the structural properties of a protein, caused by e.g. Heat, change in ph, or radiation, and in enzymes and hormones resulting in change of activity.

Dentate : With sharp teeth.

Depth factor : A depth, which would contain enough grain, so that if all the theoretical heat available for drying could be used, it would all dry to equilibrium in a period of time equal to the time of one half response.

Desiccant : Chemical drying agent (Sio, Cao etc.) With high moisture absorption capacity use to reduce humidity when stored with e.g. Seed.

Desiccation sensitive : Seeds which do not tolerate drying below certain (high) critical moisture content (*recalcitrant).*

Desiccation : To dry an object, moisture is removed up to desired level.

Desiccator : Air-tight glass or plastic container that constitutes a micro-climate to prevents absorption of moisture by seed from atmosphere.

Designated disease : The diseases specified for certification of seeds and for which certification standards must be met.

Designated weeds : Weed species specified for certification of seeds and for which certification standards must be met.

Desorption : Loss of moisture from a relatively moist hygroscopic material to a relatively dry atmosphere until the two reaches equilibrium. Ant. *Absorption*.

Detassel : To remove the tassel at the top of maize plant before pollen is released.

Deterioration (seed) : The cytological and biochemical events within a seed that ultimately lead to the death of the seed.

Determinate : Plants in which the terminal bud terminates in an inflorescence.

Deviation : Departure of an observation from its expected value.

Dew point : Temperature at which the relative humidity is 100% (saturated) and water condenses on a surface.

De-winging : Removal of fruit or seed wings during processing,

Diadelphous : In two bundles.

Diandrous : Flowers having two stamens.

Diaphanoscope : Equipment to determine the presence of Caryopsis within the glumes in certain grass seed samples.

Dichogamy : Maturation of male and female reproductive organs of a hermaphrodite flower at different times. It is of two types '*protandry*' (male first), and '*protogyny*' (female first).

Dichotomous : Branching frequently successively into two equal arms.

Diclinous : Bearing the andrcecium and gynoecium in separate flowers.

Dicliny : Presence of male and female flowers on different plants or at diffetn location on the same plant. Diclinous is of two types monoecious (on the same plant) or dioecious (on different plants).

Dicotyledons : Presence of two cotyledons as store of food with seed. Other characteristics of the plants formed by dicotyledonous seeds are broad leaves, branching veins of the leaves, persistent primary root and vascular bundles in rings.

Didynamous : In an androecium, four stamens in two pairs, one pair shorter than the other.

Diffuse flowering : Flowering throughout the year or outside the main flowering season.

Digitate : Finger like division.

Dihybrid : Heterozygous with respect to two genes.

Dimorphic : Occurring in two different forms.

Dioecious : Species with staminate (male) and pistillate (female) flowers borne on different plants. [di = two oikos = house]

Dioecy : State of being dioecious.

Diploid : Cell or organism with two basic chromosome sets, symbolized by 2n; chromosome number of somatic cells.

Diplospory : A kind of apomixis in which the embryo sac develops from megaspore.

Disc separator : A seed-grading machine that separates seeds based on length differences.

Disc : A glandular swelling below the gynoecium.

Disease cycle : The chain of events involved in disease development, including the stages of pathogen development and effects of the disease on the host.

Disease : Any disturbance of a plant that interferes with its normal structure, function or economic value.

Disinfectant : Chemicals for treatment to control pathogens established in seed parts.

Dispersal : The physical removal and displacement of the seed from the mother plant to some distance away longer than vertical drop.

Dissected : Divided into many segments.

Dissemination : Release and spread of fungus spores or disease causing agents from the site of production.

Distichous : Arranged alternately in two opposite vertical rows.

Diverticulae : The tendril like forks projecting from the ends of the adsorptive arms of the developing embryo or endosperm.

DNA : Deoxyribonucleic acid, the chemical basis of heredity.

Dockage : Total quantity of inert matter in seed samples.

Dockin's seed grader : A machine to separate long slender grains of rice varieties from thick, coarse, short rice grains.

Domestication : The process of bringing a wild species under human management.

Dominant : An allele that expresses itself to the exclusion of the contrasting allele in a hybrid.

Dormancy : Inability of a viable seed to germinate under favorable environmental conditions required for germination.

Dormancy chemical : Dormancy of seed caused by inhibitory substances that may be overcome by *leaching*.

Dormancy double : Seed coat dormancy along with embryo dormancy.

Dormancy endogenous : Dormancy due to inhibiting substances or incompletely developed embryo. Ant. *Dormancy, exogenous.*

Dormancy enforced : Failure of seeds to germinate due to certain environmental factors.

Dormancy epicotyl : When only the above part of the embryonic axis is dormant and radicle growth readily occurs.

Dormancy exogenous : Dormancy due to seed coat. It may e.g. Be mechanical, or photo-dormancy.

Dormancy mechanical : Dormancy caused by mechanical resistance of the seed covering (often the endocarp) to expansion of the embryo.

Dormancy photo : Secondary dormancy caused by high level of phytochrome Pr which must be changed to phytochrome Pfr by exposure to red light (660-760 nm) or full day light after imbibition for germination to proceed.

Dormancy physical : Dormancy caused by an impermeable seed-coat (hard seed). The embryo is quiescent (non-dormant) but is sealed inside the impermeable covering at low moisture content.

Dormancy physiological : Type of embryo dormancy in which germination is prevented by a physiological inhibiting mechanism, e.g. Chemical dormancy or thermo-dormancy.

Dormancy primary : Dormancy that exists within the seed at the time of maturity on the plant or immediately afterwards..

Dormancy secondary : Dormancy that develops in the moist seed after it has been released from the plant if subject to adverse environmental conditions. (= induced dormancy)

Dormancy seed-coat : Dormancy as a result of seed-coat conditions e.g. Impermeability (physical dormancy) or phytochrome system (photo-dormancy). Dormancy may be overcome by scarification or complete removal of the seed-coat.

Dormancy thermo : Primary or secondary dormancy in which the seed must be subjected to a high, low or fluctuating temperature prior to germination. In temperate regions thermo dormancy is overcome by pre-chilling or stratification; in tropical regions by exposing the seeds to normal diurnal fluctuations in temperature prior to germination.

Dormant mycelium : A thallus or kind of mycelium usually thick-walled resting for sometime and capable of becoming active during favourable conditions.

Dorsal : Back of an organ as to the under surface of a leaf.

Dorsifixed : Attached to the dorsal side.

Double cross : A cross between two single crosses.

Double fertilization : A phenomenon in angiosperms whereby one male nucleus unites with the egg nucleus to form a zygote developing into embryo, the second male nucleus unites with two polar nuclei in the embryo sac to form endosperm.

Double top cross : The first generation hybrid between an approved single cross and an approved open pollinated variety.

Dough stage : Seed development stage at which endosperm is pliable like dough.

Downy mildew : A type of plant disease caused by a fungus belonging to the family Peronosporaceae. E.g. Downy mildew of sunflower (*Plasmopara* sp.).

Draper separator : A machine, which separates seeds, based on their shape and surface texture.

Drift : Changes in gene and genotypic frequencies in small populations due to random processes.

Drought hardiness : The capacity of plants to survive desiccation.

Drought resistance : In plants, overall suitability for cultivation under dry conditions.

Drum mixer : A simple mixer used for dry treatment of seeds.

Drupaceous : More or less drupe-like.

Drupe : A sione fruit, the pericarp is usually differentiated into three layers, the epicarp thin and membranous, the mesocarp thick and fleshy, and the endocarp stony, inside which is the seed *e. G* . Plum.

Dryer : A unit to remove moisture generally by forced ventilation with or without the addition of heat.

Drying front : In a bin of drying grain, the advancing zone which includes all of the grain that has just reached some arbitrarily initial moisture content, the zone between the initial and final moisture content.

Drying : The removal of moisture from seeds to a point at which the rate of deterioration of chemical and biological activity is slowed down.

Dusts : Seed treating products formulated in the form of dusts.

Dyes : Colours used with seed treating formulations.

Ear : A large, dense heavy spike or spike like inflorescence.

Ebracteate : Without bracts.

Echinate : With long spines.

Ecology : The study of reciprocal relations among plants and animals and their environments.

Ecotype : A strain within a species adapted to a particular environment.

Egg : The female gamete or germ cell.

Elaiosome : Oil-rich outgrowth on seed which is eaten by ants and thus serves to disperse the seed.

Electronic colour sorter : A machine used for separating seeds by differences in colour. It views each seed with photoelectric cell.

Electrophoresis : The movement of colloidal particles through a fluid under the action of an electric field.

Electroporation : The treatment of protoplasts with high voltage electric pulse during direct gene transfer.

Electrostatic seed separator : A machine which separates seeds of different electrical properties essentially independent of differences in size, shape, weight or surface texture.

Elevators : The equipment used for moving seeds.

Elite seed : The class of Canadian pedigreed seed corresponding to the foundation seed class.

Emarginate : Notched at the apex.

Emasculation : Process of removal of male organs from flowers before the pollen is shed to prevent self-fertilization.

Embedded : Fixed in the substratum, surrounded by other matter

Embryo culture : The culturing of an immature embryo on a sterile nutrient medium.

Embryo dormancy : On removal of the seed coat of a viable seed, it fails to germinate even under favourable conditions.

Embryo sac : The mature female sacarises from the megaspore by successive mitotic division, consist of eight nucleate female gametophyte.

Embryo : The rudimentary plant in the seed, consisting of the germ and the embryonic axis.

Embryogeny : Growth and development of embryo.

Emergence : Coming out of seedlings from soil.

Endemic : (Of a plant disease). Restricted to one country or geographical region.

Endocarp : Inner most layer of the fruit wall (pericarp).

Endogenous : Originated within, or developed from a deep-seated layer of tissue. In a seed, developed from, or occurring in the *embryo*.

Endosperm : Triploid nutritive tissue with albumen in seed arising from the fusion of a sperm nucleus with the polar nucleus of the embryosac. The tissues surround the embryo in the seed that provides nutrition to the embryo. In gymnosperm it is haploid storage tissue derived from tissue associated with the female gametophyte.

Entire : Without any teeth at the margin

Entomophilous : Insect pollinated.

Environment : The physical and biological surroundings of a species or group of species.

Enzyme : A complex substance present in minute quantities, which promotes a biochemical reaction without getting itself, changed.

Epicalyx : A whorl of bracts just below the calyx.

Epicarp : The outermost layer of the pericarp of the fruit

Epicotyl : The growing point of embryo, which gives rise to the shoot and upper part of the plant. In seedlings, it is the portion of the stem between cotyledons and first leaves.

Epidermis : The outer layer of cells in plants, which protects them against drying and mechanical injury.

Epigeal germination : A type of germination in which the cotyledons are raised above the ground due to elongation of hypocotyls (epigeous)

Epigynous : Borne upon or on top of the gynoecium, with inferior ovary.

Epipetalous : Borne on the petals or corona tube, usuauy referred for stamens.

Epiphytotic : Sudden and usually widespread development of a destructive disease in plants.

Episatsis : Interaction between non-allelic genes, in which manifestation at any locus is affected by genetic phase at any other loci.

Equilibrium moisture content (EMC) : Moisture content of seeds in equilibrium with atmospheric humidity at a given temperature. The EMC is influenced by hygroscopic character of the seed storage material *viz.* Low for oily seeds and high for seeds rich in protein and carbohydrates.

Ergot : Dark, spur shaped sclerotium which develops in place of a healthy seed in a diseased inflorescence, e.g. Ergot of pearl millet, sorghum, etc.

Error variance : Variance arising from unrecognised or uncontrolled factors in an experiment with which the variance of recognised factors is compared in tests of significance.

Erumpent : Bursting through the surface of substratum.

Escape : Phenomenon of susceptible plants avoiding disease attack

Etiolation : The appearance of plants growing in the dark. The plants lack chlorophyll and are therefore pale. They develop long slender stems and have rudimentary leaves.

Evaporation : Escape of water molecules from liquid into vapour stage, e.g. During seed drying.

Evapo-transpiration : The combined loss of water from a given area and during a specified period by evaporation from the soil surface and by transpiration from plants.

Ex-albuminous seeds : The seed in which initially the endosperm was formed but all the food material of endosperm is absorbed by the cotyledonous cells during the development of seed.

Excised embryo test : A quick test for evaluating the growth potential of a root shoot axis, which has been detached from remainder of the seed.

Exerted : The stamens projecting beyond the petals

Exhaustion test : A vigour test, which measures the ability of seeds to grow rapidly under rigid controlled conditions of high temperature, relative humidity and moisture content in continuous darkness.

Exocarp : Outermost layer of fruit wall (pericarp).

Exogenous : Developed from a superficial layer of tissue. In seeds, developed from, or occurring in, the covering of the seed, either seed-coat or pericarp.

Exotic : A variety of species introduced from foreign country.

Exponential rate : Such variation in which one variable factor depends upon another resulting in an ever-increasing rate.

Expressivety : The degree of manifestation of a genetic character.

Exsertion : Complete emergence of a part of plant from an encloser.

Exstipulate : Without stipules.

Externally seed-borne : Located outside functional part of the seed, (of disease) the infection is present on the surface.

Extra-axillary : Situated away from the axil of the leaf.

Extraction : Separation of seeds from the fruit.

Extrorse : Facing outward from the centre of the flower referred for anthers. ,

Exudate : Liquid discharged from plant parts, diseased tissue, or in case of a fungal body.

Eyespot : Symptoms with oval or eye shaped areas of black dots on leaves.

Falcate : Somewhat curved.

False : Septum, not true septum.

Family : A group of individuals directly related by descent from a common ancestor.

Fascicled : Closely aggregated.

Fatuoid : A mutant commonly occurring in cultivated oats, which resembles wild oat (*Avena fatua*).

Female gametophyte : The haploid generation in a plants life cycle that produces the egg cell. In angiosperms the female gametophyte is called the *embryo sac*. It is generally composed of seven cells among them the egg cell and the central cell which after fertilization develops into the endosperm. In gymnosperms, after fertilization of the egg cell, the female gametophyte becomes gorged with reserve material and forms a storage tissue which is functionally equivalent to the angiospermous endosperm.

Female : Pistil bearing (pistiliate).

Fermentation : Anaerobic (without oxygen) decomposition of organic material.

Fertilization : The union of the nucleus and other cellular constituents of a male gamete (sperm) with those of a female gamete (egg) to form a zygote. In gymnosperms fertilization may occur months after pollination. In angiosperms there is a double fertilization in which one of the two sperm nuclei from the pollen unites with the egg nucleus to form a diploid zygote and the other sperm nucleus unites with two of the nuclei of the embryo sac to form the triploid endosperm.

Fertility restoring genes : Nuclear genes that act to restore fertility in plants with male sterile cytoplasm.

Fertility : Ability to produce viable offspring.

Fertilization : Fusion of female gamete (egg) with male gamete (sperm) to form a zygote.

Fid : Divided, lobed.

Field capacity : The percentage of water remaining in soil after having been saturated and after free drainage has practically ceased.

Field count : Number of plants/earheads, which are observed during field inspection as one count, depending upon the crop and cropping pattern.

Field desiccant : A chemical applied to dry the plant tissues in field.

Field emergence index : The difference between germination percentage and emergence in field.

Filament : The stalk of an anther or the lower part of the stamen.

Filament : Part of a stamen, which supports anther; any thread like structure.

Filiform : Thread or filament like.

Filter paper : Paper with moisture holding capacity used in seed testing for germination and health.

Fimbriate : Fringed along the edge.

Firm seeds : Grass caryopsis dormant due to seed coats that are impervious to water or gases.

First count : Number of seedlings on the fourth day after planting.

Fistular : Hollow throughout its length.

Flag leaf : The last fully expanded leaf in flowering cereal culms.

Flail : A stick used traditionally in threshing to separate the seeds from the plant material by beating.

Flesby : Thick and of leathery or succulent texture.

Floral induction : A physiological change, which occurs in plants because of external or internal stimulate that culminates in a flowering response.

Floreign : A hormone, which causes plants to change from the vegetative to the reproductive state.

Floret : The stamens, pistil and lodicules enclosed by lemma and palea.

Flotation : Method for cleaning seeds from particles with higher or lower specific density by submerging in water or other liquid. Used e.g. For separation of empty, filled and mechanically damaged seed, and weed seeds.

Flower : Angiosperm reproductive structure bearing pistils, stamens, or both, and usually also sepals and petals. So-called flowers in gymnosperms are male and female strobili before and during pollination.

Fluorescence : The property of substances that become luminous when exposed to ultraviolet and other forms of radiant energy.

Foliaceous : Leaf like.

Foliate : Referring leaflet.

Follicle : A fruit with a simple pistil, which at maturity splits open along one suture.

Foot rot : Rot involving the lower part of stem-root axis, synonymous to Collor rot.

Forages : Plants in which the top growth is harvested to use as feed for cattle.

Foreign seeds : Seeds of weeds and crops other than the kind being tested.

Formae specieales : A subdivision of a pathogen species based primarily on the host genera that are attacked. E.g. F.sp. *Batatas* of *Fusarium oxysporum* on sweet potato.

Foundation seed I : Seed produced from breeder seed.

Foundation seed II : Seed produced from foundation seed I.

Foundation seed : The progeny of breeder seed grown under supervision of certifying agency to meet the requirement of genetic and physical purity as per standard.

Fractionating aspirator : Equipment to separate seeds into fractions from light to heavy.

Frankia : Group of microsymbionts inoculating and making symbiosis with plants. Host plants are called *Actinorhizal* plants.

Free flowing seed : Non-chaffy, smooth seeds that move freely through sampling or mixing devices.

Free : Not united.

Freeze preservation : Conditioning and preservation of plant cells, tissues, or organs at extremely low temperature, usually in liquid nitrogen.

Freezing injury : Injury due to sub-zero temperature.

Frego bract : A mutant bract type that curls outward exposing buds and bolls e.g., cotton.

Fresh seeds : Seeds, which absorb water in a germination test but do not germinate.

Fructification : A fungal structure, which bears spores. Spore bearing organ like perithecium, pycnidium.

Fruit : A mature ovary and other associated parts containing seeds after fertilization.

Frutescent : Becoming shrubby.

Fruticose : Shrubby.

Fugacious : Falling off rapidly.

Full sib : Progeny from a cross between two selected plants within the population.

Fumigation : Application of a chemical compound in gas form.

Fungicide : A chemical used to kill fungi.

Fungus (pl = fungi) : A group of saprophytic or parasitic undifferentiated lower plants that lack chlorophyll and conductive tissue as mildew, smut, rusts or molds.

Funicle (funiculus) : The stalk of an ovule or seed, attaching it to the ovary placenta.

Funiculus : The stalk that connects an ovule to the placenta of the ovary wall.

Fusiform : A spike widest in the middle and tapering at tip and base. Spindle like narrow towards ends.

Fusiforme : Spindle like in shape.

GADA test : Glumatic acid decarboxylase activity test to measure the activity of a specific enzyme.

Gall : Irregular growth or Swelling of tissues.

Gamet precision divider :.Equipment for diving seed samples.

Gamete : Haploid reproductive cell produced by meiotic division for fertilization.

Gametocide : A chemical emasculant, which makes male organs defunct.

Gametophyte : Multiplication of haploid plant cell by mitotic cell division so that all the produced cells have haploid chromosome number with same alleles .

Gametophytic tissues : The nutritive tissues in conifer seeds.

Gamopbyllous : United tepals (perianth}.

GamopetaIous : United petals (corolia).

Gamosepalous : United sepals (calyx).

Gene : The unit of inheritance located at fixed loci in chromosomes.

General resistance : Non specific host plant resistance.

General seed blower : A precision seed blower used to aid in separating light seed and inert matter from heavy seed.

Genetic diversity : Variations in genotypes resulting from hybridization and selection under the law of the survival of the fittest.

Genetic engineering : Genetic manipulations that use recombinant DNA methods to change the genetic make up.

Genetic male sterility : Male sterility caused by nuclear genes.

Genetic purity : True to the type. Plants and seeds conforming to the characteristics of the variety as described by the breeder.

Genetic shift : Change in the genetic make-up of varieties, if grown over a long period in areas outside their adaptation.

Genetic sterility : A type of male sterility conditioned by nuclear genes in contrast to cytoplasmic sterility. It is transmitted by either male or female parent.

Genetic vulnerability : Pest susceptibility, when the same crop cultivar with same genes and cytoplasmic factors is planted in large geographic regions.

Genetics : The science of heredity.

Genome : A set of chromosomes corresponding to the haploid set of a species.

Genotype : The genetic constitution of an organism.

Genus : A taxonomic category that includes groups of related species.

Geotropism : The tendency of plant organs to grow in relation to the direction of gravity, either towards the ground (positive geotropism, usual for roots and radicles) or away from the ground (negative geotropism, typical for shoots).

Germ tube : The tube, which grows out from the pollen grain, usually into the stigma, down the style and into the ovary for sexual fusion.

Germ : Embryo.

Germinate : To begin to grow.

Germinating ripe : The condition of a fully developed seed, which is capable of germinating under conditions conducive to plant growth.

Germination cabinet : A type of seed germinator used for testing seeds for germination.

Germination capacity : The percentage of seeds germinated in a standard test to give normal seedlings as per ISTA rules.

Germination energy : The proportion of germination which has occurred up to the time of peak germination, or the time of maximum germination rate, or up to some pre-selected point, usually 7 test days. The critical time of measurement can be chosen by several means.

Germination percentage : Proportion of a seed sample that has germinated normally in a specified test period, usually expressed as a percentage. It should be noted that in earlier literature the term 'germination capacity' has been used to express viability.

Germination test : Standard test carried out to determine the germinability of seed under prescribed standard conditions i.e., temperature, humidity and light.

Germination : Active growth from the embryo of a seed, or a process of growth from a spore or the initiation of growth from a resting propagule. Resumption of growth by embryo and development of a young plant from seed.

Germplasm collection : A group of genotypes of a particular species from different sources to be used as source material for breeding purpose.

Germplasm : The sum total of the hereditary materials in a species. Seeds and plants used for breeding.

Gibberellins : Plant growth-stimulating substances.

Gibbous : Swollen on one side.

Gigas : Exhibiting large sized characteristics.

Glabrous : Not hairy, smooth.

Glabrous : Smooth

Glandular : Supplied with glands.

Glassine bag : A glazed, semi-transparent paper to protect emasculated flowers from cross-pollination.

Glaucous : Of a blue green colour. Covered with white or bluish waxy bloom that can be rubbed off.

Globose : Nearly spherical.

Glomerata : A compact cluster forming a round mass.

Glossy : Shinning.

Glumaceous : Resembling the glumes

Glume : One of a pair of small scale like bracts enclosing spikelet in cereals. Bracts at the base of a spikelet.

Glycolysis : It is a metabolic pathway by which a 6-carbon glucose (Glc) molecule is oxidized to two molecules of pyruvic acid (Pyr). The word glycolysis is derived from Greek glyco(sweet) and *lysis* (rupture). It is the initial process of most carbohydrate catabolism.

Goose neck : A pendent panicle with recurved peduncle.

Gossypol : A polyphenolic compound present in darkly pigmented glands in cotton plants.

Grading : The process of sorting into classes, normally according to size, colour or other factors to remove undesirable part from the lot..

Grain filling period : Duration between anthesis and physiological seed maturity.

Grain : The caryopsis of cereal plant.

Grains of moisture : The unit of measurement of the amount of moisture in air. 7000 grains are equal to one pound of water.

Gravity separator : A machine to separate seeds based on differences in specific gravity.

Green house effect : An increase in carbon dioxide in the earth's atmosphere that has a warning effect, as it restricts radiation of sunlight to the outer atmosphere.

Grow-out-test : The test to determine genuineness of seed as to species or variety or freedom from seed-borne infection.

Growth regulator : Analogues of growth hormones, which control the growth response of plants and seeds.

Gymnosperm : [Gymno=naked, sperm=seed]. Botanical classification of the group of vascular seed plants in which the ovule (and later the seed) is not enclosed in an ovary (as different from angiosperms).

Gynoecium : Female organ of plants consisting of stigma, style and ovary.

Habit : General appearance of a plant.

Habitat : The kind of locality in which plant grows.

Hair : A slender outgrowth of the epidermis.

Hand halving method : Mixing and dividing seeds for further sampling by hand.

Haploid : Cell or organism with one basic chromosome set, symbolised by n; the normal condition of gametes, i.e. Egg cell and sperm cell.

Hard seeds : Seeds, which do not absorb water owing to an impermeable seed coat at the end of prescribed test period and remain hard.

Harvest index : Grain weight as a percentage of totals above ground dry weight at maturity.

Harvest maturity : Dehydration of seed on the plant in field until the moisture comes to 16 per cent.

Head to row : Sowing the seed from an individual selected panicle.

Head : A compact (capitulum) inflorescence.

Heading : Emergence of ear from the boot.

Health test : Test for health of individual seed and seed lot that includes infection and infestation of pests viz., pathogen, insect, weeds, parasite, other crop seed etc.; and physical appearance of seed viz., shrivelled, discolored, damaged, under sized etc.

Heat injury : The injury caused to seeds due to heat.

Helobial endosperm : Development of endosperm by free nuclear division as well as cell wall formation.

Herbaceous : Not woody.

Heredity : The transmission of characteristics from parents to there off springs.

Heritability : Capability of being passed by inheritance.

Hermaphrodite (= bisexual monoclinous) : Functional male and female reproductive organs present in the same flower. Ant. *Dicliny* or *Unisexual*; a flower having both androecium and gynoecium.

Hermetic seed storage : Long term storage of seed at -18°C and 5 ± 1 per cent moisture content.

Hesperidia : Berry like fruit with papery septa and leathery separable rind.

Heterogamous : Bearing two kinds of flowers. In compositae when ray and disc florets are situated on the same head.

Heterogeneity : Variation in a seed lot.

Heterosis : Hybrid vigour such an F1 hybrid falls outside the range of the parents with respect to some characters.

Heterozygous : The state of having dissimilar alleles at corresponding loci of homologous chromosomes for one, several or all genes.

Hexapod : Having six sets of chromosomes.

Hills : Closely clustered groups of plants.

Hiltner test : Vigour test method in which the ability of germination to overcome physical stress is evaluated by germinating seeds under a layer of crushed brick stone gravel.

Hilum : The scar remaining on the seed at the place it was attached to the funiculus.

Hispid : With short scattered stiff bristles.

Homogamous : A head of Compositae bearing only one type of flowers.

Homologous : Of similar morphological significance.

Homozygous : The sate of having identical alleles at corresponding loci of homologous chromosomes for one, several or all genes.

Honeydew : The sweet liquid emitted by aphids and scale insects or secreted by fungi.

Hooded : Awns on glumes that terminate in various wings like structures.

Hormone : An organic compound produced in one part of an organism and transported to other part where it exerts profound effect.

Host range : The genera and species that are susceptible to a pathogen.

Host : The organism attacked by a pathogen.

Hull : Coat of dicot seed.

Hulling : The process to remove hulls or husk from fruits and seeds.

Humid : A geographical area where the water vapour content of the air is high.

Husk : Outer coarse or fibrous pericarp, e.g. Rice, barley etc.

Hyaline : Colourless or transparent.

Hybrid : The offspring of genetically distinctly different parents. The term is applied to progeny of mating within species (intraspecific) or between species (interspecific). Hybrids typically have morphological characters intermediate between their parents but may sometimes show exceptionally growth rate (hybrid vigour). Species hybrids often produce few seeds and of poor quality.

Hybrid variety : Commercially exploitable F1 hybrids between two different parental lines.

Hybrid vigour : The increase in vigour of hybrids over the parental types.

Hybridization : Crossing of genetically different individuals.

Hydrogen peroxide test : A quick test to determine seed viability.

Hydrogeopic : Tendency in seeds to take up moisture even as water vapour.

Hygroscopic : Sensitive to absorption or loss of water.

Hypantbium : Tubular or flask-shaped zone of the floral axis which grows above the level of the ovary and bears on its margin the floral leaves and androecium.

Hypha (pl = hyphae) : One of the filaments of a mycelium, the basic structural unit of a fungus.

Hypocotyl : The part of the embryo axis between the cotyledons and the primary root which gives rise to the stalk of the plant. In seedlings, the juvenile stem which is between the cotyledons and the root system.

Hypogeal (germination : Type of germination in which the cotyledons remain below soil level and the plumule is carried upwards by elongation of the epicotyl.

Hypogynous : Situated below the gynoecium or ovary referring to stamens petals and sepals.

Icker : Ear of corn.

IDS : Method of separating (removing) filled dead or damaged seed from sound seed. The seeds are imbibed and incubated at optimal germination conditions for approx. Two days, then dried shortly and separated by flotation. As filled dead and damaged seeds tend to lose water quicker than sound seed during desiccation, they will float in water while sound seeds will sink. (Incubation-Desiccation-Separation)

Imbibition : Absorption of moisture by a colloidal substance such as seed coats, which is accompanied by swelling of tissues.

Imbibitional injury : Damage of the dried seeds cell during the germination at low moisture.

Imbricate : A mode of aestivation in which one member of the whorl is out side au the others (*i. E.* Its margins are free) and one inside all the others (*i. E.* Both margins are overlapped), the others overlap by one margin only. Incised, deeply cut.

Immature embryo : Seed embryo which has not attained a development stage to make it capable of germination.

Immature seeds : Seeds not fully developed.

Immune : Completely free from attack by a specific pathogen.

Imperfect flower : Having either only male or only female reproductive organs in the flower, i.e. Male flowers or female flowers. (= unisexual flower)

Imperfect state : A phase in the life cycle of a fungus in which asexual spores or no spores are formed.

Impermeable : Not allowing passages to water or gases.

Improved seed : Seed of an improved variety having a very high genetic and physical purity and germination.

Inbred line : A pure line usually obtained by continued self-pollination and therefore homozygous.

Inbreeding depression : Decreased vigour in terms of growth, yield or survival following one or more generations of inbreeding.

Inbreeding : Breeding among closely related plants usually by self-fertilization.

Inclined dropper : A device of separating seeds, which are separated, based on their different tendencies to roll down the plane.

Incompatibility : Failure to obtain fertilization and seed formation after self-pollination.

Incomplete flower : A flower lacks any of the four basic parts i.e., pistils, stamens, sepals and petals.

Increase : To multiply a quantity of seed by planting it, rearing the plants grown from it and harvesting the seeds, they produce.

Incubate : (Of seed health testing). Maintaining seeds in an environment favourable to the development of symptoms or structure of pathogens.

Incubation : Maintaining seeds in an environment favourable to the development of pathogens or symptoms.

Incubator : Equipment to provide artificial atmosphere for the growth of an organism.

Incurved : With the ends curved inward.

Indefinite : Numerous.

Indehiscent : (Of anther or fruit). Not opening spontaneously at maturity to release their pollen or seeds, e.g. Some pods, samaras and nuts. Ant. *Dehiscence*.

Indent cylinder separator : A seed separator utilizing a rotating indented cylinder through which seeds are passed for cleaning.

Independent assortment : The chance distribution of two or more pairs of segregating genes to the gametes.

Indeterminate : A type of plant growth in which terminal bud remains vegetative resulting in overlapping of vegetative and reproductive stages.

Indexing : A process to test vegetatively propagated plant parts for freedom from viruses.

Indigenous : Species native to a country or an area, not introduced. Ant. Exotic.

Induced dormancy : Secondary dormancy due to unfavourable conditions.

Induplicate : Both sides rolled inwards.

Inert matter : Non seed material and seed material which is classified as inert according to rules for seed testing.

Infected : Carrying a disease organism that is associated with the host.

Infection : Process of beginning or producing disease.

Inferior : Epigynous condition.

Infested : Carried on the surface

Inflorescence : Arrangement of the flowers on the floral axis. The flowering axis or other flower bearing structure of a plant. The assembly of a few or many individual flowers into large clusters from a common axis. Inflorescences are grouped according to the manner of branching.

Infructescence : Fruit stand; cluster of fruits or arrangements of fruits on a plant.

Infundibular : Funnel shaped.

Ingestive dispersal : Seed dispersal by animals that eat (ingest) the fruit and void the seed by regurgitation or in their faeces. (= endozoochory)

Inhibition : Suppression of any function in seed due to presence of certain chemicals.

Inhibitor : A chemical substance that acts to prevent a process.

Innate dormancy : The new embryo ceases to grow when it is still attached to the parent plant.

Inoculant : Nitrogen fixing bacteria for treatment of legume seeds for nitrogen fixations.

Inoculate : To introduce pathogen propagules into a host or culture for producing infection, or for growing the pathogen in culture.

Inoculation : Application of a microsymbiont (inoculant or inoculum) to a host plant.

Inoculum : Growth of bacteria and fungi used to propagate or spread a pathogenic disease.

Insecticide : A chemical, which kills insects.

Inseparable other crop plants : Crop plants whose seed is difficult to separate from the main crop seed.

Intact seedling : Seedling with their entire essential structures well developed, complete in proportion and healthy.

Integument : The one or two layers (often fused) of tissue covering and surrounding the nucellus of an ovule. When the ovule matures, the integuments develop into the seed-coat.

Interagency certification : Certification of seed through two agencies one does field inspection and the other seed inspection.

Intermediate (seed) : In relation to storability, seeds that can be dried to a moisture content as that of orthodox seeds but are sensitive to low temperatures typically employed for orthodox seeds.

Internally seed borne (of fungus) : When pathogen is located inside the functional part of the seed and at least some of the infected seeds will give rise to diseased seedlings.

International Seed Analysis Certificate : A certificate issued by the authorized seed laboratories as per ISTA rules.

Internode : Space between two successive nodes. The portion of a stem or other structure between two nodes.

Introgression : Transfer of a few genes from one species into the full diploid chromosome complement of another species.

Introrse : Facing inwards usually referred for anthers.

Inversion : A rearrangement of a chromosome segment so that its genes are in reversed linear order.

Invigoration : The infusion of bioactive chemicals into the seed through some solvent.

Involucre : Whorl of bracts situated below the flower or inflorescence.

Irradiation : Exposure of plants or plant parts to X-rays or other radiation to increase mutation rate.

Irregular : Unsymmetrical, zygomorphic.

Isoalleles : Indistinguishable alleles.

Isolation distance : A gap that should be maintained between the seed crop and the contaminant or other varieties of the same crop to avoid cross-pollination.

Isolation requirement : The minimum isolation required avoiding cross-pollination or contaminant as per seed certification standard.

Isolation : Separation of seed fields from fields of other varieties of the same crop as per rule to avoid genetical and physical contamination.

Isoline : Lines that are genetically similar except for one gene.

Isotherm : Graph showing the relationship between relative humidity of the surrounding air and moisture content of e.g. A seed or an embryo at a given temperature.

ISTA (International Seed Testing Association) : International association which issues rules and regulations for standard seed testing on agricultural, horticultural and forest tree seed. Revised and updated rules and regulations are issued every three years. ISTA rules are mainly followed in Europe, Africa and Asia, while a slightly different system is used in North and part of South America.

Jointing : The period when nodes and internodes develop during stem elongation as in cereals.

Joints : The solid or nodal portion separating the internodes of a cereal stem from which leaf sheath, adventitious roots or tillers arise.

Juvenile phase : An early stage in the development of a plant.

Juvenile : Young or non-mature stage of a plant.

Karnal bunt : A bunt disease that infects immature seeds. It is caused by *Tilletia indica*, named for the town Karnal (Haryana), India.

Keel : The two fused anterior petals of a legume flower.

Kernel : A cereal caryopsis as in maize.

Key : Abbreviated statement of contrasting diagnostic characters.

Kiln : Drying device in which fruits are exposed to above ambient temperature by solar or artificial heat source to make them open. The most common kiln forms are stationary kilns and rotating drum kilns.

Klendusity : The ability of a susceptible plant to escape disease by changing date of planting to allow rapid growth and early maturity.

Kneeing ability : A positive geotropic response whereby the floating leaves and stems of deep water or floating rice becomes erect.

Labiate : Lipped.

Lactophenol : A cleaning solution consisting of 20.20.40.20 parts of lactic acids phenol, glycerine and water respectively.

Lag phase : In connection with germination, the phase after imbibition and before radicle elongation. During the lag phase there is no or very little water absorption and no or very little sign of development.

Laid crop : Crop flattened in the field due to weather conditions.

Lamina : The blade or expanded part of a leaf.

Lamination : Sealing of seed in polythene bag air tight with heat sealer.

Lanceolate : Lance-head shaped, a leaf broad below the middle and tapering towards apex.

Land races : Traditional cultivars with sufficient genetic integrity to be morphologically identifiable and differing in adaptation and to cultural practices, but genetically variable.

Land requirement : Required information about preceding crop of a field in which seed certification programme has to be taken up.

Larva : An immature but active stage of an insect.

Lateral : Borne on the sides.

Latex : Milky juice.

Lax ears : Ears with long segment.

LD-50 : The dose of any agent, which kills 50 per cent of the treated individuals.

Leaching : Removal of chemical inhibitors in fruits or seeds by washing in running water.

Leaf spot : Circular, irregular spots on leaves due to infection caused by a pathogen.

Leaflet : One of the blades of compound leaf.

Legume : Generally referring to fruits of the family Leguminosae, here used equivalent to 'pod'. Some Leguminosae have, however, modified fruit types such as samaras or follicles. The term should be reserved to the prevalent fruit type in Leguminosae viz. A multi-seeded dry fruit that may or may not open at maturity.

Legume seed : Seed of plant from family Leguminosae. The seeds are often hard and impermeable. Seed-coat consists of three distinct layers of which the outer two, the cuticle and the palisade layer provide impermeability to water and protect against desiccation.

Leguminous : Resembling legume

Lemma : One of the two bracts of the grass floret which is located on the side nearest the embryo and opposite the rachilla.

Lesion : A well marked localized diseased area.

Light line : In legume seed-coats, a distinct line across the cells in the palisade layer. Impermeability is generally believed to be connected to the light line.

Ligulate : Furnished with a ligule.

Ligule : A membranous outgrowth at the junction of leafblade and leaf sheath in grasses.

Limb : The broader or expanded part of a gamopetalous corolla.

Line : A group of individuals from a common ancestry.

Linear : Long and narrow with parallel margins.

Linkage : Association of genes in proximity on the same chromosome resulting in simultaneous inheritance.

Lint : Long fibres of cotton separated from seed.

Linters : Short fibres of cotton that generally remain attached to seed.

Lip : One of the two divisions of a bilabiate corolla.

Lobed : Cut less than half way.

Local variety : Mixture of different types well adapted to the local environment.

Locular : Chambered.

Locule : Compartment of the ovary containing seed.

Loculicidal : Fruits which open by the dorsal suture of the carpels.

Locus : The location of a specific gene on a chromosome.

Lodging : The bending or breaking of plants before harvest.

Lodicule : One of the small sacs at the flower base which takes up water to force apart the glumes at anthesis.

Lomentum : A pod or legume constricted between the seeds.

Long day plants : Plants which require short dark period for reproductive development in comparison of vegetative phage..

Longevity seed : The period of time seed will maintain viability in storage under a given set of storage conditions. Often used equivalent to *storability*.

Looped structure : A seedling, which instead of being straight forms a loop or circle.

Loose smut : A smut disease cause by *Ustilago tritici* that infects ovaries only e.g. In wheat.

Loosely twisted : A seedling that becomes twisted instead of remaining straight.

Lost capital : Tillers, which fail to reach maturity.

Lot numbers : The number given to a seed lot.

Lot : A specified quantity of seed, which is physically identifiable.

Low- land rice : Rice grown in fields flooded with water.

Lowest Safe Moisture Content (LSMC) : Moisture content below which freshly collected seeds die when the seed lot desiccates, same as critical moisture content (CMC).

Maceration : The process of softening (fleshy fruits) by steeping in a liquid, with or without heat; to wear away or separate the soft parts, by steeping.

Macro-nutrient : An essential element required by plants in relatively large quantities.

Macrosclereids : Layer of elongated cells perpendicular to the seed-coat surface. (= palisade cells)

Maggot : A leg less larvae of certain two winged flies of the order Diptera.

Magnetic separator : A machine to separate seeds based on magnetic property.

Maintainer line (B-line) : Line with all the characters of male sterile line (A Line) except fertility, used for maintaining a cytoplasmic male sterile line.

Male fertile : A plant having staminate flowers producing functional pollen.

Male flower : Androecium bearing (staminate) flower.

Male sterile : A plant devoid of staminate flowers or producing non-functional pollen.

Male sterility : A condition in which pollen is absent or non-functional which may be genetic, cytoplasmic or genetic cytoplasmic.

Mass selection : A type of selection in which individual plants with desirable phenotypes are selected from a large population of plants grown in a field and the next generation propagated from the aggregate of their seeds.

Mast year : Year with large seed production. (= seed year)

Masting : Ecological term used to describe the strategy of some plant species to produce large crops of seeds at long intervals, e.g. Some dipterocarps and araucarias.

Mating system : A method by which individuals are assorted in pairs leading to sexual reproduction.

Mature : Of plants. The stage of individuals after having passed the juvenile stage; often considered synonymous with reproductive age. Of fruits. The stage of the fruit when seeds are ready to be dispersed; usually, but not always, coinciding with germinability of the seeds. Often visible by e.g. Colour change, desiccation and opening structures. Of seeds. Seeds which are fully germinable after extraction from their fruits and possible breakage of dormancy.

Maturity index (pl indices) : Changes of fruit and seed appearance indicating maturity e.g. Change of colour, odour, drying, dehiscence, softening of fruit pulp and hardening of the seed-coat and endosperm. Maturity indices are used to schedule seed collection.

Mechanical ventilation : Process of supplying or remaining air with a mechanical device to or from any space.

Medium term storage : The period of storage not more than five years.

Megagametogenesis : Production of female gametophyte (embryo sac) from megaspore.

Megaspore mother cell : Diploid cell in ovary that gives rise to four haploid megaspores after meiosis.

Megaspore : One of the four haploid spores originating from the meiotic divisions of diploid megaspore mother cell in the ovary.

Megasporogenesis : Production of the female spore from megaspore mother cell through meiosis.

Megasporophyll : Leaf or modified leaf bearing macrosporangia. In angiosperms, the carpel producing ovules. In gymnosperms the megasporophyll is called ovuliferous scale or cone scale. (= macrosporophyll)

Meiosis : Two successive nuclear divisions, in the course of which the diploid chromosome numbers, is reduced to haploid.

Membrane integrity : Increase in permeability of seed membrane due to ageing.

Mericarp : One half of a two-sectioned fruit called schizocarp as in Umbellioferae.

Meristem regions : Areas in plant where cells rapidly divide.

Meristem : Undifferentiated tissue located at the tips of growing points of vegetative or reproductive organs capable of undergoing cell division.

Mesocarp : Middle layer of the fruit wall (pericarp); the pulp of berries and drupes.

Mesocotyl : The portion of embryo present in between epicotyl and hypocotyl.

Mesophyll : The photosynthetic parenchyma of leaf.

Metabolism : The chemical changes within a cell that provide the energy required by a plant or animal.

Metaxenia : Influence of pollen on material tissues of the fruit.

Microgametogenesis : Production of male gametophyte from microspore mother cell.

Micro-gametophyte : A mature pollen grain or male gamete.

Micron (µ) : A unit of length equal to 0.001 mm.

Micronutrient : An essential element required in relatively small quantities of plants.

Micropyle : Minute opening in the integuments of an ovule through which the pollen grain or pollen tube passes to reach the embryo sac. On mature seeds the micropyle is sometimes visible as a small pore on the seed-coat. In seeds derived from orthotropic ovules the micropyle and hilum lie distant to each other, from anatropous ovules close to each other and from campylotropous ovules intermediate. The radicle of the embryo always faces the micropyle.

Microspore mother cell : One of many cells in the anther (microsporangium) which produce four microspores.

Microspore : The male spore in anther from which male gametes develop.

Microsporogenesis : The development of microspores from the microspore mother cell.

Microsporophyll : A leaf or modified leaf bearing the male reproductive part. In angiosperms the stamens, in gymnosperms the scales of the male strobili.

Microsymbionts : Micro-organisms living in close association (symbiosis) with a host plant.

Middling : The seed from the middle portion of gravity separator, which have a mixture of light and heavy seed.

Midrib : The central or main rib of leaf.

Milk stage : The stage in grain formation when endosperm is like whitish liquid like milk.

Mist-O-Matic treater : A machine to treat seeds as mist.

Mitochondria : Microscopic rod shaped or spherical bodies present in living cells containing enzymes for respiration.

Mitosis : Cell division in which each daughter cell has exactly the same chromosome numbers as the mother cell.

Mobilization efficiency : The capability of a seedling to draw food from endosperm or cotyledons for establishment.

Modifying genes : Genes that affect the expression of a non-allelic gene or genes.

Moisture content (mc) : The amount of water present in a material, e.g. Wood, soil or seed. It is normally expressed on a weight basis, either as the weight of the water in % of the material's oven-dry weight ('dry-weight basis') or, preferably in the case of seeds and fruits, as a % of the material's fresh weight including water ('wet-weight' or 'fresh-weight basis').

Moisture content of seed : The quantity of moisture in the seed, which is expressed in per cent.

Moisture meter : Instrument for quick measurement of moisture content of seed without drying. The instrument measures the electrical properties of the seed tissue, which is correlated with the content of water (moisture content). Moisture meters must be calibrated for each species with standard method of moisture content measurement i.e. Oven-drying.

Mold (Mould) : Grayish to black fungi grow on seeds at higher moisture content.

Monitoring : Inspections of breeder seed plots by a team to verify different parameters as per rules.

Monoadelphous : United into one bundle by the filaments.

Monocotyledons : Subclass of angiosperms characteristic by having only one cotyledon in the embryo. Differ from dicotyledons in a number of morphological characters like leaf, root and stem structure. Woody monocotyledons are e.g. Bamboos and palms. Ant. *Dicotyledons*.

Monoecious : A species with male and female sexual reproductive organs borne on the same individual at different locations. [mono = one oikos = house]

Monogerm seed : A seed containing one ovule.

Monohybrid : Heterozygous with respect to one gene.

Morphology : The form and structure of plant.

Mottled : Disease symptom characterised by an irregular pattern of light and dark areas.

Mould : Fungi that produce distinct mycelium or spore mass on the surface of their host, e.g. Mucor and certain *penicillium* species.

Mucilage : The gummy complex carbohydrate substance, which covers the seed and other plant, parts.

Mucronate : Tipped with an abrupt point.

Multigerm seed : An aggregate fruit containing several ovules.

Multiline variety : It is a mixture of two or more near isogenic lines of normally self fertilizing plants.

Multiline : A composite population of several genetically related lines of a self pollinated crop.

Multilocular : Ovary with two or more locules each containing one or many seeds.

Multiple allele : Character with more than two alternative forms of a gene and their expression.

Multiple genes : Two or more genes at different loci, which produce complementary or cumulative effects on a single quantitative genetic trait.

Multiple seed unit : One fertile floret attached with one or more fertile or sterile florets.

Multiplication ratio : Number of quantity of seed harvested per unit number or quantity of seed sown.

Multiseptate : (Of spores). Having several septa (cross walls) e.g. In *fusarium* sp.

Mutant : Acquisition of a heritable variation because of mutation.

Mutation : A sudden heritable change in a gene or in chromosome structure other than those due to normal segregation and recombination..

Mycelium : A mass of thread like filaments of a fungus comprising the main body.

Mycorrhiza : Fungi living in symbiosis with plant roots. The fungus provides the plant with mineral nutrients and gets in return sugar and other organic compounds. [Myco = fungus; rhiza = root].

Myreaochory : Dispersal of seeds by ants.

Naked stratification : Pre-chilling of seeds without the use of a moisture-holding medium.

Natural selection : Selection due to natural forces i.e. Environments.

Necrosis : Death of a plant cells.

Nectary : A nectar secreting gland.

Nectariless : Devoid of extrafloral nectaries in cotton.

Nematodes : Thread likes round shaped worms causing diseases of plants.

NES - pistillate : Femaleness in monoecious plants due to homozygous recessive gene.

Net assimilation rate : Total rate of photosynthesis minus respiration.

Nick : Synchronous or simultaneous flowering of both the parents in hybrid seed production.

Nicking : Synchronization of the receptivity of the male sterile plant to the maximum pollen load of the pollinator for cross-pollination in hybrid seed production.

Nitrate reductase : An enzyme in tissues involved in the first step of nitrate fertilizer reduction and influences the levels of amino acids and proteins in plants.

Nobbe trier : A pointed tube, long enough to reach the centre of the bag, with an oval hole near the pointed end to draw seed samples.

Nodal bud : The lateral shoot bud located at the node.

Node : The place on a stem, which normally bears a leaf or whorl of leaves. Solid constrictions on the culm of grasses.

Non-recurrent apomixis : A type of apomixis in which a haploid embryo sac is produced giving rise to haploid embryo and haploid plants which are usually sterile.

Non-restorer line (B line) : A male parent that cannot restore male fertility and when crossed to a male sterile line (A- line) produced a hybrid which is male sterile.

Normal seedlings : Seedlings possessing essential structure and show the capacity for development into normal plants.

Notified variety : A variety notified under section 5 of seeds Act. 1966.

Noxious weed seed : Seed of weed species considered harmful in agriculture.

N-pistillate : Femaleness in monoecious plants due to heterozygous genes for pistillate nature.

Nubbin : A small poorly developed corn ear.

Nucellar embryony : Seeds resulting from nucellus rather than the zygote.

Nucellus : In angiosperm ovules, the tissue in the central part of the ovule inside the integument, in which the embryo sac is embedded. During maturation the nucellus is usually absorbed but develops in some species into a nutritive tissue, surrounding the endosperm. The nucellus is of entirely maternal origin and hence diploid.

Nucleus seed : A group of progenies of individual plants taken at random from a variety for purifying from mixtures and off-types.

Nullisome : A 2n plant which lacks both members of one specific pair of chromosomes making 2n-2 chromosomes.

Nurse crop : A rapidly growing non-smothering crop, sown and harvested in the first season of a perennial seed crop.

Nursery : A place, where plant offspring and introductions are grown for early crop establishment.

Nut let : A small dry indehiscent fruit composed of one half a carpel enclosing a single seed.

Nut : A dry, indehiscent, one seeded fruit with a hard woody shell.

Nymph : Immature insect that resembles the adult but has no fully developed wings.

Obconical : Inversely conical.

Objectionable weed seeds : Seeds of weed plants, which are difficult to be separated from crop seeds, when mixed and has smothering, effect on the main crop.

Obligate parasite : A parasitic fungus that can develop only in living tissues e.g. Downy mildew of pearl millet, sunflower.

Oblong : Longer than broad with parallel sides.

Obovate : Inversely ovate.

Observation nursery : A nursery in which a row of selections and progenies are examined on a preliminary basis in one or more rows each without replication or randomization.

Obtuse : Rounded and blunt.

Obvoid : Egg shaped with narrow end at the base.

Official seed sample : He sample collected by seed control or seed-law enforcement officers.

Off-type : A plant in a seed crop, which deviates from the norm of the cultivar, but does not belong to another cultivar or species.

OLDA : Category of storage behaviour intermediate of that of orthodox and recalcitrant seed. (Orthodox with Limited Desiccation Ability)

Oospore : A thick walled resting spore produced by the union of an oogonium (female) and an antheridium (male) e.g. Downy mildew of sunflower, pearl millet.

Ooze : Movement of fungal spores or bacterial cells through apertures of micro or macroscopic dimensions e.g. Bacterial ooze in cowpea, French bean, cabbage, cauliflower etc.

Open pollinated variety : A heterogeneous variety of cross-pollinated crops, which is allowed to inter-pollinate freely during seed production.

Open pollination : Normal cross-pollination.

Operculum : Dehiscent 'cap' or 'lid' structure of e.g. Some anthers and capsules.

Opposite : Arranged on different sides of the axis but on the same level.

Optimum leaf area : The leaf area at which maximum dry matter is produced.

Orthodox seeds : Seeds, which withstand dehydration without damage.

Orthodox : Term used to describe seeds which can be dried down to a low moisture content of around 5% and successfully stored at low or sub-freezing temperatures for long periods. There is practically no metabolism in dry, cooled orthodox seeds, but they may deteriorate by general ageing and thus ultimately lose their viability. Ant. *Recalcitrant*.

Orthotropous (= atropous) : A straight ovule with the micropyle opposite to the chalaza.. Ovule orientation in which the ovule develops in an upright position from the placenta, consequently with hilum and micropyle located at opposite ends of the developing seed.

Osteosclereids : Layer of specialised cells between the macrosclereids and the parenchyma cells in legume seed-coats.

Other crop seed : Seeds of domesticated plants, which are grown as crops, other than the main crop.

Other distinguishing variety : Seeds or plants of other varieties, which can be easily differentiated, based on morphological characters.

Other seeds : Seeds and seed like structures of any plant species other than the pure seeds.

Out cross : Cross-pollination usually by natural means with a plant differing in genetic constitution.

Oval : Broadly elliptical.

Ovary : Part of gynoecium containing the ovules. The enlarged female reproductive structure of the pistil containing an ovule or egg in which seeds are borne.

Ovate (ovoid) : Egg shaped with basal broader end.

Oven-dry : In connection with moisture content, the standard ISTA / AOSA method of drying seeds at 103° C for 20 hours to reveal weight loss, which is an indication of moisture content of the seed.

Over winter : Survival of a fungus (organism) from one season to the other.

Ovule orientation : Position of the ovule within the ovary, which may be upright (orthotropous), completely inverse so that the micropyle faces the placenta (anatropous) or at an angle to the placenta (campylotropous). Ovule orientation is visible on the seed as the position of the micropyle in relation to the hilum and chalaza.

Ovule : A structure in seed plants consisting of the nucellus which contains the female gametophyte, one or two integuments and the funiculus. After fertilization of the egg cell the ovule develops into the seed.

Ovuliferous scale : The ovule- (later seed-) bearing scales in the cones of conifers (contrast bract scales).

Ovum : An egg cell.

Prostrate : Lying horizontally on the ground.

Packaging : The process of filling, weighing and sewing seed bags.

Palea : The tiny upper bract, which encloses the flower in grasses.

Palindrome : A word with same spelling when written from back or reverse side

Palisade cells : Same as *Macrosclereids.*

Palisade layer : In seeds a layer, which is also called Malpighian layer.

Pallet : Wood frame structures used to support a stack of bags.

Palmate : With palm-like spreading segments.

Panicle : A compound recemose inflorescence; An inflorescence in which the lateral branches arise from the peduncle, which produce flower-bearing branches.

Paniculate : Bearing panicle type inflorescence.

Paper piercing test : A stress test for seedling vigour.

Papilionaceous : Butterfly-like as corolla of Pea family.

Pappus : A tuft of delicate fibres or feathery appendages or scales on seed. Modified calyx into bristles *e. G.* In many Composites.

Parasite : An organism living upon another organism and drawing the food from it, with or without fatal effect on the host.

Parenchyma : Soft tissue consisting of undifferentiated, living cells. In seed-coats a layer of parenchyma is often present beneath the hard seed-coat cells.

Parietal cell : The sister cell of megaspore mother cell, which is nonfunctional and degenerates.

Parietal placentation : The placentation in which the seeds are attached in the ovary near the outer ovary wall.

Parthenocarpy : Development of fruit without viable seed. The fruits may be either seedless or seeds may lack embryos. May result from a failure of pollination, a failure in fertilisation, or a failure in embryo development.

Parthenogenesis : Reproduction from an unfertilised egg, a type of apomixis.

Partial hybrid : Seed lots with 75 to 95% hybrid seed.

Partials : Presence of fertility in A-line.

Pathogen : A microorganism capable of causing disease

Pathogenicity : The ability of an organism to cause a disease.

Pathology seed : The science of seed diseases or seed-borne pathogens.

Peak germination : A loose term which describes the point in time when rate of germination is highest; it can be calculated in several ways.

Pearson square : A method to determine the proportion of different lots that can be blended together to give a final lot of a desired quality.

Peat : Organic planting medium manufactured from bogs or swamps. Peat has a high water holding capacity. Also used as a carrier of rhizobium inoculates.

Pedicel : Stalk of flower.

Pedicellate : The flower with pedicel.

Pedigree : A record of the ancestry of an individual, family or strain.

Peduncle : Stalk of an inflorescence.

Peduncle : Stalk of an inflorescence.

Pelleted seed : Seeds with a coating of some inert matter along with other substances.

Pelleting : Procedure by which individual seeds are provided with an envelope of adhesive material containing e.g. Nutrients, microsymbiont inoculant and/or pesticides. In addition to providing these beneficial compounds, pelleting facilitates mechanical sowing because of the more uniform seed size.

Pentamerous : With five members in each whorl.

Pepo : A fruit with a hard rind without internal separations.

Percentage germination : The proportion of the number of seeds, which have formed normal seedlings within the specified period and conditions.

Perennial : Can live for more than two years; A plant, which survives and produces vegetative growth and flowers year after year without being replanted.

Perfect flower : Flower possessing both stamens and pistils.

Perfect state : Sexual part of the life cycle of the fungus during which spores are formed as a result of some sort of sexual process; sexual stage in which a fungus is capable of sexual reproduction.

Perianth : The two floral envelopes, no differentiation into calyx and corolla..

Pericarp : Wall of a ripened ovary i.e. Fruit wall. The pericarp is homogeneous in some genera and in others it is composed of three distinct layers. Exocarp, mesocarp and endocarp.

Perigymy : The ovary is enclosed in a cup shaped thalamus.

Perigynous : Sepals, petals, and stamens inserted above the hypanthium

Periodicity : The tendency, in individuals, stands or species, to produce seed crops at long but often more or less regular intervals, e.g. Many dipterocarps. Periodicity is closely connected to gregarious flowering and masting. Periodicity indicates the interval (in years) between good seed or cone crops. Compare *seasonality*.

Perisperm : The endosperm like storage tissue in a mature seed, which is formed from the nucellus of the parent plant.

Perithecium : A flask shaped sexual fruiting body of a fungus with a pore at the top.

Perlite : An inorganic, very light planting material manufactured from volcanic lava. It has anvery high water holding capacity. Used primarily to improve the aeration of a planting medium.

Permeable : Allowing passage of motile molecules, e.g. A gas or liquidto pass through. A seed-coat is permeable when water can be absorbed for the seed to imbibe. A barrier may be selective or semi-permeable if it allows only passage of small molecules, e.g. A polyethylene sheet which is permeable to gas but impermeable to water. Ant. *Impermeable.*

Persistent : Plant parts, which are not shed until after maturity.

Pest : Organisms, which harm plants.

Pesticide : Substances or chemicals, which kill pests.

Petal : One of the floral leaves of the corolla.

Petaloid : White or coloured.

Petiole : Stalk of leaf.

pH : The negative logarithm of hydrogen ion activity (potential of hydrogen ion).

Phage plaque : A clear area caused by a bacteriophage in a bacterial colony.

Phenology : Study of the relations between seasonal climatic changes (e.g. Temperature, day length and precipitation) and periodic biological phenomena such as flowering, fruiting, leaf flushing and dormancy.

Pheno-period : Abbreviation of phenological period i.e. A period of a phenological stage such as flowering or fruiting.

Phenotype : External appearance of an organism.

Photochemical : A chemical reaction activated by light.

Photoperiodism : Response to the relative length of light and dark periods in a day.

Photosynthesis : The process of transforming carbon dioxide and water into carbohydrate.

Photothermal : Combined effect of light and temperature.

Phyllody : Change of floral parts into foliage.

Phylogeny : The evolutionary development of a species or genus.

Physiological dormancy : Seed dormancy caused by internal physiological conditions, which prevents germination.

Physiological maturity : The maturity of seed when it reaches its maximum viability and vigour capability and its physiological activities cease.

Physiological race : Pathogens of the same species with identical morphology but differing in pathogenic capabilities.

Physiology of seed : The study of the functions and activities of seed.

Physiology : The study of processes taking place in living organisms.

Phytase : An enzyme, which catalyzes the breakdown of phytin, the source of inorganic phosphorus in seed metabolism.

Phytochrome : The bluish photo-reversible protein pigment responsible for the photoperiodic control of flowering and seed germination.

Phytosanitary certificate : A certificate issued by a legally constituted authority stating that the seed lot has been inspected and found substantially free from pathogens.

Phytosanitary : [Phyto = plant, sanitary = cleaning]. A term applying to the condition of potential transfer of *seed-borne pathogens.*

Phytotoxic : A compound poisonous to plant tissue. Applies to e.g. Side effects of pesticides.

Phytotoxic : Poisonous to plants.

Picking : Harvesting mature ears, pods, or fruits from plants.

Pigment : Substances that appear coloured by virtue of differential absorption of radiant energy.

Pinna : The leaflets of a bipinnate leaf.

Pinnate : A compound leaf in which the leaflets are arranged on the two sides of the rachis.

Pistil : A collective term for ovary, style and stigma; Ovule-bearing organ of angiosperms, composed of ovary, style and stigma. A pistil is made up of one (simple pistil) or more (compound pistil) carpels.

Pistillate flower : A flower bearing pistil but no stamens.

Placenta : The organ through which ovules are attached to the ovary wall.

Placentation : The method of attachment of ovule or seed within the ovary.

Plant architecture : The number, size, shape, structure, arrangement and display of particular parts on plants.

Plant quarantine : Official precautions used to prevent the introduction of plant pathogens and pests from one geographical area to other usually non-infested area.

Planting ratio : The recommended ratio in which male and female parental lines are planted in hybrid seed production.

Planting stakes : Stem cuttings used for planting.

Planum : An air chamber from which air is distributed or collected in a seed drying system.

Plastid : Small cytoplasmic organelles containing chloroplasts giving green colour to leaves.

Pleiotropy : A single gene affecting two or more different characters.

Pleurogram : A distinct horse-shoe formed line occurring on either side of certain legume seeds

Plumaose : Feathered,

Plumule : The embryonic shoot derived from the epicotyl. In dicotyledons situated between the cotyledons.

Pneumatic conveyor : A method of transferring seeds in a processing plant utilizing air as the driving force.

Pneumatic separator : A seed processing machine to separate seeds by using air.

Pod : One or many-seeded dehiscent or indehiscent dry fruit formed from a single carpel. Resembles the follicle but differs from it in splitting along both sides. Prevailing fruit type in Leguminosae. Also found in some Sterculiaceae.

Polar nuclei : Two nuclei of the female gametophyte, which unite with one of the sperm cells to form endosperm.

Polisher : The machine used to polish seeds to improve their lusture.

Pollen chamber : In gymnosperms, a modification of the apex of the megasporangium which receives the pollen, and where the pollen remains dormant until fertilisation can occur.

Pollen grains : The male gametophyte originating from a microspore.

Pollen parent : The parent, which furnishes pollen for fertilization of ovule.

Pollen sac : Locule in the anther containing the pollen grains.

Pollen shedder : The plants of B-line present in A -line in hybrid seed production programmes.

Pollen tube : Tubular protuberance of maturing male gametophytes in seed plants, occurring during fertilisation. In angiosperms the pollen germinates on the stigma and grows through the style to reach the ovule. In gymnosperms the pollen germinates at the tip of the ovule.

Pollen vector : The vehicle by which pollen is transferred from the male flower or strobilus to a female flower or strobilus, e.g. Wind, water, or animals. In animal pollinated species usually called a *pollinator*.

Pollen : The male gametophyte of seed plants produced in the anther or in strobili. Wind pollinated species have a large production of light (in conifers usually winged) pollen. Animal pollinated species often have sticky pollen.

Pollination droplet : Droplet of fluid exuded from the micropyle of gymnosperms at pollination, the function of which is to catch the pollen and later upon evaporation draw the pollen into the *pollen chamber*.

Pollination : The transfer of pollen from anthers to stigma.

Pollinator : A living organism transferring pollen, e.g. Insects, birds or bats. (= pollination agent)

Polyadelphous : Stamens united through filaments into several bundles.

Polyandry : Infinite number of stamens.

Polycross : Random inter-pollination.

Polyembryony : An ovule with more than one embryo.

Polygamaous : Bearing uni and bisexual flowers on the same plant.

Polygenes (Multigenes) : Numerous genes that influence a single character, each have a small effect.

Polymaorphic : With many forms.

Polymer : A larger molecule formed by joining many small identical molecular units.

Polymorphism : The occurrence of two or more distinct forms in the same population.

Polypetalous : With free petals.

Polypoid : A plant with two or more complete sets of homologues chromosomes.

Polysepalous : With free sepals.

Polythene : Abbreviation of polyethylene, a material used for plastic sheet and pots (slang. 'polypot').

Pomae : Most of the edible part develops from the thalamus *e g*. Apple.

Pome : A fruit in which the floral cup forms a thick outer fleshy layer and which lay a papery inner pericarp layer forming a multi-seeded core.

Population : A community of individuals, which share a common gene- pool.

Poricidal capsule : A capsule that at maturity splits opens at pores near the top releasing mature seed.

Porphyllum : The first leaf or protective scale of a lateral shoot.

Portmanteau : A word formed by fusing two or more words or word parts to give a combined or loaded meaning.

Post control : The process of taking a sample from a certified seed lot, sowing it in a control plot the following year and examining the plants to confirm cultivar purity.

Pre control : The process of taking a sample from a seed lot, which is to be multiplied under a certification scheme. Sowing it in a control plot and examining the plants concurrently with the farm multiplication crops growing from that lot.

Pre-basic seed : Seed from which basic seed is produced.

Prechilling : The practice of exposing imbibed seeds to cool (5-10°C) temperature for a few days prior to germination at warmer conditions.

Precision grader : A machine to separate seeds based on width and thickness.

Precocious germination : Same as *vivipary*.

Precuring : The deliberate storage and slow air drying under shade of fruits and contained seeds in order to ease extraction and after-ripen the seeds.

Predation C : Onsumption of a living organism by an animal predator, e.g. An insect eating a seed.

Pre-emergence : The stage between germination of seed and coming out of seedling from soil.

Preheating : Exposure of seed at higher temperature (30-35°C) to break dormancy.

Pressure system : Method of air movement under pressure for seed drying.

Pre-treatment : Any kind of treatment applied to seeds to overcome dormancy and hasten germination.

PREVAC (pressure-vacuum) : Method of separating mechanically damaged seed from sound seed in a seed lot. Dry seeds are exposed to vacuum in water. When the pressure is released, damaged seeds absorb water quicker than undamaged seeds. During subsequent flotation, damaged seeds tend to sink while undamaged seeds float.

Preventive : A precautionary measure to prevent the occurrence of infection (fungicide).

Previous crop : The crop grown in the season immediately preceding the one in which the seed crop is grown.

Primary dormancy *: See* Innate dormancy.

Primary inoculum : The first inoculum of a pathogen available for infection to the host.

Primary leaf : The first leaf or pair of leaves developed after the cotyledon.

Primary root : The root developing directly from the radicle.

Primary sample : A small portion taken from one point of a seed lot during sampling.

Priming (osmotic) : Pre-treatment method to promote rapid and uniform germination. The seeds are soaked in a liquid solution (e.g. Polyethylene glycol (PEG), sugar or salt) of sufficiently low water potential to regulate moisture content at a level where the germination process initiates but radicle protrusion is prevented.

Primordia : Organs in their earliest stage of development.

Processing of seed : Pre conditioning, cleaning, size grading and upgrading of harvested seeds.

Procurement seed : Practical methods of seed handling from collection to sowing including e.g. Methods of collection, processing, storage and pre-treatment.

Proembryo : The young embryo in its early stages of development.

Progeny testing : A test of the value of a genotype based on the performance of its offspring produced in some definite system of maturing.

Progeny : Offspring, plants grown from the seeds produced by a parent plant.

Propagule : A plant part, e.g. Root, bud or shoot, used for vegetative propagation. Ant. *Seed.*

Prophylactic spraying : Protection of crop by spraying pesticides before possibility of being infected.

Protandry : Maturation of anthers before pistils.

Protein : Complex substance built up from amino acids.

Protogyny : The female reproductive structures mature before the male structures.

Protoplasm : The essential complex living substance of cells on which all-vital functions depend.

Provenance : The region where seed was harvested.

Provisional certificate : A certificate issued before the completion of a test or tests, which includes a statement that finals certificate, will be issued on completions.

Pseudo self-compatibility : Partial seed setting following self-pollination in an otherwise self-incompatible plant.

Pseudo-carpic fruit : A fruit consisting of one or more ripened ovules attached or fused to modified tracts or other non-floral structures.

Pseudo-gamous apogamy : A type of diplospory (apomixis) in which the seed develops from some cell of the diploid embryo sac other than the egg after pollination.

Pseudogamy : A type of apomixis in which the diploid egg cell develops into the embryo without fertilization of the egg cell.

Psychrometric chart : The chart for the measurement of air properties.

Puberulent : (Glabrous) without hairs.

Pubescent : Densely clothed with hairs.

Pulp : Fleshy part of fruits which may make up the entire pericarp (berries) or only the exo- and mesocarp (drupes). Also including aril and sarcotesta in some species.

Pulvinus : The swollen base of the petiole.

Pupa : A resting stage in insect metamorphosis between larva and young adult.

Pure line : A genetically pure population of plants descended from an ancestral single plant by pollination in every generation.

Pure live seed (PLS) : The percentage of pure seeds in a seed lot, which have the ability to germinate.

Pure seed content : The percentage each crop species, which comprise five per cent or more of a seed lot by weight.

Pure seed : That component of a seed lot which consists of seeds of the designated species. According to ISTA rules, it includes not only mature, undamaged seeds but also undersized, shrivelled, immature and germinated seeds provided they can be positively identified as the designated species, and pieces of seed resulting from breakage which are more than half their original size.

Purity test : Identification of seeds of various species/varieties and inert matter in a seed sample.

Purity : Proportion of clean, intact seed (according to pure seed definition) of the designated species in a seed lot, usually expressed as a percentage by weight.

Pycnidium : The sporocarp of the group of fungi belonging to the order Sphaeropsidales, frequently globose or flask shaped e.g. *Macrophomina* sp., *Phoma* sp.

Pyrene : Seed enclosed by the hard endocarp of a drupe.

Pyriform : Pear shaped.

Quadrate : One of the small sample areas of a seed crop, which are examined in detail to determine the cultivar purity of the crop as a whole.

Qualitative character : Character showing distinct classes and little or no effect of environment.

Quantitative character : A character showing continuous variance, which can be measured, and its expression is influenced by a group of genes at different loci, which are cumulative in their effect.

Quarantine : Period during which seeds are kept under observation and examination for possible seed-borne pests and pathogens. Transfer over international borders are often subject to quarantine regulations.

Quiescence I : Nability to germinate even in the presence of faovurable environmental conditions.

Rachilla : Little rachis of a spikelet.

Rachis : The central axis of an inflorescence of a cereal crop plant.

Radicle : The embryonic root, i.e. The part of the seed embryo that develops into the primary root. In seeds, the radicle is always facing the micropyle.

Radiograph : Image created on a film or photographic paper after radiation e.g. By X-rays.

Ramose : A lax inflorescence having lateral branches.

Random cup method : Small cups (6-8) are arranged in a tray and seeds are poured on the tray, the seeds which fall in the cup are considered as working sample.

Random deviation : Uncontrolled, non-directed variation occurring by chance.

Random : Arrived at by chance without discrimination.

Randomization : Process of making assignments at random.

Raphe : Ridge formed on the seed-coat if the funiculus is fused with the integument in part of its length in anatropous or campylotropous ovules.

Recalcitrant : Term used to describe seeds that cannot survive drying below a relatively high moisture content (30-40%) and, for tropical species, do not tolerate low temperature. The seeds rapidly lose their viability and cannot be successfully stored for long periods. Ant. *Orthodox.*

Receptacle : End of a flower stalk on which the floral organs are borne. In epigynous flowers the receptacle enlarges and encloses the ovary.

Receptivity : Condition of the female flower or cone that permits effective pollination.

Recessive : The member of an allelic pair, which is not, expressed when the other (dominant) member occupies the homologous locus.

Recurved flag leaf : Flag leaf at 180° angle to the stem.

Reduction process : The addition of hydrogen to or removal of oxygen from a substance is known as reduction process It is a process of lowering the positive valence state of an element

Registered seed : the progeny of breeder's seed or foundation seed handled under procedures acceptable to the certifying agency to maintain satisfactory genetic purity and identity

Re-inspection : If the seed crop does not meet the prescribed standards, another chance is given to rectify according to the suggestions made by the inspectors.

Reinvigoration : Improving germination percentage of old stored seed lot.

Rejection of the field : When the field of seed production is not found up to the standard of the class, the case is dropped.

Rejuvenation of seed : Maintenance of seed quality even after normal storage period to satisfy the requirement of high quality seed.

Relative humidity : The amount of water present in air at a given temperature in proportion to its maximum water holding capacity at that temperature.

Release : A crop variety which is permitted to be reproduced, marketed, and made available as seed for public use.

Reniform : Kidney shaped.

Renovation : Removal of plants from a very dense, unproductive stands to revitalize the productivity.

Repeated halving method : A procedure of sampling by dividing seed quantity into two parts several times.

Replications : Repetition of treatments in experiments that is needed for statistical analysis.

Reproductive age : The age at which the tree produces its first fruit crop, i.e. The demarcationfrom juvenile to mature in respect to reproduction.

Resistance : The genetic characteristic of a plant that prevents the pathogen from disease development.

Resistant : Capability of host plant not to allow infection and development of a pathogen.

Respiration : Normally distinction is made between external and internal respiration. External respiration is taking oxygen from the environment and giving off carbon dioxide. Internal respiration are the chemical reactions in which energy is released by the breaking down of organic compounds. If oxygen is consumed during this process it is called aerobic respiration, if not, anaerobic respiration. External and internal respiration occur in both plants and animals.

Respiratory quotient (RQ) : The ratio of volume of CO_2 evolved and the volume of O_2 consumed per unit time.

Restorer line (R-line) : An inbred, pollen- parent, containing fertility restorer genes, when crossed with a male sterile strain (A-line) produce male fertile hybrid. Use for the production of hybrid seed.

Reticulation : The net-like appearance of teliospores wall of smut fungi e.g. In *Teletia indica*.

Rhizobium : Bacterial root symbiont that infects roots of plants of the family Leguminosae. During infection the plant forms root nodules, in which the bacteria fix atmospheric nitrogen which benefits plant growth.

Rhizome : A horizontal underground stem.

Rib : A primary or prominent vein of a leaf.

R-line : *See* restorer line.

Rodewald apparatus : A devise to test seed for germination under controlled conditions.

Rogue : An undesirable offtype plant.

Rouging : The act of removing off type plants.

Roll mill : A machine to clean smooth seeds contaminated by rough seeds.

Rolled paper towel test : Seed germination test between moist paper towels then rolled and kept in germinator at desirable temperature.

Room germinator : The required conditions for germination provided in a room where the workers can enter to work.

Root slips : The plants arising from portion of a root, when bedded for propagation.

Rough cleaning : A process to clean larger as well as smaller material from crop seed.

Rudimentary : Incompletely developed.

S_0 : Symbol used to designate the original plant produced by self- pollination.

S_1 S_2 and S3 : Symbols to designate first, second, third etc. Selfed generations from an ancestral plant.

sacs : It consists of filament and anthers

Sagittate : Arrow shaped with the basal lobes directed backwards.

Saline soil : A soil containing sufficient soluble salts to impair crop productivity.

Salver-shaped : With a long tube and horizontal spreading limb usua.lly referred to corolla.

Samara : An indehiscent, winged fruit in which the seed coat is loose inside the pericarp.

Sample : In the context of seed testing, a small representative quantity drawn from a seed lot. The different types of samples in seed testing are *primary sample*, *composite sample*, *submitted sample* and *working sample*.

Sampling error : Deviation of a sample value from the true value owing to the limited size of sample.

Sampling intensity : The required number of primary samples to represent a seed lot or population.

Sampling : The method by which a representative sample is taken from a seed lot or a population.

Sand medium : Sand particles, which pass through 0.8mm diameter hole sieve used for seed germination.

Saprophyte : An organism, which is incapable of photosynthesis and feeds on dead plant or animal tissues.

Sarcotesta : Soft, fleshy outer layer of a testa.

Saturated air : Air that contains the maximum amount of water at a given temperature. The relative humidity of saturated air is 100%.

Scabrid : clothed with small and stiff hairs

Scabrous : Rough

Scale : A thin scarious membranous outgrowth usually a degenerate leaf.

Scalper : A screening air blast type machine to separate very large, very small or very light contaminant from a seed lot.

Scaly : with scale

Scapoletin : Growth promoting substance to stimulate seed germination of some plant species.

Scarification : The process of mechanically roughing or abrading seed coat to make it more permeable.

Scheduled weed : Objectionable weed.

Schizocarp : A dry, two seeded fruit, which separates into two mericarps.

Sclerenchymatous : The tissue composed of cells with thick and hard walls.

Sclerotium (pl = sclerotia) : A hard, dark, pigmented body of a fungus capable of remaining dormant for a long period, comprising of fungal tissues e.g. *Sclerotium rolfsii*.

Scorching : Brief exposure to high temperature, e.g. Open fire or kiln, applied for extraction of seeds from serotinous fruits or cones.

Scorpioid cyme : A determinate inflorescence in which the lateral buds on one side are suppressed during growth, resulting in a curled or coiled arrangement.

Scorpioid cyme : A determinate inflorescence in which the lateral buds on one side are suppressed during growth, resulting in a curled or coiled arrangement.

Screen dams : Strips of wood etc. Fastened across the screen to retard flow of seed down the screen.

Scutellum : The cotyledon of an embryo of a cereal present in between embryo and endosperm, which absorb food from endosperm.

Seasonality : Phenological periods of seed production occurring tightly connected to distinct climatic season over the year, e.g. Rainy season, dry season, summer, winter. Compare *periodicity*.

Secondary dormancy : *See* induced dormancy.

Secondary infection : Spreading of disease again, due to inoculum from a source, other than the crop seed under test.

Seed act : An act to regulate the seed quality established in 1966.

Seed analysts : The person supervising or carrying out the seed analysis work and appointed under section 13 of seed Act 1966.

Seed blower : The equipment to separate, light-weight material from heavier seeds.

Seed certification : Verification of field and seed standards based on official inspection and testing of seed with regard to its quality and origin as per seed rule. It maintains genetic and physical purity as well as seed health of the produced seeds.

Seed coat dormancy : Checking of germination of seed in spite of imbibition of water, until the seed coat is removed or embryo excised.

Seed coat : The outer protective cover of the seed.

Seed crop : The total seed production during one defined fruit/seed season.

Seed deterioration : Reduction in seed quality which starts from physiological maturity and continuous till the seed is dead.

Seed disinfection : Making seed free from pathogens established in or on the seed.

Seed disinfestation : Making seed free from pathogens present on the surface as contaminant.

Seed dressing : Chemical treatment of seed.

Seed drill : A machine used to sow seeds in rows.

Seed germination : Resumption of active growth in a seed which results in its emergence from the seed and development of those structures essential to normal plant development.

Seed grading : The process of removing inert matter and smaller seeds from a seed lot.

Seed handling : Practical methods of seed procurement from collection to sowing. Sometimes used synonymously with seed procurement but generally it covers a wider range e.g. Sowing techniques.

Seed hardening : Alternate hydration and dehydration cycles prior to completion of germination

Seed health : Presence or absence of disease causing organisms in or on the seeds.

Seed inspector : Persons appointed under Section 13 of the Seeds Act to take seed samples get them analysed and initiate prosecution against offenders, if necessary.

Seed leaf : Cotyledon.

Seed lot : A specified quantity of seed of the same species, provenance, date of collection and handling history, and which is identified by a single number in the seed documentation system.

Seed mates : Broad sheets of degradable material for spreading seeds.

Seed moisture : Loss in weight of seeds when dried.

Seed mycoflora : Refers to the fungi associated with seeds which may or may not have the potential of causing diseases of the plant or seed

Seed packaging : Packing of seed in a way to prevent absorption of water and to avoid contamination.

Seed parent : The line (female parent) from which seed is harvested in the hybrid-seed production plot.

Seed pellets : Seed enclosed in inert matter certain times with nitrogen fixing bacteria as well as pesticides.

Seed piece : Pieces of stem, root, and leaf to be used for vegetative multiplication.

Seed potato : Small sized tubers of potato used as seed.

Seed priming : Pre sowing treatment of seed with osmotic solution or salts or polymers based on controlled hydration of seed to such a level that only pregerminative metabolic activity is initiated without active emergence of the radicle.

Seed processing : Upgrading the seed quality by removing under sized, undesirable seed, inert matter etc. And improving the planting value.

Seed protection : Treating the seed to prevent infection by pathogens.

Seed quality : The seeds meeting required standards as per rule.

Seed repository : Storing of seeds after drying to 5 per cent moisture content sealed in laminated aluminium foils and stored at -20°C

Seed rot : Seed decay caused by pathogenic fungus.

Seed stalk : The erect stalk on a plant of root crop that produces flowers and seeds.

Seed stock : Seed, which is used as source of germplasm for maintaining and increasing seed of crop varieties.

Seed technology : The science, which deals with production, harvesting, processing, testing, packaging, storing and marketing of seeds.

Seed testing : Examination of physiological seed quality of a seed lot, normally consisting of four standard parameters *viz.* Purity, seed weight, moisture content and viability or germination percentage. Sometimes also including special parameters such as vigour or health test. Standard rules and procedures for seed testing are issued by ISTA (Europe) or AOSA (America).

Seed transmission : An act of perpetuation of a pathogen from one generation to another through the seed.

Seed vigour : Seed property, which determines the potential for rapid, uniform emergence and development of normal seedlings under a wide range of field conditions.

Seed weight : Seed weight is indicated in number of seeds per weight unit (usually kg) of the seed lot including impurities. Another measure is the thousand seed (grain) weight indicating the weight in grams of 1000 pure seeds.

Seed year : In respect of any species, particularly trees of irregular or infrequent seed production, a year in which it produces an adequate amount of seed, either as an individual or a crop, Many periodic seeders produce heavy ('bumper') seed crops during their seed years.

Seed : A mature ovule consisting of an embryonic plant and a store of food all surrounded by a coat The grain or mature ovule of a plant used for sowing

Seed-bearing organ : Common term for all seed-enclosing structures in seed plants, which comprise fruits in angiosperms (true fruits and multiple fruits), and cones and other structures in gymnosperms

Seed-borne : An organism or an infectious agent carried in, on or with the seed, but not necessarily transmitted.

Seed-borne diseases : Plant diseases caused by pathogens carried through seeds.

Seed-borne pathogen : An infectious agent associated with seed and having the potential of causing a disease.

Seed-coat : Protective outer layer(s) on a seed derived from the integuments. When two layers of the seed-coat are distinguishable, the terms testa for the outer coat and tegmen for the inner is often used.

Seedling : Plant produced from a seed as different from plants produced by vegetative propagation. The latter are called cuttings, stumps, or plantlets depending on mode of propagation.

Seed-transmitted pathogen : Infectious micro-organism carried in, on or with the seed with the potential of causing disease of the seedling or the plant, but harmless to the seed itself. Compare *seed-borne pathogen.*

Segmented seed : Seeds of certain root crops borne in clusters of fruits each containing a single true seed.

Segregation : Separation of paternal from maternal chromosomes at meiosis and consequent separation of alleles.

Seizure : Legal court acts to take off the seeds from the market.

Select seed : The approved progeny of breeder's seed in Canadian certification system.

Selection intensity : The percentage of the population selected.

Selection : A plant, line or strain originating by the process of preference.

Selenium : Is a chemical element with atomic number 34, with the chemical symbol Se. Selenium occurs only rarely in the free state in nature. It is a nonmetal that is chemically related to sulfur and tellurium. It is toxic in large amounts, but trace amounts of it, forming the active center of certain enzymes, are necessary for the function of all cells.

Self sown plant : A plant which grows from seed scattered from the previous year crop plant

Selfed : Fertilization of pistil with pollen from the same plant which bears the pistil

Self-fertility : Capability of producing seed upon self-fertilization.

Self-fertilization : Fusion of male and female gametes from the same individual.

Self-incompatibility : Inability to set seed from application of viable pollen produced on the same plant due to Genetically controlled physiological hindrance.

Self-sterility : Failure to complete fertilization and set seed after self-pollination.

Semigany : Abnormal fertilization in which the male gamete fertilizes the egg, but does not fuse with the egg nucleus.

Seminal : Not arising from the original place.

Semi-permeable : Allowing water to pass through, but not dissolved substances.

Sepal : a segment of calyx

Sepaloid : Green.

Septicidal capsule : A capsule splitting along the septa by ventral sutures and releasing the seeds.

Septum : Partition between locules of a fruit

Sequential sampling : A technique for monitoring viability of seeds in Seed Bank based on sequential probability ratio test.

Serotinous : Coming late, particularly applied to plant species or individuals that flower or fruit late in the season and to fruits or cones that remain on the tree without opening for one or moreyears e.g. *Pinus contorta*. Serotinous fruits normally need high temperature for opening and seed extraction.

Serrate : toothed with teeth inclined upwards

Service sample : The sample submitted to the central or state seed laboratory for testing.

Sessile : Without stalk.

Set : A vegetative propagule consisting of one or more stem nodes with buds and root primordia.

Seta (pl = setae) : Stiff hair, usually dark and thick walled e.g. In *Colletotrichum* sp.

Setae : Bristle like structures.

Sexual reproduction : Reproduction involving germ cells and union of gametes.

Shattering : Bursting of fruit and dispersing the seeds before harvest.

Sheath : A tubular envelope forming lower part of the leaf in grasses

Sheathing : The leaf having expanded base of the petiole encircling partially or.completely the stem.

Shedding tassel : The tassels in female parent rows shedding pollen in hybrid maize plot.

Sheller : A machine or device used for shelling.

Shelling : The process of separating the fruit-wall from its enclosed seeds.

Shoot apex : Terminal part of the shoot with main growing point.

Short day plants : Plants, which initiate flowers best under, short day and long night regions.

Short term storage : Storage period between one year and one and a half-year.

Sib : Progeny of the same parents derived from different gametes.

Sigmoid : curved in zigzag fashion

Sign : Any indication of disease other than reaction of the host plant i.e. Spores, mycelium, exudate or fruiting bodies of the pathogen.

Significance test : Statistical test designed to distinguish differences due to sampling error and discrepancy between observation and hypothesis.

Silk : The long style and stigma of the maize ovary.

Silky : Covered with close pressed soft and straight pubescence.

Sillique : A fruit with two locules that at maturity split away from the base and disperse the seed.

Silo : Storage structures whose height is more than the diameter or width.

Simple cyme : A determinant inflorescence in which lateral flowers develop later than the terminal flower.

Simple fruit : Fruit developed from single pistil or ovary.

Single cross hybrid : A hybrid between two inbred lines.

Single seed descent : Selection procedure in which F_2 plants and their progenies are advanced by single seed until genetic purity is virtually attained.

Sizing : The process of removing seeds smaller and larger than the required dimensions in order to have a lot of uniform size.

Sleeve trier : A hollow bass tube inside a closely fitting outer shell, which has a solid pointed end for drawing seed samples.

Sling Psychrometer : A type of dry and wet bulb thermometer.

Slot seeding : A method of reseeding without destroying all existing plants.

Slurry treater : A machine used for slurry seed treatment.

Slurry : Semi-liquid suspension of wettable powder of fungicide or insecticide in small quantity of water.

Smut : A disease characterized by a mass of dark spores caused by members of Ustilaginales.

Sod binding : A condition of grass fields where seed production is decreased due to high plant population.

Soil seed bank : Dormant, viable seeds of one or several species accumulated in the soil over one to several years.

Solanad embryo type : A type of embryo in which the terminal cell of the proembryo divides by a transverse wall and the basal cell plays a minor or no part in subsequent embryo development.

Solitary flower : The simplest expression of determinant inflorescence by producing only one flower.

Somatic cells : Cells of plant body other than reproduction tissues.

Somatic embryo : Embryo developed synthetically without surrounding nutrient tissues and protective seed coat.

Somatic embryogenesis : The production system of somatic embryos.

Somatoplastic sterility : Sterility due to collapse of zygote, during embryonic stages because of disturbances in embryo endosperm relationships.

Sorus (pl = sori) : A fruiting structure in rusts and smuts, mass of spores.

Sound seed : Healthy, viable seed without major damages.

Source-identified seed : Seeds collected from natural stands where geographical origin (source and elevation) is known and specified, or from seed orchards or plantations of known provenance, specified by a seed-certifying agency.

South Dakota blower : A type of seed blower used in purity testing of seeds based on their specific gravity.

Spadix : A special type of spike with a fleshy inflorescence axis possessing male and female flowers.

Spathe : A large bract surrounding an inflorescence especially a spadix.

Species : A taxonomic division of a genus.

Sperm : A male gamete.

Spike : An unbranched inflorescence with elongated axis bearing sessile spikelets.

Spikelets : A unit of the inflorescence in grasses composed of glumes, rachilla and florets.

Spine : a sharp pointed hard body

Spiral separator : A type of seed separator with no moving parts but a spiral running.

S-pistillate : Femaleness in monoecious.

Spoon method : Sampling by the help of spoon.

Spore ball : An aggregate of many spores or a group of spores clustered together usually formed in smut fungi. A structure made up of sori and peripheral cells.

Spore plants : Group of lower plants whose main characteristic is reproduction by wind-borne spores. The spores are haploid and germinate into haploid gametophytes, which in these groups are independent plants. Spore plants include e.g. Mosses and ferns.

Spore : A general term for a reproductive unit in fungi.

Sporodochium : A cushion shaped fruiting. Body consisting of conidiophores and conidia.

Sporogenesis : Production of mega and microspores from mega and microspore mother cell respectively after meiosis.

Sporophyte : An individual of the diploid generation of the life cycle of a plant. It is the dominating part of all higher plants where the gametophyte (the haploid generation) is reduced to a dependent organ in of the sporophyte (embryo sac in angiosperms). In some lower plants, the sporophyte is reduced and the dominating generation is the gametophyte.

Sprout : Emergence of shoot, root, bud etc. Before planting spud.

Spud : Potato tuber.

Square : Bracts of cotton.

Squash test : Simple, indirect test of viability, by which seeds are first allowed to imbibe water and are then squashed with a pair of forceps to reveal the condition of the embryo. The number of seeds appearing fresh and healthy per unit weight of seeds plus chaff (eucalypts) or per 100 seeds (larger seeds) provides a rough estimate of viability.

Stack : Storage of seed bags by placing them one over the other.

Stacking : The arrangement of seed bags in a manner to prevents their falling.

Staggering : Difference in date of planting of A and R or A and B lines for synchronization of flowering.

Stalk : A stem like supporting structure.

Stamen : The pollen-bearing organ of a flower composed of anther and filament.

Staminate flower : A flower bearing stamen but no pistil.

Staminode : Rudimentary stamens which do not bear fertile pollen.

Standard deviation : A measure of variability.

Standard error : A parameter, which is the estimated value of the standard deviation.

Standard : The broad upper petal of Papilionatae.

Steckling : Small sugar beet stored over winter and planted for seed production.

Stele : the stele is the central part of the root or stem containing the vascular tissue and occasionally a pith

Stem rot : Decay of the stem caused by a pathogenic fungus.

Sterile : Unproductive or infertile.

Sterility : Failure to complete fertilization.

Steriobinocular microscope : A kind of microscope with lenses arranged so that the object in view is seen in three dimensions

Stigma : The part of the pistil on which pollen must be deposited in order to germinate and reach the ovule

Stipulate : Having stipules.

Stipule : a leaf-like outgrowth at the base of the petiole usually in pairs

Stone : Common term for the hard endocarp of a drupe containing the seed; same as *pyrene*.

Stoner : A modification of gravity separator to separate stones from crop seeds.

Storability : Potential life time of seed under optimal storage conditions, cf. *Longevity*.

Storage fungi : Fungi, which grow on seeds under storage conditions.

Strain : A group of similar individuals within a variety.

Stratification : Scratching or rupturing of seed coat to facilitate absorption of water to hasten germination.

Stress : Sub-optimal growth conditions like shortage of water, too high or too low temperature, shortage of light or fertiliser etc.

Strobilus (pl strobili) : A spiral arrangement of modified leaves bearing the reproductive organs in conifers and certain other plants, functionally equivalent to angiosperm flowers. Male strobili bear microsporophylls, female strobili megasporophylls. As female strobili grow after pollination, they are called conelets, then as reaching their full size, cones.

Stroma (pl = stromata) : A matrix of vegetative hyphae, with or without host on which spores are produced.

Strophiole : Arilloid (type of outgrowth on a seed) occurring on the raphe. In legume seeds the strophiole is often the weakest spot on the seed-coat and the place where e.g. Hot water treatment has impact and imbibition begins.

Stubbing root : Root becoming and club shaped with intact root tip.

Stubble : Part of the crop plants left standing in the field after harvesting.

Style : The stalk of the pistil between stigma and ovary.

Subepidermal : Occurring just beneath the epidermis.

Submitted sample : A representative sample taken from the seed lot and sent to seed testing laboratory for testing purpose.

Sub-sampling : The procedure for obtaining representative working sample from the submitted sample.

Substratum : The material upon which seeds are placed for germination test.

Succulent : Fleshy.

Sucker : A vegetative shoot of subterranean origin.

Sulcate : grooved

Superficial : Occurring on the surface of the structure.

Supplementary pollination : Extra pollination for production of hybrid seed.

Surge Bin (*Holding bin*) : The bins mounted above the processing machine

Susceptible : Incapability of a plant to suppress or retard an injurious pathogen or other factor.

Suspensor : The group of cells produced from zygote, which pushes the developing proembryo towards the centre of the ovule having nutrients.

Suture : the line of dehiscence of dry fruits

Swarm : Simultaneous movement of large numbers of insects in a group.

Swollen seeds : Healthy looking seeds imbibing water but not germinating during the prescribed test period.

Symbiosis : Literally 'living together'. Applies to two organisms of different species forming an association with one another. If the symbiosis is of mutual benefit, it is properly called mutualism; if one of the symbionts benefits at the expense of the other, it is called parasitism.

Symptom : The indication of a disease in a plant.

Syn 0 Syn 1 Syn 2 etc : Symbols for designating the original, first and second synthetic generation.

Syncarpous (syncarpy) : Fruit composed of two or more united carpels.

Syngamy : Sexual fusion of sperm and egg cells.

Syngenesious : with anthers cohering in a ring

Synthetic seed : Somatic embryos engineered for commercial plant production.

Synthetic variety : A variety composed of an interbreeding population of several cross-pollinated plant lines.

Tailings : Partly threshed material, which has passed through the coarse shakers and eliminated.

Tap root : Primary root, deeply penetrating the soil.

Tape : A collapsed tube of soluble plastic material containing seeds spaced at regular intervals for planting.

Tapering ear : Upper most spikelet of ear pointing.

Tassel : The inflorescence of male flowers produced at the apex of maize plant.

Taxon : Any unit used in the science of biological classification, or taxonomy. Taxa are arranged in a hierarchy from kingdom to subspecies, a given taxon ordinarily includes several taxa of lower rank.

Tegmen : A thin layer in between testa and cotyledon of seeds.

Teliospore : A dikaryotic thick walled, black resting spore produced by fungi, with two haploid nuclei (n+n), which become diploid (2n) before germinating.

Temperate recalcitrant : A group of recalcitrant seed that are sensitive to desiccation and do not tolerate storage with low moisture content but are tolerant to low temperature, a difference from tropical recalcitrant seed. Includes many temperate forest species such as *Quercus* and *Acer* species.

Temperature alternation : During germination test, lower and higher temperatures are maintained in seed germinator as per need of the seed.

Temperature constant : A specific temperature, which should not vary more than 1°C in germination test of seeds.

Temperature sensitive male sterility : Line of rice showing sterility when temperature exceeds 30°C in day and 24°C in night, fertility is exhibited when day and night temperatures are 24 and 18°C respectively.

Tempering : To bring grains to a desired moisture or temperature for processing.

Tendril : They are either modified branches or leaves. They are filament like usually coiling round the support.

Tepal : A segment of perianth.

Terete : Cylindrical.

Terminal : Borne at the apex.

Terminator technology : The seed produced by the plant with terminator gene will be sterile that means in next generation seed will not germinate so the farmer has to purchase new seed.

Test cross : A cross of a double or multiple heterozygote to the corresponding multiple recessive to test the homozygosity or linkage.

Test weight : Seed weight expressed as the weight of 1000 pure seeds; a standard measure during seed testing.

Testa : Outer most layer of the seed coat.

Tetradynamous : With four long and two short stamens.

Tetraploid : A plant with four sets of identical chromosomes.

Tetrazolium test : Quick test to determine seed viability.

Tetrazolum : A class of chemicals with ability to undergo reduction due to dehydrogenase enzymes during respiration process in viable seeds changing from colourless to red.

Thermal induction : The change in growth and development of plants brought about by exposure to a given temperature resulting in flowering of biennial plants.

Thermoperiod : A period of the proper temperature to elicit a specific growth or developmental response for flowering or seed germination.

Thermo-periodism : Plant responses to temperature cycles for floral induction.

Three line system : The exploitation of hybrid vigour by the use of cytoplasmic genetic male sterile system involving male sterile line, restorer line and maintainer line.

Three way cross : A cross between a single cross (F_1 hybrid) and an inbred line.

Threshing : Separation of seeds from other plant material by mechanical impact.

Tightly twisted seedling : Completely turned seedling.

Tiller : A branch arising from the base of a monocot plant (in Gramineae).

Tilth : Physical condition of the top soil after tillage.

Time of one half response : Time required to obtain a change in moisture content from the original moisture content half of the way to equilibrium, usually expressed in hours.

Tissue culture : The development of an entire organism from plant cells and tissue *in vitro* or artificial media.

Tissue : An aggregate of cells usually similar in structure.

Tolerance : 1.The maximum acceptable difference between the results of two tests of samples taken from the same seed lot. 2. Ability of a host to avoid the loss in productivity although it has been infected by a pathogen.

Top cross : A cross between selections, clones, lines or inbreds with a common pollen parent.

Top dressing : Application of feritlizer at a definite gap after crop establishment.

Top of paper : Conduction of seed germination test on one or more layers of paper towel or filter paper.

Totipotency : The potential of an undifferentiated plant cell to develop into a plant when cultured *in vitro*.

Trait : A synonym of character with respect to function and performance.

Transgenic plant : A plant regenerated from transformed cell /protoplast.

Transgressive segregation : Appearance of individuals falling outside the parental range in respect to some characters in segregating generations

Treatment : Any process, physical or chemical to which a seed lot is exposed.

Trichome : A leaf hair arising from an epidermal cell.

Trier : A tube-like devise to take out samples from a seed lot. The trier consists of two tubes, one fitting into the other. Openings in the two tubes permit seed to flow into the tube when the holes are at the same position; the holes are closed by turning the inner tube.

Trifoliate : With three leaflets.

Trihybrid : A hybrid resulting from a cross between parents that differ in three specific genes.

Triploid : An organism with three basic sets of chromosomes.

Tripping : Pollen dispersal mechanism in alfalfa in which the stanminal column is sprung free of the keel and exposed.

Truncated glume : A glume with medium nerve elongation.

Truthfully labelled seed : The seed of notified variety which is not up to the standard of certification but have met the seed standard up to good level. Genetic purity is not guaranteed.

Truucate : Cut off at the end, square or so.

TTZ : Tetrazolium or topographical tetrazolium. Chemical used for examination of living tissue. TTZ stains living cells red by the reduction of a colourless tetrazolium salt to form red formazan. The reduction is caused by dehydrogenases, a group of oxidising enzymes present in living cells.

Tuber : A short underground stem with starch.

Tumbling : The operation by which cones or fruits are placed in a rotating drum, which rolls and tosses them around to induce the release of the contained seeds. Tumbling may be used in connection with artificial heat in a rotating kiln.

Turgid : Swollen.

Twiner : Climber, which climbs by spirally coiling stems.

Umbel : An inflorescence having radial branches of equal length from a common point therefore flower units are arranged in a flat or umbrella shaped heads

Umbellate : Umbel like.

Unicellular : One celled

Uniculum : A plant having only one shoot and lacking secondary tillers.

Uniformity : Absence of significant variation in a seed lot.

Unilocular : One chambered ovary.

Unisexual flower : A flower bearing only staminate or pistillate part.

Unitegmic testa : A seed coat made up of only one integument.

Unsaturated fatty acid : A fatty acid that has a double bond between the carbon atoms at one or more places in the carbon chain and hydrogen can be added at the site of the double bond.

Upgrading : Improving the average quality or performance by removing inferior individuals. In connection with seeds, the increase of viability and vigour of a seed lot by removal of small, immature, empty and otherwise inferior seeds from a seed lot. In connection with seed sources, the culling or roguing of inferior phenotypes to improve the genetic quality, e.g. Used in connection with establishment of seed production areas.

Upland rice : Rice grown in non- bunded fields like wheat.

Utricle : A small, thin walled, one seeded fruit in which the seed is loosely attached to the pericarp.

Valvate : An aestivation, when the segments of calyx or corolla are so placed that their edges touch each other, not overlap.

Vapour proof storage : Sealed storage.

Variance : The average of the squared deviations about a mean. The variance may be environmental genetic or phenotypic.

Variate : A single observation or measurement.

Variation : The occurrence of differences among individuals due to differences in their genetic composition and/or the environment in which they are raised.

Variety (Cultivar) : A subdivision of species. An agricultural variety is a group of similar plants that by structural features and performance can be identified from other varieties within the same species.

Variety blend : Mechanical mixture of seed of two or more varieties.

Vector : 1. An agent for transferring a gene from one plant to another. 2. Insects responsible for transmission of diseases specially virus.

Vegetative reproduction : Reproduction by vegetative means.

Veins : Threads of fibro-vascular tissue in a leaf.

Venation : The arrangement of veins in a leaf.

Ventral : Inner face of an organ

Ventricose : Suddenly bulged.

Vermiculite : Mineral with high water holding capacity; used as medium for germination tests.

Vernalization : The treatment of seeds before sowing to hasten flowering.

Versatile : Can move in any way.

Vertical resistance : Resistance conferred by few genes specific to few known races of pathogens.

Verticellate : Whorled.

Viability : Capability of living, growing and developing.

Viable seed : A seed which can germinate under favourable conditions, provided that any dormancy that may be present is removed.

Viable : Alive, ability to live, grow and develop.

Vibrator separator : Seed separator run by an electromagnetic vibrator.

Vigour test : Germination test usually carried out under sub-optimal conditions (stress) or after accelerated ageing in order to reveal deterioration of seed lots, not evident during normal germination test.

Vigour : It is the sum total of all seed attributes which favours rapid and uniform stand establishment in the field.

Virus : Nfectious nucleoprotein.

Viscid : With sticky secretion.

Visual : Examination a method of examining seeds with or without the help of a magnifying lens.

Vitascope : A commercial devise for speeding up the tetrazolium test.

Vitreous : A hard, transparent, glasslike structure.

Vitrification : In the dry state, the cytoplasm of seed exists in glassy state. This glass is a liquid of high viscocity, which slowdown all chemical reactions requiring molecular diffusion. Thus prevents damaging interactions between cell components. The glass formation is known as vitrification.

Viviparous : (Vivipary = precocious germination). Seeds germinating while still attached to the parent plant, e.g. *Rhizophora* spp. Viviparous seeds are extremely difficult to store.

Volunteer plants : Unwanted plants growing from residual seeds of previous crops remaining in the field.

Water holding capacity : The amount of water that can be retained by a substrate after free drainage. Gravel and sand have low water holding capacities while peat and other organic substrates have high ones.

Water potential : The energy status of water in the soil, root, seed etc. Water will flow from a place with high water potential to a place with low potential. Water potential is, depending on type, influenced by e.g. Osmotic potential, gravitational potential, turgor potential. Measured in negative pressure units (- bars).

Weed seed : Seeds of weeds mixed with seed lot whose percentage is determined in purity test.

Weed : An unwanted plant appearing in a cultivated crop.

Weevil : Small beetles, which cause damage to seed in storage.

Whorl : An arrangement where the leaves are circled around the stems.

Windrow : Cut or uprooted plants gathered loosely into a long row to facilitate drying and subsequent collection.

Wing : A membrane, or thin, dry expansion or appendage of a seed or fruit.

Winnower : A machine used for winnowing.

Winnowing : Removal of light inert matter from the seed lot by flowing it away.

Withering : (Weathering). Impact of environmental fluctuation on seed quality.

Working sample : A reduced seed sample taken from the submitted sample in the laboratory on which the seed quality tests are made.

Xenia : Effect of pollen on embryo and endosperm.

Xerophytic : Capable of living under extremely dry soil conditions.

X-radiography : A non-destructive method of examining seed quality by exposing them to X-ray and capturing the images on photographic film on which the condition of the embryo can be assessed. (= X-ray radiography)

X-ray : Penetrating electromagnetic radiation having wavelength shorter than those of ultraviolet or visible light.

Zombie seed : It is an alternative form of Internationally banned terminator technology, the zombie type seed is sterile seed that means it will not germinate. To get it germinate, farmer has to purchase chemical from the company that has formed that seed.

Zygomorphic : Asymmetrical, irregular.

Zygote : The fertilized egg.